Mahler · Weberruß   Quantum Networks

# Springer

*Berlin*
*Heidelberg*
*New York*
*Barcelona*
*Budapest*
*Hong Kong*
*London*
*Milan*
*Paris*
*Tokyo*

Günter Mahler
Volker A. Weberruß

# Quantum Networks

## Dynamics
## of Open Nanostructures

With 172 Figures

 Springer

Professor Dr. rer. nat. Günter Mahler
1. Institut für Theoretische Physik und Synergetik
Abteilung Festkörperspektroskopie
Universität Stuttgart
Pfaffenwaldring 57
D-70569 Stuttgart, Germany

Dr. rer. nat. Volker A. Weberruß
V.A.W. scientific consultation
Im Lehenbach 18
D-73650 Winterbach, Germany

Working out of the text, graphics, parts of the numerical calculations,
softbook, and production of the camera-ready manuscript
by V.A.W. scientific consultation.

ISBN 3-540-58850-7 Springer-Verlag Berlin Heidelberg New York

CIP data applied for

Typesetting: Camera-ready copy from the authors using a Springer TeX macro package
SPIN 10478302      56/3144 - 5 4 3 2 1 0 - Printed on acid-free paper

# Preface

This book grew out of lectures on density matrix theory given by one of us (G. Mahler) at the Universität Stuttgart in the academic years 1989/90 and 1993.

Basic properties of the density matrix are covered in many books and from a number of different points of view; however, we felt that an extensive treatment on coupled few-level quantum objects is missing. This may not be too surprising as the necessity to understand such systems has emerged only in the 1980s. In Stuttgart the main motivation has originally been a special research project on molecular electronics, funded by the Deutsche Forschungsgemeinschaft.

However, there are many other areas like photon optics, atom optics, cavity electrodynamics, and combinations thereof, which typically can be mapped onto the same mathematical framework: the density matrix theory cast into an $SU(n)$ lattice description.

Though there are powerful alternatives, we think that this approach is particularly useful to "see quantum dynamics at work": quantum dynamics is often believed to be "counter-intuitive" (which may simply mean that we have no or the wrong intuition). In fact, from experiments and detailed modelling of individual quantum objects one almost gets the opposite impression. Though part of the game is non-deterministic, the type of events, the alternatives, are controlled by an amazingly strict "logic", which derives from the embedding of the quantum object into a classical environment. The measurement protocol ("information dynamics") feeds back into the system dynamics. The resulting visualization of quantum dynamics in terms of "clockwork of pointers", moving and jumping, disappearing and reappearing, may provide us with some kind of experience, which we are used to getting for free in the classical world.

This book is intended to attract not only specialists, but also students trying to gain some working knowledge of quantum mechanics. For this purpose all calculations are given fairly explicitly; furthermore, they do not require more than a decent understanding of vector analysis and vector algebra. A quantum-mechanical background will be provided in the introduction.

We have focussed on the discussion of so-called nanostructures, as these seem to be of great importance for technical products. In order to write a book not only aimed at a scientifically interested public, we have added an extensive introductory part dealing with technical aspects of nanotechnology. Moreover, this part is to convince those not interested in technology that even the most complicated formulae and trains of thought presented in this book control objects which are becoming – though often invisible for us – a matter of course for all our lives: electronic devices such as micro-chips.

*Stuttgart and Winterbach*                          *Günter Mahler*
*January 1995*                                  *Volker A. Weberruß*

# Acknowledgements

The authors wish to thank Dr. Matthias Keller, Dipl. Phys. Rainer Wawer, Dipl. Phys. Jürgen Schlienz, and Dipl. Phys. Holger Hofmann (1. Institut für Theoretische Physik und Synergetik, Universität Stuttgart) for many valuable discussions and for supplying us with numerical simulation data. Furthermore, it is a pleasure to thank Dr. Heinz Schweizer, Dipl. Phys. Uwe Griesinger, and Dipl. Phys. Renate Bergmann (4. Physikalisches Institut, Universität Stuttgart) for making available measurement data on nanostructures and for valuable discussions. We thank Prof. Dr. Wolfgang Eisenmenger and Dr. Bruno Gompf (1. Physikalisches Institut, Universität Stuttgart) for the kindly released video scanning tunneling microscope images of molecular structures. It is a pleasure for the authors to thank Springer-Verlag, especially Dr. Hans J. Kölsch, Jacqueline Lenz, Gisela Schmitt, and Dr. Victoria Wicks, for the excellent cooperation. This cooperation has guaranteed a fast and smooth passing of the project. Last but not least we would like to thank Dorothee Klink, Klink translations (English/German and German/English translations), for helpful proofreading and for translations of parts of the text.

# Table of Contents

# List of Symbols

In this book, operators are indicated by $\wedge$ (e.g. $\hat{A}$). The symbol $*$ denotes conjugate complex quantities, and $\dagger$ denotes adjoint quantities. For vectors of physical quantities (or of operators), bold italic letters are used (e.g. $\boldsymbol{A}$ or $\hat{\boldsymbol{A}}$, respectively), where the components are indicated by indices (e.g. $A_i$ or $\hat{A}_i$ are components of $\boldsymbol{A}$ or $\hat{\boldsymbol{A}}$). In the case of matrices (or tensors), letters without serifs are used (e.g. A). Greek letters are exceptions: in this case, matrices (tensors) are indicated by parantheses (e.g. $(\Omega_{ij})$). Components of $SU(n)$ vectors (matrices) are specified by calligraphic letters (e.g. $\mathcal{A}_i$ or $\mathcal{A}_{ij}$, respectively). $SU(n)$ vectors and matrices are also indicated by parentheses (e.g. $(\mathcal{A}_i)$ or $(\mathcal{A}_{ij})$). The $SU(n)$ vector $\boldsymbol{\Gamma}$ with components $\Gamma_i$ represents an exception. The corresponding $SU(n) \otimes SU(n)$ matrix is written as $(\Gamma_{ij})$.

For the Heisenberg picture the symbol (H) is used as an upper index (e.g. the operator $\hat{A}^{(\mathrm{H})}$ is an operator in the Heisenberg picture). The interaction picture is indicated by the additional index (i) (e.g. the operator $\hat{A}^{(\mathrm{i})}$ represents an operator in interaction picture).

Blackboard bold letters are used if a vector space is considered, e.g., a Hilbert space is written in the form $\mathbb{H}$.

The main symbols are presented in the following list. The Greek symbols are at the end of the list.

| | |
|---|---|
| $\hat{1}$ | Unit operator |
| $\hat{a}^+, \hat{a}$ | Creation and annihilation operator |
| $\boldsymbol{a}$ | Direction of polarization measurement |
| $\hat{A}$ | Operator |
| $\boldsymbol{A}, A_i$ | Vector of eigenvalues, eigenvalue |
| $|A_i\rangle, |i\rangle$ | Eigenstates (discrete) |
| A, $\left\langle i \left\| \hat{A} \right\| j \right\rangle = A_{ij}$ | Matrix, matrix element |
| $\left\langle \hat{A} \right\rangle = \left\langle \psi \left\| \hat{A} \right\| \psi \right\rangle = \overline{A}$ | Expectation value |
| $(\mathcal{A}_i), \mathcal{A}_i$ | $SU(n)$ vector, element of |
| $(\mathcal{A}_{ij}), \mathcal{A}_{ij}$ | $SU(n) \otimes SU(n)$ matrix, element of |
| $\hat{A}^{(\mathrm{H})}(t)$ | Heisenberg operator |
| $\left( \mathcal{A}_i^{(\mathrm{H})}(t) \right), \mathcal{A}_i^{(\mathrm{H})}(t)$ | $SU(n)$ vector (Heisenberg picture), element of |

| | |
|---|---|
| $\hat{A}^{(i)}(t)$ | Operator in interaction picture |
| $\boldsymbol{b}$ | Direction of polarization measurement, vector quantity |
| $C(n,q)$ | Trace of $\hat{\rho}^q$ in $SU(n)$ |
| $\hat{C}$ | Casimir operator |
| $C_{\mathrm{F}}, C_{\mathrm{R}}$ | Coulomb coupling constants |
| $e$ | Electron charge |
| $\boldsymbol{e}_i$ | Unit vector |
| $\boldsymbol{E}$ | Electric field vector |
| $E_i$ | Eigenvalues of the Hamilton operator |
| $E_{ij} = E_i - E_j$ | Transition energy |
| $f_{ijk}$ | Structure constants |
| $F$ | Free energy |
| $g$ | Coupling to the electromagnetic field |
| $g_{\mathrm{s}}$ | Landé $g$ factor |
| $\hat{G}$ | Super operator |
| $G_{ij}(\tau)$ | 2-time-correlation function |
| $G_{ij}(\omega)$ | Spectral density |
| $\hbar = h/2\pi$ | Planck's constant |
| $\hat{H}$ | Hamilton operator |
| $(\mathcal{H}_i)$ | Vector representation of a Hamilton operator in $SU(n)$ |
| $(\mathcal{H}_{ij})$ | Matrix representation of a Hamilton operator in $SU(n) \otimes SU(n)$ |
| $\hat{H}^{(\mathrm{H})}(t)$ | Hamilton operator (Heisenberg picture) |
| $\hat{H}^{(i)}(t)$ | Hamilton operator (interaction picture) |
| $\mathbb{H}$ | Hilbert space |
| $\mathrm{i}$ | Imaginary unit ($\mathrm{i} = \sqrt{-1}$) |
| $I$ | Intensity |
| Im | Imaginary part |
| $k, k_{\mathrm{B}}$ | Proportional constant, Boltzmann constant (do not confuse the index B with the index B used in the context of operators of a "bath"!) |
| $\boldsymbol{k}$ | Wave vector |
| $\mathsf{K}, K_{ij}(1,2)$ | Correlation tensor of second order, component of |
| $K\begin{pmatrix} 1 & 2 \\ i & j \end{pmatrix}$ | Alternative notation for components $K_{ij}(1,2)$ |
| $\hat{K}$ | Statistical operator |
| $\hat{L}_i$ | Angular momentum operator |
| $\hat{\boldsymbol{L}} = (\hat{L}_x, \hat{L}_y, \hat{L}_z)$ | Vector operator of angular momenta |
| $\mathsf{L}, L_{ijk}$ | Correlation tensor of third order, element of |
| $\hat{\mathfrak{L}}$ | Lindblad operator |
| $m_0, m_{\mathrm{e}}$ | Particle mass, electron mass |
| $M_i$ | Magnetic moment |

| | |
|---|---|
| $M_{ij}$ | Element of correlation tensor proper |
| $n$ | Dimension of Hilbert space |
| $p_i$ | Probability for state $i$ |
| $\boldsymbol{P}, P_i$ | Polarization vector, component of |
| $\hat{P}_{ij} = \lvert i \rangle \langle j \rvert$ | Projection operator |
| $\hat{\boldsymbol{p}} = (\hat{p}_x, \hat{p}_y, \hat{p}_z)$ | Momentum operator |
| Q | Transformation matrix |
| $r$ | Rank of a group |
| $\boldsymbol{r}, \boldsymbol{R}$ | Position vector, position vector of centre of mass (COM) |
| $\mathsf{R}_\phi$ | Rotation matrix |
| $\hat{R}, \hat{R}(\nu)$ | Reduced density operator, reduced density operator of subsystem $\nu$ |
| $\mathsf{R}, R_{ij}$ | Matrix representation of reduced density operator, element of |
| $\tilde{\mathsf{R}}, R_{ijkl}$ | Relaxation matrix, element of |
| $\hat{R}^{(i)}(t)$ | Reduced density operator (interaction picture) |
| Re | Real part |
| $s = n^2 - 1$ | Number of generating operators |
| $S$ | Entropy |
| $t$ | Time coordinate |
| $T$ | Absolute temperature, time period |
| $T_1, T_2$ | Longitudinal and transverse relaxation time |
| Tr | Trace |
| $\text{Tr}_i$ | Trace operation in subsystem $i$ |
| $\hat{u}_{12}, \hat{v}_{12}, \hat{w}_1$ | Generating operators in $SU(2)$ |
| $u_{12}, v_{12}, w_1$ | Expectation values of generating operator in $SU(2)$ |
| $U$ | Internal energy |
| $\hat{U}$ | Operator of unitary transformation |
| $\mathsf{U}, U_{ij}$ | Unitary matrix, matrix elements of |
| $\hat{V}, \hat{V}^{(i)}(t)$ | Interaction operator, interaction operator in the interaction picture |
| $\mathbb{V}$ | Euclidean vector space |
| $W_{ij}$ | Transition probability from state $j$ into state $i$ |
| $Z$ | Partition function |
| $\alpha$ | Angle, coefficient in a superposition, phase, index |
| $\beta$ | Angle, coefficient in a superposition, index |
| $\beta_1 = 1/(k_B T)$ | Thermodynamic parameter, $T$ = temperature |
| $\beta_\nu \ (\nu = 1 \dots b)$ | Lagrangian parameter |
| $\gamma$ | Damping parameter (off-diagonal) |
| $\Gamma_{ijkl}^{\pm}$ | Damping parameters in the relaxation matrix |
| $\Gamma_i = \mathcal{H}_i / \hbar$ | Element of the $SU(n)$ vector $\boldsymbol{\Gamma}$ |

| | |
|---|---|
| $\Gamma_{ij} = \mathcal{H}_{ij}/\hbar$ | Element of the $SU(n) \otimes SU(n)$ matrix $(\Gamma_{ij})$ |
| $\delta = \omega_{ij} - \omega$ | Detuning parameter of laser with frequency $\omega$ |
| $\delta_{ij} = \begin{cases} 1: & i = j \\ 0: & i \neq j \end{cases}$ | Kronecker delta |
| $\nabla$ | Nabla |
| $\Delta E = E_{i+1} - E_i$ | Energy level spacing |
| $\epsilon_0$ | Dielectric constant |
| $\epsilon_{\mathrm{r}}$ | Relative dielectric constant |
| $\epsilon, \boldsymbol{\epsilon}$ | Electric field, vector of |
| $(\varepsilon_{ijk}), \varepsilon_{ijk}$ | $\varepsilon$ tensor, elements of |
| $\boldsymbol{\eta}, \eta_i$ | Damping vector, component of |
| $\eta_F$ | Measure for the indeterminacy of the experiment $F$ |
| $\eta\begin{pmatrix} \nu \\ i \end{pmatrix}$ | Alternative notation for components $\eta_i(\nu)$ |
| $\Theta$ | Angle in the context of a polarization measurement |
| $\boldsymbol{\lambda}, \lambda_i$ | Coherence vector (generalized Bloch vector), component of |
| $\lambda\begin{pmatrix} \nu \\ i \end{pmatrix}$ | Alternative notation for components $\lambda_i(\nu)$ |
| $\nu$ | Subsystem index, common index |
| $\mu$ | Subsystem index, common index |
| $\mu_{\mathrm{B}}$ | Bohr magneton |
| $(\xi_{ij}), \xi_{ij}$ | Damping matrix, component of |
| $\xi\begin{pmatrix} \nu & \nu \\ i & j \end{pmatrix}$ | Alternative notation for components $\xi_{ij}(\nu)$ |
| $\hat{\rho}$ | Density operator |
| $(\rho_{ij}), \rho_{ij}$ | Density matrix, matrix element |
| $\hat{\rho}^{(\mathrm{H})}$ | Density operator (Heisenberg picture) |
| $\hat{\rho}^{(\mathrm{i})}(t)$ | Density operator (interaction picture) |
| $\hat{\sigma}_i$ | Spin operator |
| $\hat{\boldsymbol{\sigma}} = (\hat{\sigma}_x, \hat{\sigma}_y, \hat{\sigma}_z)$ | Vector of spin operators |
| $(\sigma^x), (\sigma^y), (\sigma^z)$ | Pauli matrices with components $\sigma^i_{jk}$ |
| $\sigma_{ij}(\tau)$ | 2-time-correlation function |
| $\hat{\sigma}^+, \hat{\sigma}^-$ | Creation and annihilation operator (of spin states) |
| $\Sigma$ | System |
| $\tau_{\mathrm{c}}$ | Correlation time |
| $v$ | Subsystem index with $v = 1, 2$ |
| $\varphi$ | Phase |
| $\phi, \boldsymbol{\phi}/\phi$ | Angle of rotation, unit vector of rotation |
| $\vert\phi\rangle, (\vert\phi\rangle)^* = \langle\phi\vert$ | Wave function (vector in Hilbert space) |

| | |
|---|---|
| $\lvert\psi\rangle, (\lvert\psi\rangle)^{*} = \langle\psi\rvert$ | General wave function (vector in Hilbert space) |
| $\lvert\psi^{(i)}(t)\rangle$ | General wave function (interaction picture) |
| $\Psi$ | Scalar field |
| $\boldsymbol{\Psi}$ | Column matrix |
| $\omega = 2\pi/t_0$ | Circular frequency ($t_0 =$ time period), driver frequency |
| $\omega_{ij} = (E_i - E_j)/\hbar$ | Transition frequency |
| $\boldsymbol{\omega}$ | Vector of rotation |
| $(\Omega_{ij}), \Omega_{ij}$ | Rotation matrix (rotation of coherence vector), matrix elements of |
| $\Omega_R$ | Rabi frequency |

# 1. Introduction

## 1.1 Motivation

How do classical properties emerge within a quantum world? How does genuine quantum behaviour survive in a classical environment? While the first, in some sense more fundamental problem has intensively been addressed, for example, within so-called *decoherence schemes* (cf. [155]), the latter question has not yet enjoyed that much attention, let alone systematic inquiry. Shouldn't there exist *design principles* by which quantum phenomena would tend to be enhanced or suppressed, respectively? From design of structure to design of dynamics: is this the way in which R. Feynman's dreams might be coming true: "There is plenty of room at the bottom" (cf. [47])?

There was a time when quantum mechanics, despite disturbing interpretation problems, appeared to be well-established to the extent that exciting new developments would no longer be anticipated. Quantum mechanics was put to work as the obvious tool of choice to solve standard (though often quite demanding) problems like energy spectra, eigenstates, and dynamical properties in terms of transition probabilities. This was the time when condensed matter physicists turned, in part, to potentially more rewarding subjects like *phase transitions* (cf. Fig. 1.1 and [64, 134, 150]), nonlinear classical properties including *soliton theory*, *chaos* (cf. Fig. 1.2), and *pattern formation* (cf. Fig. 1.1 and [64]). One subject of these investigations has been "control", a central theme also of this book.

Quantum mechanics constitutes a theoretical scheme mathematically quite different from its classical counterpart: *operators* acting on wave vectors in *Hilbert space* have no direct classical correspondence. Nevertheless, at the heart of non-classical properties is typically just one specific feature: the *superposition principle*, known also in the classical domain of linear waves. Here, however, combined with *quantization*, it gives rise to *quantum coherence*. In the case of composite quantum systems, coherence within and between the various *quantum subunits* may result. (Following Schrödinger, this coherence is called *entanglement*.)

Coherence as a relational property does not necessarily lead to observable effects. Any pure quantum state can be represented as a superposition of an appropriate set of basis states and, vice versa, any superposition specifies

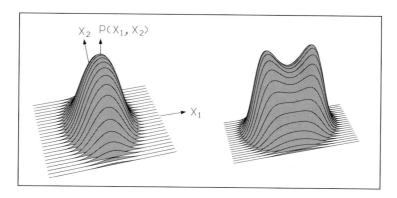

**Fig. 1.1.** Control of macroscopic systems: near critical points the behaviour of physical systems changes dramatically as a function of external control parameters (like temperature or energy supply). Then more or less complicated structural patterns (like ferromagnetic states in ferromagnetic materials or lasing states in laser systems) emerge. The correlated fluctuating behaviour of specific variables (like components of a magnetization or amplitudes of laser modes) is described by a statistical distribution function. On this statistical level, the emergence of structural patterns is represented by the occurrence of new maxima. Here, a special (2-dimensional) probability density $P(X_1, X_2)$ before (l.h.s.) and after phase transition (r.h.s.) is shown (cf. [150])

a pure state. Entanglement between subsystems may alternatively be interpreted as a "simple" superposition in the total state space. It is the way in which the quantum system is embedded into a classical environment which defines the actual reference frame. The necessity of such frames is obvious already in classical mechanics: inertial frames, e.g., are not just abstract math-

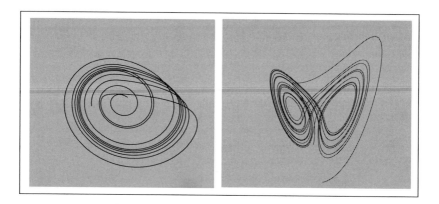

**Fig. 1.2.** Limits of control: we show 2-dimensional projections of a Rössler attractor (l.h.s.) and a Lorenz attractor (r.h.s.). Such chaotic trajectories represent highly complicated but totally deterministic motions which are sensitive with respect to initial conditions

ematical constructions but involve "real" length scales and clocks. Here the frames are more elaborate: they include external fields and contacts as well as controlled or uncontrolled measurements implying *dissipative state changes* and thus tend to limit quantum coherence to finite time scales. This is the price we have to pay for having access to this strange world at all. Thus to enhance quantum phenomena means to look on sufficiently small time scales and length scales, i.e. to study individual quantum objects as long as they are almost isolated. The engineering challenge is to increase pertinent coherence times and coherence lengths in optimized scenarios.

Novel experimental techniques allowing for *high temporal resolution* thus offer new insight. In this way coherence phenomena become accessible even within large ensembles of (non-interacting) quantum objects. *Femtochemistry*, e.g., is able to monitor the transient *coherent dynamics* within molecules after excitation (wave packet motion in the femto-second regime), and even in bulk semiconductors, despite their short *coherence time* (ps-regime), coherent properties are now being detected. Experiments with entangled photon pairs (generated by down-conversion in nonlinear crystals) require only short time scales even if the travelling distance is macroscopically large: the speed of light is conveniently high.

Since the late 1970s, improved structuring techniques (*epitaxy, lithography*) have become available, in particular for technologically important materials like semiconductors: we have been witnessing the gradual change from microfabrication to nanofabrication in the late 1980s. Presently we are in the midst of novel approaches including *scanning tunneling microscopy, lithography* based on *"natural" masks, self-assembly techniques* and *electron-holography*, to name but a few. These techniques essentially make possible synthetic *nanostructures* consisting of hierarchies of structurally well-defined subsystems. In the extreme case these subsystems might appropriately be termed *artificial molecules* in so far as they contribute localized states connected with discrete subspectra. The spectral jumps of individual defects in a matrix, which are induced by uncontrolled micromotion within their respective environment, explicitly demonstrate the importance to study *individual objects* also for dynamics.

*Cavities for photons* can now be built with high enough quality $Q$ to store photons for many milliseconds: superpositions of *photon number states* can be prepared at will. A trailblazing experiment has been the investigation of single ions in so-called *electrodynamic traps* (1986). About the same time single electrons have been prepared in a vacuum chamber. It was only a few years later when the first observation of single defects (*SMD = single molecule detection*) in solid-state matrix was reported, followed in 1994 by the *near field scanning optical spectroscopy* of individual organic molecules and clusters on substrate. Single subsystems are being investigated also in *nanobiology*: single viruses and bacteria, single proteins, motor cells (*quantum biomechanics*), and single-ion channels. Single *electron tunneling* through "islands", controlled by voltage gates, is another pertinent example; also current *fluctuations* induced by the individual or even *cooperative stochastic motion* of near-by defects have been reported. This latter example lends support to the expectation that *cooperativity* does not necessarily end up in well-known bulk properties. (The *molecular design* of bulk properties like nonlinear optical response can be an interesting target in its own right, though.)

These considerations show that quantum mechanics today is just as up-to-date as in the early days, but now with a completely different appeal than in the days of atomic spectra and Planck's radiation law: we are about to learn how to manipulate those tiny quantum objects, by design and by the interface to their environment. Nobody can predict right now how far this fascinating enterprise will take us. The modest observer might expect that those experiments and the supplementing computer simulations will give us a more intuitive understanding of how (and how strangely!) quantum mechanics actually works, also, how classical ensemble properties might emerge. The more daring observer might foresee possibilities to realize versions of *quantum computation*, a recent and very challenging prospect, which would bring *information theory* and physics in close relation already on the very fundamental level (and, at the same time, on the level of applications).

The general theoretical model to cover all these scenarios should be a *network of interacting quantum objects* (e.g., individual particles). In Fig. 1.3 the fundamental structure of such a *quantum network* is sketched. As will be discussed in detail, the quantum nature derives from the *local properties* (described by a special algebraic structure rather than a simple discrete state space which could also be realized by classical attractors) and the complicated *non-local correlations* (which also differ from their classical counterparts).

This book is aimed at the theoretical study of small (coherent) *quantum networks* by means of the *density matrix theory* (cf. [20]): the network node $\nu$ is taken as a finite *local state space* of dimension $n(\nu)$. The network might be a regular lattice or an irregular array of $N$ nodes. We restrict ourselves, in the coherent case, to $n, N \leq 4$. This basic model will already exhibit a large dynamical repertoire when coupled to external (e.g. optical) driving fields and to dissipative channels, which are, *inter alia*, required for measurement. These scenarios include many of the experimental and theoretical modes of dynamical operation, which are currently discussed in the field of nanosystems. Our approach should provide a useful *system-theoretical tool* easily adaptable to situations in which a finite quantum mechanical state space is controlled by a classical environment. This feature becomes apparent if one recognizes that the well-defined parameters entering the network model can be interpreted by any type of physical implementation of that given structure.

For $n$ and/or $N > 4$ we will restrict ourselves to networks in the incoherent ("classical") limit: otherwise the number of equations and the numerical efforts involved increase very rapidly. In fact, the state space (represented by the number of independent components) increases like $n^{2N}$, i.e. exponential

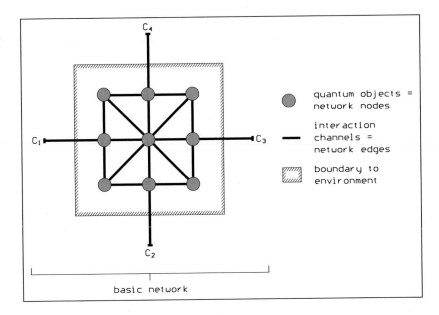

**Fig. 1.3.** Quantum network: the subsystems of the network are special quantum objects denoted as "network nodes". Some interaction channels are indicated. $C_i$ ($i = 1, 2, 3, 4$) denotes local interactions with the "outside world" or other networks

in the network size $N$. As already noted by R. Feynman, the simulation of quantum systems is thus computationally a hard problem, at least on this microscopic level. A method to describe high-dimensional networks (quantum networks and "classical" networks) on the basis of Feynman path integrals and self-similar power series functions was considered in [150]. A pathway to include high-dimensional correlations was studied. However, this requires a difficult mathematical treatment which would go beyond the scope of this book. Moreover, the emphasis is here on the density matrix theory and not on Feynman's approach to quantum systems.

Chapter 2 introduces the *SU(n) algebra* as an appropriate language for dealing with networks (cf. [69, 39]). *Ensemble dynamics* is covered in Chap. 3 for closed as well as open systems. Chapter 4 is devoted to a study of individual *small networks* with emphasis on *stochastics* and *information retrieval*. This approach goes beyond the conventional density matrix theory (*master equations*); it necessarily rests on quantum measurement models. Chapter 5 finally presents a short summary with special consideration of the main topics.

We continue this chapter with some introductory remarks on the implementation of $SU(n)$ *networks*. These may consist of localized spins proper, but more typically of artificially confined systems such as electrons, phonons, photons, atoms, etc. (One should note that for the communication with the outside world we also need unconfined systems like atom beams or photon beams.) We then close with a tutorial on fundamental quantum mechanics.

## 1.2 Confined Electrons: Nanostructures

Confinement is a classical concept: it refers to the pertinent parameter fields ("structure") controlling a particular (dynamical) quantum field (like electrons). It is by no means obvious that such parameter fields exist at all (cf. [136]): fortunately, it is very often a reasonable approximation to start from such static (symmetry breaking) parameter fields (neglecting any fluctuations or treating them as perturbations: phonons).

A really fascinating field of modern research is *nanostructures* (cf. [79]). Nanostructures can be characterized as *assemblies of quantum objects* in the *nanometer region* ($1\,nm = 10^{-9}\,m = 10\,\text{Å}$). On the basis of such structures, *technical devices* with extremely small extension can be constructed. Examples are *quantum dots* in semiconductor materials (cf. [120, 132]), *quantum corrals* (cf. [30]), or *quantum wires*. In Figs. 1.4, 1.5 electron microscope images of $300\,nm$ wide gallium indium arsenide (GaInAs) wires – surrounded by

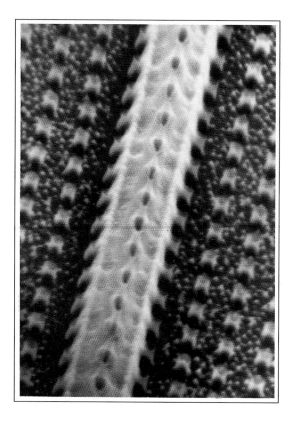

**Fig. 1.4.** Electron microscope image of a $300\,nm$ wide wire with a chain of antidots in the middle of the wire. Kindly provided by H. Schweizer, R. Bergmann (cf. [14])

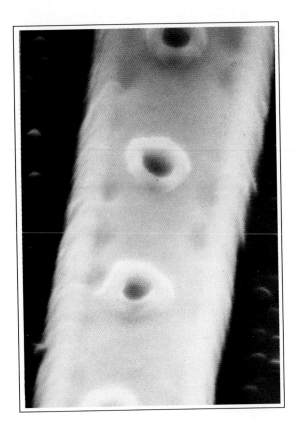

**Fig. 1.5.** Electron microscope image of individual antidots in a 300 nm wide wire. Kindly provided by H. Schweizer, R. Bergmann (cf. [14])

indium phosphide (InP) – are shown. The visible surface of these wires consists of InP. Periodically occurring antidots create an additional structuring of the wire (cf. [14]). Such a wire allows well-defined one-dimensional electrical conduction (within the GaInAs part) near helium temperature (4.2 K).

Nanostructures can be fabricated starting from *organic molecules* (a type of *bottom-up approach*) or from bulk semiconductor materials (*top-down approach*). Either way has its advantages and disadvantages; examples of the former are shown in Figs. 1.6, 1.7. In this section we focus on the semiconductor top-down scheme. This scheme is based on spatial confinement in one dimension (layers), two dimensions (wires), and three dimensions (dots).

The further reduction of size of *semiconductor devices* (like transistors, diodes, and complete *integrated circuits*) is an important technological aim, because the reduction of the size of such devices supports a faster and denser type of electronics. Modern nanotechnology allows such a reduction almost down to the nanometer region.

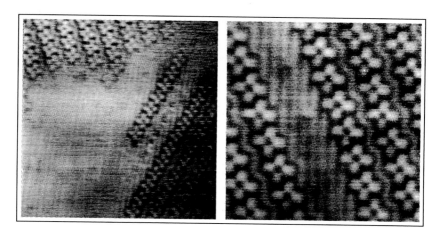

**Fig. 1.6.** Rows of copper-phthalocyanine molecules (CuPHTH molecules) on molybdene-disulfide ($MoS_2$), 200 Å × 200 Å. A video-STM (STM = scanning tunneling microscope) image produced by a working team at the Universität Stuttgart (W. Eisenmenger, B. Gompf, Ch. Ludwig, et al., 1993, cf. [88, 89])

## 1.2.1 Fabrication

The fabrication of semiconductor nanostructures is, by itself, an example for controlling essentially quantum dynamical processes by macroscopic constraints. It typically involves *self-organization* (like crystal growth) and direct manipulation. Such steps will now be discussed.

After epitaxial growth of layers on a substrate, and after applying an additional electron sensitive resist, a resist mask can be created with the help

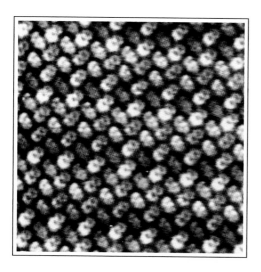

**Fig. 1.7.** PTCDA monolayer on graphite (100 Å × 70 Å). This image was produced by a video-STM working team at the Universität Stuttgart (W. Eisenmenger, B. Gompf, Ch. Ludwig, et al., 1993, cf. [88, 89])

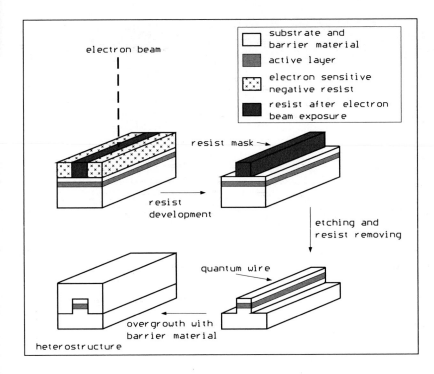

**Fig. 1.8.** The fabrication of a quantum wire by electron beam lithography

of an electron beam (*electron beam lithography*): following the electron beam exposure, the sample with the resist is baked and afterwards removed by a *solvent*. After this process, only the part exposed to the electron beam (negative resist) or the part not exposed to the beam (positive resist) remains, i.e. a resist mask is created. A subsequent etching process then leads to a structuring defined by the resist mask. After overgrowth with barrier material, one obtains a compact *heterostructure* with enclosed active parts. (In the wires shown in Figs. 1.4, 1.5 an active part is represented by the inner – in the images invisible – GaInAs part of a resulting wire.) In Fig. 1.8 the negative resist principle is detailed for the fabrication of a single quantum wire with such an *internal substructure*. In the same way, quantum-dot arrays can be fabricated, and also antidots in quantum wires.

### 1.2.2 Characterization Methods

Methods to locally examine the structure of quantum objects in the nanometer region include in particular electron microscopy and atomic force microscopy. For example, Figs. 1.4, 1.5 show electron microscope images of single quantum wires, and Figs. 1.9–1.11 represent *atomic force microscope* (AFM) images of various kinds of nanostructures.

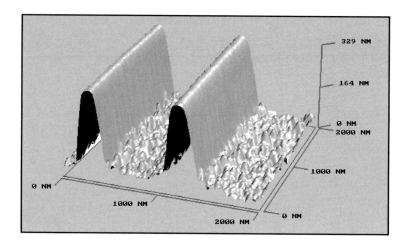

**Fig. 1.9.** Two parallel wires. AFM image of the surface. Kindly provided by H. Schweizer, U. Griesinger (cf. [142])

In Fig. 1.9 a first example of such a structure is shown, in which the surface of two parallel nanoscale wires is scanned by an AFM. Electron microscope images of single wires of this kind – though, with additional antidots – have been given in Figs. 1.4, 1.5.

It is remarkable that ordered structures with various inherent length scales ("hierarchical structures") can thus be generated. An example with effective *lattice constant* below 460 nm is shown in Fig. 1.10. Each of the wave maxima specifies the position of a wire. Also networks of quasi-zero dimensional subsystems ("boxes" or "dots") can be fabricated: an example is represented

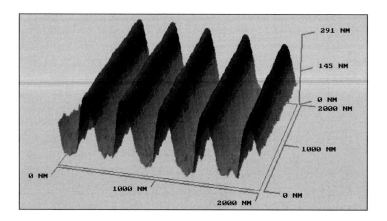

**Fig. 1.10.** Structure of wires with lattice constant < 460 nm. AFM image kindly provided by H. Schweizer, U. Griesinger (cf. [142])

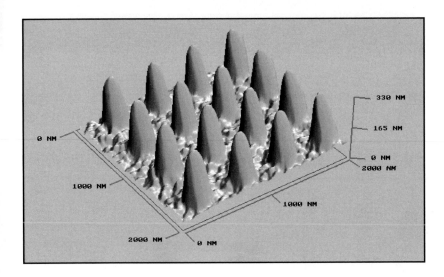

**Fig. 1.11.** Dot array observed with an AFM. Kindly provided by H. Schweizer, U. Griesinger (cf. [142])

by a so-called *quantum-dot array* (cf. Fig. 1.11). Such an array represents a lattice of artificial dots (the network "nodes"). The dots can have further internal substructures (again a hierarchy of various inherent length scales). The properties of the individual nodes or of node clusters can be adjusted (for example, by external fields or contacts).

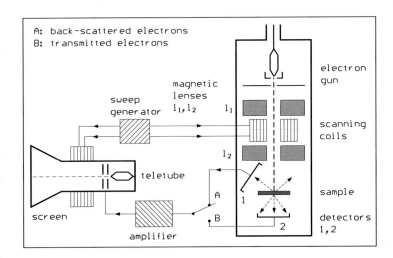

**Fig. 1.12.** A schematic representation of an electron microscope

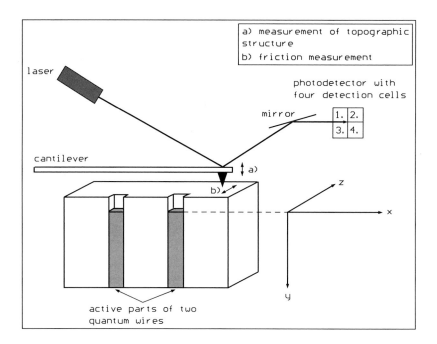

**Fig. 1.13.** A schematic representation of atomic force microscopy applied to a 2-wire system

In Fig. 1.12 a schematic representation of an electron microscope is shown. As illustrated, an electron beam is used to scan over the sample. The back-scattered (or transmitted) electron intensity is measured so that one obtains information about the sample surface. Figures 1.4, 1.5 show images obtained by a *scanning electron microscope*, i.e. images produced by back-scattered electrons.

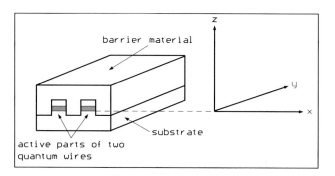

**Fig. 1.14.** Side-view of two burried wires

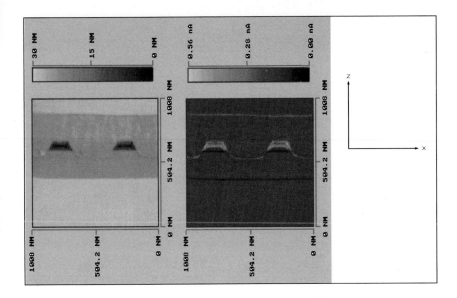

**Fig. 1.15.** Structure of 2 wires observed by an AFM (side view). L.h.s.: structure as seen by an AFM topography measurement. R.h.s.: structure as seen by an AFM friction measurement. Kindly provided by H. Schweizer, U. Griesinger (cf. [142])

In Fig. 1.13 the principle of atomic force microscopy is illustrated by a sectional view of the heterostructure shown in Fig. 1.14. With an AFM, the topographic structure as well as the friction of a surface can be measured. In the example shown in Fig. 1.13 two geometric levels (the level of the cross-section of the active wire parts and the level of the barrier material) are considered. A respective measurement result is shown in Fig. 1.15. There, topography and friction of the sectional view is illustrated. While the image in Fig. 1.10 shows wire surfaces, the image in Fig. 1.15 represents the internal structure in side view. In contrast to Fig. 1.10, Fig. 1.15 shows wires overgrown with barrier material.

## 1.2.3 From Structure to Dynamics: Energy Spectra

The motivation for all those structuring procedures is to implement new dynamical modes, if possible, to enhance quantum behaviour. Basic to the understanding of the (electron) dynamics is the energy spectrum and the localization of the respective eigenstates, as they result from the underlying structure ("size quantization").

Bulk semiconductors are characterized by a quasi-continuum of states separated by a *band gap*: this gap reaches from the *valence band edge* $E_{VB}$ (the highest occupied states at zero temperature and zero doping) to the *conduction band edge* $E_{CB}$ (the lowest unoccupied electron state). The position of these two energy markers varies from material to material so that

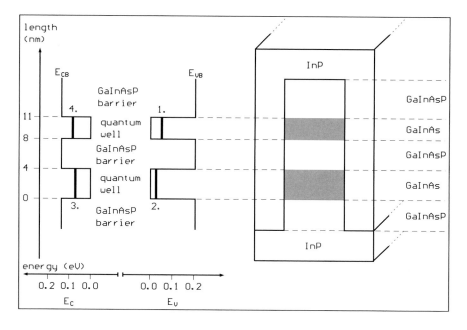

**Fig. 1.16.** Bound states of 2 different semiconductor quantum dots. On the r.h.s the structure is sketched. On the l.h.s. the corresponding valence band edge, $E_{VB}$, and the conduction band edge, $E_{CB}$, with quantum wells and GaInAsP barriers are shown. 1., 2., 3., 4. represent electronic eigenstates

in a heterostructure they form *potential wells* and *barriers*, which, in general, will be modified by *charge redistribution*. In the so-called *effective mass approximation* the electrons in either band are treated as *free particles* with an effective mass; in the electron–hole picture the occupied *valence band* is interpreted as an empty *hole band* (with energies increasing downwards from the valence band edge) so that excitations can be specified as the creation of an *electron–hole pair*. The electron states (and hole states) of the heterostructure then follow from an effective *one-particle Schrödiger equation* for effective mass-particles moving in the potential $E_{CB}(\boldsymbol{r})$ (and $E_{VB}(\boldsymbol{r})$). This simplified picture works surprisingly well and demonstrates how the known bulk properties are used to derive novel properties of the nanostructure.

It is a textbook example to show how potential wells lead to discrete bound states. In the present case, however, we have to consider all three orthogonal spatial directions. Confinement in one direction then gives rise to bound states which are superimposed by the continua arising from the still unconfined directions, so-called *subbands*. If the subband splitting is large enough to address only the lowest subband, we speak of a *quantum layer* (or of a "quantum film"), for a system confined in 2 directions of a *quantum wire* (cf. [14]). Only confinement in all 3 directions leads to a discrete spectrum

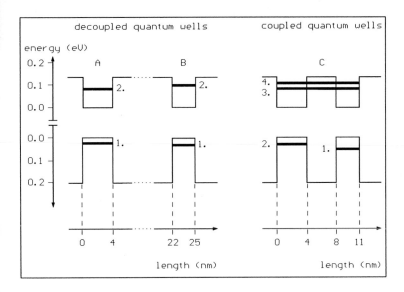

**Fig. 1.17.** Bound states of coupled (r.h.s) and uncoupled (l.h.s.) dots. This may be seen as a transition from two 2-level nodes (A, B) to one 4-level node (C). 1., 2., 3., 4. represent electronic eigenstates

(below a certain limiting energy) (cf. [24, 120, 132]). Two such *quantum dots* are sketched in Fig. 1.16.

By geometrical design the nature of the bound states can be changed: as the barrier width is reduced, the wave functions within either dot may start to overlap forming states delocalized over the whole double-well potential. The resulting system is then better seen as one quantum object with differently localized states. This is exemplified in Fig. 1.17. We may speak of an effective 4-level node. A similar situation may be obtained with a wire structure. Noncrystalline materials also offer size-quantization effects. The preparation of colloidal semiconductor material has been reported with particle diameters in the range of 5 nm. Even composite particles of HgS and CdS, for example, can be manufactured (cf. [151]). Surface states of metals have been shown to become "pinned" within so-called *quantum corrals* (radius ≈ 7 nm) made of individually positioned atoms (cf. [30]).

## 1.2.4 Optically Driven Nanostructure: DFB Laser

A technologically relevant application of nanostructures is represented by the so-called *distributed feedback laser (DFB laser)*. Based on coupled 2-layer (or multi-layer) quantum wires, a resonator is constructed, which, for example, can be pumped into the continuum by an excitation laser to create, after relaxation, a *population inversion* between the upper and the two lower subbands (cf. Fig. 1.17). Then, between the active parts of a single wire

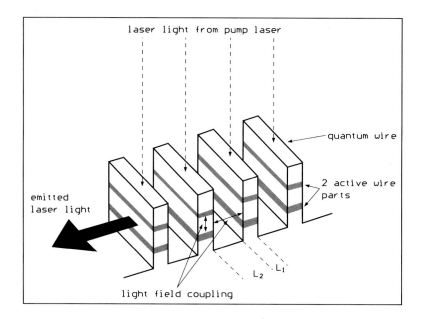

**Fig. 1.18.** A system of quantum wires homogeneously pumped by an external laser field. $L_1 \approx 60\,\mathrm{nm}$, $L_2 \approx 120\,\mathrm{nm}$. Each quantum wire consists of 2 active parts giving rise to a 3-subband system with a density of transition frequencies peaked at $\omega_{13}$, $\omega_{23}$. Using a suitable lattice constant $G = L_1 + L_2$, the structure can be made to act as a distributed feedback resonator (DFB resonator) for $\omega_{13}$ and/or $\omega_{23}$

and between the different wires, interactions via the light field evolve. If the pump energy reaches a critical value, a coherent light field emerges, i.e. a *lasing state*. In Fig. 1.18 this situation is depicted. (Also see Fig. 1.1: the phase transition would consist here in the transition from a non-lasing to the lasing state.)

## 1.3 Confined Photons: Cavity Electrodynamics

The manipulation of a mode structure of a given field by changing its boundary conditions is by no means restricted to the Schrödinger field describing electrons in vacuum or effective particles within condensed matter. In fact, there are many analogies between the various *geometries* encountered in the low-dimensional electron physics in semiconductor nanostructures and states on *ring-shaped molecules* with modes of the approximately *confined electromagnetic field*. Qualitatively, all the various effects known for the electrons (like modified density of states, various localization patterns of the modes) find their counterparts in *confined photon physics*. Confinement, of course, requires material walls specified by definite dielectric properties. We will briefly discuss some typical scenarios and their relations to network theory.

### 1.3.1 Mirror Gaps

It was recognized long ago that the modes of the electromagnetic field confined by metallic boundaries (plates) change dramatically and become dependent on *geometrical shape* and *size* of this confinement (cf. [96, 152]). The simplest geometry is the 2-dimensional layer geometry, approximately realized by 2 parallel mirrors. The mode density $\rho(\omega)$ becomes highly anisotropic (with respect to field direction $\epsilon$) and furthermore depends on distance from the mirrors. Only the density of *photon modes* with electric field parallel to the mirrors is strongly affected by confinement. The properties of that anisotropic field can be tested by letting excited atoms interact with the resulting vacuum state. By applying an additional magnetic field to the atoms, one can change their radiation characteristics: if the excited atoms are made to couple mainly to the photon modes with the electric field parallel to the mirrors, the decay can be suppressed.

### 1.3.2 Ring Cavities

For a *ring cavity* (as the analogue of a ring molecule) the photon modes can be classified, inter alia, by their propagation direction (clockwise (cw) and counter-clockwise (ccw)) and/or polarization. Both constitute effective 2-state subspaces, equivalent to a pseudospin 1/2, for every mode $k_i$ fitting the ring circumference. Due to this analogy, driven ring cavities have been termed *optical* (2-level) *atoms*. (This system is described in detail in [133].)

So-called *whispering gallery modes* (WGMs) with high quality factors $Q$ have been demonstrated in quartz microspheres. (The term *whispering gallery mode* originates in the analogy with acoustic modes in large auditoriums.) In these optical modes, light circulates in a thin annular region near the equator, just inside the surface of the sphere. The evanescent component can be used for interactions which reach into the outside world, even to induce trapping (cf. [90]).

### 1.3.3 Box Cavities

For microwave fields, sophisticated *superconducting cavities* exist which allow the "trapping" of photons up to the $10^{-1}$ s scale. Unfortunately, a superconducting cavity does not exist at optical frequencies: a metal-clad optical cavity suffers from a large absorption loss. Here other designs are being discussed which make use of semiconductor nanostructures (e.g. planar dielectric *microcavity structures*). A dielectric 3-dimensional periodic structure has been proposed to realize a stop-band at optical frequencies.

Of special interest are weak fields consisting of few photons. Such fields are too small to be detected directly. They are injected and measured by excited atoms being sent through the cavities (atom beams). These beams thus play a similar role as light beams for the excitation of localized matter systems.

The dramatic progress in experiments has been triggered by the introduction of *frequency-tunable lasers*, which can prepare large populations of highly excited atomic states, so-called *Rydberg states* (cf. [148]). As the induced transition rates between neighbouring levels scales as $n^4$, these high-$n$ Rydberg states are very strongly coupled to the radiation field. They have long life-times with respect to spontaneous decay.

If one can restrict the field states to zero-photon and 1-photon states (*Fock states*), we, again, have an effective 2-state subspace. It has been shown that coherent superpositions are possible, and even entangled states between a cavity mode and a 2-level atom. The sharing of a single photon between two cavities is also possible: this is a first step towards *cavity networks*.

## 1.4 Confined Atoms: Electrodynamic Traps

Precision spectroscopy on atoms is limited by *Doppler effects* and by *atomic collisions*. The observation of single atoms localized in space would therefore be of considerable interest. This has become possible by placing ions in electromagnetic or electrodynamic traps (*Paul traps*) (cf. [72]). Such a trap is shown in Fig. 1.19. The most useful types are based on an axially symmetric electric field with the shape of a quadrupole. In the Paul trap an rf electric field between the cap electrodes and cylindrical ring electrodes drives the ions in small orbits ("micromotion"). These orbits go to zero at the trap centre, where the electric field has a node. The corresponding kinetic energy may be reinterpreted as a pseudo-potential, which gives rise to a radial harmonic force towards the centre. Tight confinement requires fast and efficient removal of kinetic energy: this is accomplished by "laser cooling".

A single atomic Barium ion was prepared and observed for the first time in 1979. In principle, this scenario is similar to the "trapping" of defects within a solid-state matrix (see Fig. 1.19 and [3, 108]). This situation has also been verified experimentally. Single optically driven 3-level atoms may exhibit a luminescence signal jumping between light and dark periods. Experiments of this type have significantly increased the interest in stochastic modelling of single quantum objects.

Even though the trap is certainly macroscopic (because the ion is – via the electromagnetic field – eventually bound to the earth; this setup has sometimes been called *geonium*), the local properties of the single quantum objects need to be described by quantum dynamics in a finite space.

This trap can also be used to store several ions. While these ions do not interact in terms of their internal degrees of freedom (excitation spectrum), their translational motion is coupled via Coulomb forces. This gives rise to equilibrium configurations. (The average ion–ion distance is orders of magni-

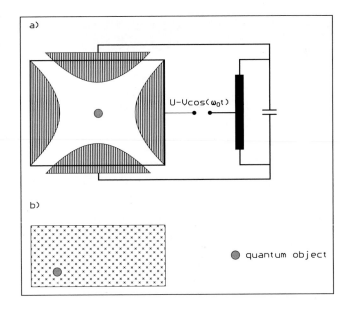

a)

$U-U\cos(\omega_0 t)$

b)

● quantum object

**Fig. 1.19.** Quantum object in a Paul trap (part a)) and in solid matrix (part b))

tude larger than that in an ion crystal, say.) Elongations from this equilibrium state can be described by phonon modes (like in the solid-state counterpart).

## 1.5 Applications: Present and Future

There are already a number of applications based on confined quantum fields; in particular, devices using semiconductor microstructures and nanostructures. For example, DFB lasers can be used as energy sources in the context of *information transmission* via *optical waveguides*. In quantum cascade lasers electrons are streaming down a staircase sequentially emitting photons at each step. The steps consist of coupled quantum wells, in which population inversion is achieved by control of tunneling. The wavelength is entirely determined by quantum confinement ([43]). Semiconductor structures with internal lengths down to the nanostructure scale allow the implementation of electronic circuits: an example of such a *micro-chip* is shown in Fig. 1.20.

Electronics based on the manipulation of single electrons is being discussed: it typically involves tunneling controlled by appropriate point contacts. Kinds of "wave electronics" can be envisioned (cf. [32]), which would try to exploit the interference of two single-particle paths as does, for example, a Mach-Zehnder interferometer in optics. As, contrary to photons, electrons are fermions, and as there is a strong Coulomb interaction between them, this optical analogue is perfect only on the single-electron level. On

**Fig. 1.20.** An example of a nanostructure micro-chip fabricated by R. Bergmann, 1994

the other hand, requiring that only one electron be active within a coherent subsection of a network appears to be a very severe and impractical condition. Alternative architectures would involve network modes with a fixed electron number, i.e., excluding charge transfer. In this way one can include many-body effects in the local spectrum (cf. [140]), while interactions would be restricted to Coulomb forces. Such quantum-dot arrays might be used as prototype structures for nanoscopic *parallel computing architectures* like *cellular automata*, with local rules based on quantum mechanical processes ([17, 81, 104, 140]).

Pseudospin networks ("spin glasses") have long since been under intense investigation as a kind of representation of so-called *neural nets*. Up to now these systems have been studied exclusively in the classical limit (*Ising spin limit*) so that there was no need for an actual quantum mechanical implementation. Under appropriate conditions their relaxation dynamics can be interpreted as *pattern recognition* ("associative memory").

A completely new type of application based on the superposition principle of quantum mechanics is the so-called *quantum cryptography*. Schemes involving photon pairs as well as single photons have been proposed. Meanwhile there are already experimental realizations (cf. [137]). Further challenging applications could lead towards quantum computation and quantum communication (cf. [22, 36]).

## 1.6 Fundamentals

In this section, a brief overview of the fundamental methods of *quantum theory* will be presented. Fundamental aspects of *group theory* and their relation to quantum theory are given. Special terms which are often used in the book in hand are considered in detail.

### 1.6.1 Operators in Hilbert Space

The fundamental mathematical terms relevant for the book in hand will now be explained.

**1.6.1.1 Hilbert Space.** A set of elements $f, g, \ldots$ "spans" a *Hilbert space* $\mathbb{H}$ if the following conditions are fulfilled:

*Condition 1.* $\mathbb{H}$ is a linear space, i.e. if $f$ and $g$ are elements of $\mathbb{H}$, then also $f + g$ and $\zeta f$ ($\zeta \ldots$ any complex number) are elements of $\mathbb{H}$.

*Condition 2.* Between two elements $f, g \in \mathbb{H}$ a *scalar product*

$$(f, g) = c \quad (c \ldots \text{ complex number}) \tag{1.1}$$

is defined. This scalar product has the following properties:

$$(f, \zeta g) = \zeta (f, g) , \tag{1.2}$$

$$(f_1 + f_2, g) = (f_1, g) + (f_2, g) , \tag{1.3}$$

$$(f, g_1 + g_2) = (f, g_1) + (f, g_2) , \tag{1.4}$$

$$(f, g) = (g, f)^* , \tag{1.5}$$

$$(f, f) \geq 0 . \tag{1.6}$$

($*$ denotes a conjugate complex number, and $\zeta$ represents any complex number.)

*Condition 3.* $\mathbb{H}$ is complete, i.e. the completeness relation

$$(f, f) = \sum_i (f, g_i) (g_i, f) \tag{1.7}$$

holds for any $f$.

**1.6.1.2 Operators.** Consider elements $f$, $g$ of an $n$-dimensional Hilbert space $\mathbb{H}^n$. Any rule $\hat{A}$ which relates an element $f$ to another element

$$g = \hat{A} f \tag{1.8}$$

is called an *operator* (in this Hilbert space).

*Linear Operators.* An operator $\hat{A}$ is *linear* if the relation

$$\hat{A}\left(\zeta_1 f + \zeta_2 g\right) = \zeta_1 \hat{A} f + \zeta_2 \hat{A} g \tag{1.9}$$

holds. ($\zeta_1$, $\zeta_2$ are complex numbers.)

*Inverse Operators.* Consider the relation

$$\hat{A} f = g \ . \tag{1.10}$$

An operator is called the *inverse operator $\hat{A}^{-1}$* with respect to $\hat{A}$ if

$$\hat{A}^{-1} \hat{A} f = \hat{1} f = f = \hat{A}^{-1} g \tag{1.11}$$

($\hat{1}$ is the unit operator).

*Adjoint Operators.* The adjoint operator $\hat{A}^\dagger$ of an operator $\hat{A}$ is defined by

$$\left(\hat{A} f, g\right) = \left(f, \hat{A}^\dagger g\right) \ , \tag{1.12}$$

where the parantheses indicate scalar products.

*Hermitian (self-adjoint) Operators.* The operator $\hat{A}$ is a *Hermitian (self-adjoint) operator* if

$$\hat{A}^\dagger = \hat{A} \ . \tag{1.13}$$

*Unitary Operators.* An operator $\hat{U}$ is *unitary* if

$$\hat{U} \hat{U}^\dagger = \hat{1} \tag{1.14}$$

and thus

$$\left(\hat{U} f, \hat{U} g\right) = (f, g) \tag{1.15}$$

holds, i.e. a *unitary transformation* leaves scalar products unchanged.

**1.6.1.3 Eigenvalue Equations.** Consider elements $f_i$ and operators $\hat{A}$ defined in Hilbert space $\mathbb{H}$. Then the equation

$$\hat{A} f_i = a_i f_i \tag{1.16}$$

is called *eigenvalue equation* of $\hat{A}$ (in $\mathbb{H}$), where $a_i$ represents possible *eigenvalues*, and $f_i$ *eigenvectors*. Hermitian operators have real eigenvalues.

Eigenvectors are *orthonormal* if

$$(f_i, f_j) = \delta_{ij} \tag{1.17}$$

holds. Orthonormal systems of eigenvectors consist of eigenvectors which are both orthogonal (i.e. $(f_{i \neq j}, f_j) = 0$) and normalized (i.e. $(f_i, f_i) = 1$).

**1.6.1.4 Dirac's Bra and Ket Notation.** In the book in hand, state vectors and scalar products are formulated on the basis of *Dirac's bra and ket notation*, i.e. state vectors of a Hilbert space $\mathbb{H}$ are denoted by $|f\rangle$, $|g\rangle$, etc., and $\langle f|$, $\langle g|$ denote conjugate state vectors. The respective scalar products are written as

$$(f, g) = \langle f | g \rangle \; , \tag{1.18}$$

$$\left(f, \hat{A}g\right) = \left\langle f \left| \hat{A} \right| g \right\rangle \; , \tag{1.19}$$

where

$$\langle f | g \rangle = \langle g | f \rangle^* \; . \tag{1.20}$$

Based on such a notation, operators can be represented in terms of *projection operators*. For example, the operator $\hat{P}_{ff}$ in

$$\hat{P}_{ff} | g \rangle = | f \rangle \langle f | g \rangle \tag{1.21}$$

can be represented by

$$\hat{P}_{ff} = | f \rangle \langle f | \; , \tag{1.22}$$

and the completeness relation (1.7) by

$$\hat{1} = \sum_i | g_i \rangle \langle g_i | = \sum_i \hat{P}_{g_i g_i} \; . \tag{1.23}$$

### 1.6.2 Basic Group Theory

An essential tool for the analysis of quantum systems is provided by *group theory*. In this section some aspects of group theory are summarized (cf. [93]).

**1.6.2.1 Group.** A set $G$ of elements (like operators or matrices) is called a *group* if following conditions are fulfilled:

*Condition 1.* For any pair of elements $A, B \in G$ there exists a product $AB$, where this product, again, is an element of $G$. In general, the multiplication is not commutative, i.e. $AB \neq BA$.

*Condition 2.* The multiplication is associative, i.e. $(AB)C = A(BC)$.

*Condition 3.* A unit element $1 \in G$ exists so that $1A = A1 = A$ for any element $A \in G$.

*Condition 4.* Every element $A \in G$ has an inverse element $A^{-1} \in G$ so that $AA^{-1} = 1$ or $A^{-1}A = 1$.
     If conditions 3 and 4 are not fulfilled, the resulting structure is called a *semigroup*.

**1.6.2.2 Isomorphic Mapping.** Consider two groups $G$ and $G'$ with elements $A, B \in G$ and $A', B' \in G'$, and a mapping of $G$ into $G'$. The mapping is called *homeomorphic* if the group operations are conserved, i.e. $(AB)' = A'B'$. In the case of a reversible unambiguous mapping the term *isomorphic mapping* is used.

**1.6.2.3 Matrix Representation of a Group.** A group of quadratic matrices to which the group $G$ is homeomorphic is termed *matrix representation of $G$*. Such matrices can be considered as transformation matrices defining transformations within a vector space.

**1.6.2.4 Unitary and Special Unitary Group.** The group of matrices (or of operators) $\mathsf{U}$ (with elements $U_{ij}$) which transforms vectors $v = (v_1, v_2, \ldots, v_n)$ of an $n$-dimensional linear vector space $\mathbb{V}^n$ by

$$v_i' = \sum_{j=1}^n U_{ij} v_j \tag{1.24}$$

is called a *unitary group* $U(n)$ if the relation

$$\mathsf{U}\mathsf{U}^\dagger = \mathsf{U}^\dagger \mathsf{U} = 1 \tag{1.25}$$

and thus

$$\sum_{j=1}^n U_{ij} U_{kj}^* = \sum_{j=1}^n U_{ji}^* U_{jk} = \delta_{ik} \tag{1.26}$$

($\delta_{ik} \ldots$ Kronecker delta) holds. Due to this relation, the modulus of the determinant is equal zero:

$$|\det(\mathsf{U})| = 1 . \tag{1.27}$$

If

$$\det(\mathsf{U}) = 1 , \tag{1.28}$$

the unitary group is called *special unitary group* $SU(n)$. The special unitary group $SU(2)$ is isomorphic to the group of rotations in a 3-dimensional space.

**1.6.2.5 Lie Group.** A continuous group containing uncountable many elements, which depend analytically on one or several continuous parameters, is also called a *Lie group*. A pertinent example from physics is the so-called *Poincaré group* (inhomogeneous group of *Lorentz transformations*). This group reflects the fundamental space-time property of (microscopic and macroscopic) physical systems.

**1.6.2.6 Algebra.** A set $E$ of elements $X, Y, Z \ldots \in E$ is called *algebra* if $E$ is a linear space in which a multiplication is defined which is associative and distributive:

$$\zeta(XY) = (\zeta X)Y = X(\zeta Y) , \tag{1.29}$$

$$(XY)Z = X(YZ) \,, \tag{1.30}$$

$$(X+Y)Z = XZ + YZ) \,, \tag{1.31}$$

$$X(Y+Z) = XY + XZ \,, \tag{1.32}$$

with $\zeta$ being any number.

Due to the fact that a language (mathematical or non-mathematical) is nothing but a set of elements with additional linking rules, an algebra represents a special kind of (mathematical) language.

**1.6.2.7 Generating Operators.** Every element of a group can be represented by combinations of corresponding *generating operators* (*generators*), i.e. a corresponding set of such operators "generates" the group. This set of generators is said to "span" the algebra of the group.

### 1.6.3 Quantum Systems

In this section the fundamental mathematical terms and relations introduced above are applied to quantum systems (cf. [16, 75, 127]).

**1.6.3.1 Measurement Operators.** *Eigenstates* of a microsystem (i.e. quantum systems like phonon systems, photon systems, or molecular systems) can be described by orthonormal *state vectors* $|i\rangle$ of a Hilbert space $\mathbb{H}$. Corresponding operators $\hat{A}$ allow the calculation of observable values $a_i$ of a measurement quantity $A$ by using eigenvalue equations:

$$\hat{A}\,|i\rangle = a_i\,|i\rangle \,. \tag{1.33}$$

$\hat{A}$ will be called *measurement operator* (or *observable*). Here, and in the following, we specify the eigenstate of a specific operator $\hat{A}$ by its corresponding eigenvalue $a_i$ or, if confusions are excluded, simply by its index $i$. A general, unspecified state will be denoted by $|\psi\rangle$.

If different measurement quantities represented by operators $\hat{A}$, $\hat{B}$ are measureable at the same time, the corresponding *commutator* is identically zero:

$$\left[\hat{A}, \hat{B}\right]_{-} = \hat{A}\hat{B} - \hat{B}\hat{A} = 0 \,. \tag{1.34}$$

In this case, $\hat{A}$ and $\hat{B}$ have a common set of eigenvectors.

**1.6.3.2 Spectral Representation of Measurement Operators.** The measurement operator $\hat{A}$ can be represented by its eigenvalues and eigenvectors:

$$\hat{A} = \sum_{j} |j\rangle\, a_j\, \langle j| \,. \tag{1.35}$$

*Proof.* Multiplying (1.35) with a state vector $|\psi_i\rangle$, and using the orthonormality relation

$$\langle j\,|i\rangle = \delta_{ji}\,, \tag{1.36}$$

one obtains, again, the eigenvalue equation (1.33):

$$\begin{aligned}
\hat{A}\,|i\rangle &= \sum_j |j\rangle\, a_j\, \langle j\,|i\rangle \\
&= \sum_i |i\rangle\, a_i \delta_{ji} \\
&= a_i\,|i\rangle\,.
\end{aligned} \tag{1.37}$$

As values of measurement quantities are real, measurement operators have to be Hermitian.

**1.6.3.3 Schrödinger Equation and Schrödinger Picture.** A very important measurement operator is the *Hamiltonian* $\hat{H}$ with the corresponding eigenvalue equation

$$\hat{H}\,|E_i\rangle = E_i\,|E_i\rangle\,, \tag{1.38}$$

the so-called *stationary Schrödinger equation*. (State vectors which correspond to the same energy eigenvalue, i.e. *degenerate states*, are not considered.) On the basis of this equation, observable eigenvalues $E_i$ of the total energy of the quantum system can be calculated. A well-known example is represented by the (discrete) energy spectrum of the hydrogen atom which can be calculated exactly.

The dynamics of a ("closed") quantum system can be described by the time-dependent Schrödinger equation

$$\hat{H}\,|\psi(t)\rangle = i\hbar \frac{\partial}{\partial t}\,|\psi(t)\rangle\,. \tag{1.39}$$

The time-evolution is thus contained in time-dependent state vectors $|\psi(t)\rangle$. In the case of time-independent forces, the Hamiltonian $\hat{H}$ and other measurement operators $\hat{A}$ are time-independent. This "picture" of description is called *Schrödinger picture*.

If $\hat{H}$ does not explicitly depend on time, (1.39) is solved by

$$|\psi(t)\rangle = |E_i\rangle\, \mathrm{e}^{-\mathrm{i}E_i t/\hbar} \tag{1.40}$$

with the *initial condition* $|\psi(0)\rangle = |E_i\rangle$. A general state can be written as a *superposition*:

$$|\psi(t)\rangle = \sum_i b_i\,|E_i\rangle\, \mathrm{e}^{-\mathrm{i}E_i t/\hbar} \tag{1.41}$$

or, using any complete *basis set* $|i\rangle$, as

$$|\psi(t)\rangle = \sum_i c_i(t)\,|i\rangle\,. \tag{1.42}$$

### 1.6.3.4 Heisenberg's Equation of Motion and Heisenberg Picture.
Applying the unitary transformation

$$\hat{U} = e^{-i\hat{H}t/\hbar} \tag{1.43}$$

to (time-dependent) state vectors $|\psi(t)\rangle$ and (time-independent) measurement operators $\hat{A}$, one obtains time-independent state vectors $|\psi^{(H)}\rangle$ and time-dependent operators $\hat{A}^{(H)}(t)$, i.e. after this transformation the time-evolution of the quantum system is described by time-dependent operators $\hat{A}^{(H)}(t)$:

$$\left|\psi^{(H)}\right\rangle = \hat{U}^\dagger |\psi(t)\rangle = |\psi(0)\rangle , \tag{1.44}$$

$$\hat{A}^{(H)}(t) = \hat{U}^\dagger \hat{A} \hat{U} . \tag{1.45}$$

The *Heisenberg operators* $\hat{A}^{(H)}(t)$ obey *Heisenberg's equation of motion*

$$\frac{d\hat{A}^{(H)}(t)}{dt} = \frac{\partial \hat{A}^{(H)}(t)}{\partial t} + \frac{i}{\hbar} \left[\hat{H}, \hat{A}^{(H)}(t)\right]_- . \tag{1.46}$$

This kind of description defines the *Heisenberg picture*.

### 1.6.3.5 Interaction Picture.
For systems composed of interacting subsystems with

$$\hat{H} = \hat{H}_0 + \hat{V} \tag{1.47}$$

($\hat{V}$ denotes the interaction operator, $\hat{H}_0$ the Hamiltonian of the non-interacting systems), a third picture, the *interaction picture*, can be introduced by application of the unitary transformation $\hat{U}_0 = e^{i\hat{H}_0 t/\hbar}$:

$$\hat{V}^{(i)}(t) \left|\psi^{(i)}(t)\right\rangle = i\hbar \frac{\partial}{\partial t} \left|\psi^{(i)}(t)\right\rangle , \tag{1.48}$$

$$\hat{V}^{(i)}(t) = \hat{U}_0 \hat{V} \hat{U}_0^\dagger , \tag{1.49}$$

$$\left|\psi^{(i)}(t)\right\rangle = \hat{U}_0 |\psi(t)\rangle , \tag{1.50}$$

$$\hat{A}^{(i)}(t) = \hat{U}_0 \hat{A} \hat{U}_0^\dagger . \tag{1.51}$$

We see that the operators $\hat{A}^{(i)}(t)$ obey Heisenberg's equation of motion with respect to $\hat{H}_0$, while the dynamics of $|\psi^{(i)}\rangle$ is generated by $\hat{V}^{(i)}$.

### 1.6.3.6 Expectation Values.
The general formulation (1.42) enables us to calculate expectation values $\overline{A}$ of the physical quantity $A$. Taking eigenvectors of the measurement operator $\hat{A}$ as basis vectors, one obtains

$$\overline{A} = \left\langle \hat{A} \right\rangle = \left\langle \psi(t) \left| \hat{A} \right| \psi(t) \right\rangle = \sum_{i,j} c_i^*(t) c_j(t) \left\langle i \left| \hat{A} \right| j \right\rangle$$

$$= \sum_{i,j} c_i^*(t) c_j(t) a_j \left\langle i | j \right\rangle$$

$$= \sum_{i,j} c_i^*(t) c_j(t) a_j \delta_{ij}$$

$$= \sum_{i} c_i^*(t) c_i(t) a_i . \tag{1.52}$$

$c_i^*(t) c_i(t)$ represents the probability of finding the measurement value $a_i$ in a state $|\psi\rangle$.

**1.6.3.7 Density Matrix and Density Operator.** If an ensemble of quantum systems can be described by a single state vector $|\psi\rangle$ (such as the state vector defined by (1.42)), the terms *pure population* and *pure state* are used. In the case of macroscopic statistical perturbations, statistical effects due to the influence of the surrounding have to be taken into account. Then, various state vectors $|\psi_\nu\rangle$ occurring with the probability $\rho_\nu$ have to be taken as a basis. In such a case the terms *mixed population* and *mixed state* are used. The description must then resort to the *density matrix*.

In the book in hand the *density matrix theory* is a main subject; it will therefore not be discussed here in detail. Nevertheless, some essential facts should be appropriate. Introducing the density matrix $\rho$ with elements

$$\rho_{ij} = c_i^*(t) c_j(t) , \tag{1.53}$$

expectation values defined by (1.52) and possible eigenvalues $a_i$ are connected by the relation

$$\left\langle \hat{A} \right\rangle = \sum_{i} \rho_{ii} a_i = \left\langle \psi \left| \hat{A} \right| \psi \right\rangle \tag{1.54}$$

(inserting (1.53) into (1.52), one obtains (1.54)).

Introducing the density operator

$$\hat{\rho} = \sum_{ij} \rho_{ij} |i\rangle \langle j| = |\psi\rangle \langle \psi| , \tag{1.55}$$

such expectation values can be represented by

$$\left\langle \hat{A} \right\rangle = \mathrm{Tr} \left\{ \hat{A} \hat{\rho} \right\} = \mathrm{Tr} \left\{ \hat{\rho} \hat{A} \right\} , \tag{1.56}$$

where Tr denotes the trace-operation:

$$\mathrm{Tr} \left\{ \hat{\rho} \hat{A} \right\} = \sum_{i=1}^{n} \left\langle i \left| \hat{\rho} \hat{A} \right| i \right\rangle \tag{1.57}$$

with $|i\rangle$ representing a complete basis in the Hilbert space of dimension $n$.

The quantities $\left\langle \hat{A} \right\rangle$ given by (1.54) with (1.53) define expectation values in pure populations. In the case of mixed populations, expectation values can be defined by

$$\left\langle \hat{A} \right\rangle = \sum_{\nu} \rho_{\nu} \left\langle \psi_{\nu} \left| \hat{A} \right| \psi_{\nu} \right\rangle , \qquad (1.58)$$

where $\rho_{\nu}$ denotes the probability of finding state $|\psi_{\nu}\rangle$. In this case (1.56) still holds.

**1.6.3.8 Liouville Equation.** The dynamics of the density operator is defined by the *Liouville equation*

$$i\hbar \frac{\partial}{\partial t} \hat{\rho} = \left[ \hat{H}, \hat{\rho} \right]_{-} , \qquad (1.59)$$

where $\hat{H}$ specifies the Hamiltonian of the quantum system. This equation of motion will be considered in more detail in the book in hand. Then it will be shown that the dynamics of a time-dependent expectation value $\overline{A}$ can be determined by

$$i\hbar \frac{\partial}{\partial t} \overline{A} = i\hbar \frac{\overline{\partial A}}{\partial t} + \text{Tr} \left\{ \left[ \hat{H}, \hat{\rho} \right]_{-} \hat{A} \right\} . \qquad (1.60)$$

### 1.6.4 Groups in Quantum Theory

**1.6.4.1 Symmetries and Symmetry Groups.** Physical systems on any level of description (in terms of elementary particles, atoms, or phenomenological properties) are characterized by structural patterns. Very often, structural patterns are *covariant* with respect to a class of transformations (covariance = invariance with respect to the form). Such properties are usually called *symmetries*: one finds that the symmetry elements of the considered system form a group, the *symmetry group* of the considered system.

On a mathematical level of consideration, symmetry elements are represented by symmetry operators $\hat{\chi}_{\alpha}$. Applying $\hat{\chi}_{\beta}$ to fundamental equations (which contain the symmetry described by the operators $\hat{\chi}_{\beta}$), the equations remain unchanged (i.e. these equations are covariant with respect to the symmetry transformations). For example, considering a quantum system described by the Hamiltonian $\hat{H}$, the group of symmetry operators $\hat{\chi}_{\beta}$, which represents the symmetry elements (implicitly contained in $\hat{H}$), commutes with $\hat{H}$ so that the corresponding Schrödinger equation remains unchanged:

$$\left[ \hat{\chi}_{\beta}, \hat{H} \right]_{-} = 0 , \qquad (1.61)$$

$$\hat{\chi}_{\beta} \hat{H} \left| \psi_i^{(\kappa)} \right\rangle = \hat{H} \underbrace{\hat{\chi}_{\beta} \left| \psi_i^{(\kappa)} \right\rangle}_{\left| \psi_i^{(\kappa')} \right\rangle} = E_i \underbrace{\hat{\chi}_{\beta} \left| \psi_i^{(\kappa)} \right\rangle}_{\left| \psi_i^{(\kappa')} \right\rangle} . \qquad (1.62)$$

(In this context degenerate states are considered. Therefore, the additional index $\kappa$ is used.)

Pertinent examples of symmetry groups are *Lie groups*: in quantum theory (in particular, in the theory of elementary particles) unitary groups $U(n)$ and special unitary groups $SU(n)$ are relevant. Those reflect internal symmetries of the quantum systems. Such unitary groups will be considered in the book in hand.

**1.6.4.2 Matrix Representations and Vector Basis.** Symmetry operators of fundamental equations can be represented by matrices which are strongly correlated with solutions of these fundamental equations. For example, consider the degenerate states

$$\left| \psi_i^{(\kappa)} \right\rangle = \left\{ \left| \psi_i^{(1)} \right\rangle, \left| \psi_i^{(2)} \right\rangle, \ldots, \left| \psi_i^{(m)} \right\rangle \right\} \tag{1.63}$$

as solutions of the Schrödinger equation (1.62) with respect to the energy eigenvalue $E_i$. Every superposition of those state vectors is also a solution with respect to $E_i$. Applying the symmetry operators $\hat{\chi}_\beta$ (representing the symmetry elements of $\hat{H}$) to the state vectors represented by (1.63), one obtains new state vectors $\left| \psi_i^{(\kappa')} \right\rangle$. Due to (1.62), these new state vectors, again, are solutions of the Schrödinger equation (1.62) with respect to the same eigenvalue $E_i$. Thus, the relation

$$\hat{\chi}_\beta \left| \psi_i \right\rangle = \chi_\beta \left| \psi_i \right\rangle \tag{1.64}$$

with

$$\left| \psi_i \right\rangle = \begin{pmatrix} \left| \psi_i^{(1)} \right\rangle \\ \cdot \\ \cdot \\ \cdot \\ \left| \psi_i^{(m)} \right\rangle \end{pmatrix} \tag{1.65}$$

holds, where $\chi_\beta$ represents matrices which coincide with the symmetry operators $\hat{\chi}_\beta$. The group of matrices $\chi_\beta$ is a representation of the group of symmetry operators $\hat{\chi}_\beta$. The state vectors (1.65) define the *basis* of the representation.

**1.6.4.3 Generating Operators.** As mentioned already, every operator of a group of symmetry operators can be expressed by combinations of fundamental operators, the (infinitesimal) generators of the group. In quantum theory the generating operators essentially are identical with the operators defining observable values of measurement quantities. For example, in the case of the special unitary group $SU(2)$, angular momentum operators are the generating operators (where the Pauli matrices specify the non-trivial matrix representation of lowest dimension). Such generating operators will be extensively used in the following.

# 2. Quantum Statics

## 2.1 Introduction

*Quantum statics* is meant to deal with time-independent aspects of quantum mechanics. In this chapter, *static properties* of *quantum networks* consisting of few subsystems ("nodes") will be discussed. The nodes, specified by local $SU(n)$ *algebras*, can be thought to stand for real spins (spin 1/2, spin 1, etc.) or any *pseudospins* like few-level systems such as molecules or number states of a quantized field (e.g. photons). Based on the *generating operators* of $SU(n)$, *Hamiltonians* specifying such networks will be considered.

In the same way, a convenient representation of the density operator follows, which allows for a natural description of *1-node* and *more-node coherence*. *Coherence vectors* which characterize 1-node coherence and *correlation tensors* characterizing few-node coherence will be discussed.

In order to reconstruct a state of a quantum system, a *measurement* is necessary. In this chapter, the influence of a measurement on a quantum system will be discussed in terms of the projection postulate (dynamical measurement models will be taken up in Chap. 4). In particular, *direct measurements* on single nodes and *indirect measurements* in composite systems are studied.

Few-node coherence ("entanglement") is basic to the violation of *Bell inequalities*, to *quantum cryptography*, and to mind-boggling schemes like *quantum teleportation*.

*Entropy* will be introduced as a measure to describe the lack of information about a network or a node state.

## 2.2 Quantum Mechanical Systems

Contents: algebra of angular momentum, representations in Hilbert space, Pauli matrices, spin polarization, atomic polarization, projection operators, generators of the $SU(n)$ algebra, representations of operators in $SU(n)$, Hamilton models in $SU(n)$.

### 2.2.1 Quantum Mechanics of Angular Momentum

While many operators in quantum mechanics have a continuous spectrum (similar to position or momentum), this is not so for the angular momentum: there are even well-defined finite subspectra, the eigenvectors of which span finite Hilbert spaces. The physics of *angular momenta*(cf. [119]) is used to motivate the introduction of a special operator basis for the representation of observables in such finite spaces. In this section fundamental properties and basic definitions relevant in this book are introduced, in particular, *commutator relations, anticommutator relations, eigenvalue equations, eigenvectors, projection operators*, and *representations*.

**2.2.1.1 General Definitions.** Classically, the angular momentum of a single point mass with position $r$ and linear momentum $p$ is defined by $L = r \times p$. Quantum-mechanically, the angular momentum $\hat{L} = \left( \hat{L}_x, \hat{L}_y, \hat{L}_z \right)$ is introduced via the commutator relations for its components $\hat{L}_i$ ($i = 1, 2, 3$ or $x, y, z$):

$$\boxed{\left[ \hat{L}_i, \hat{L}_j \right]_- = i\hbar \sum_{k=1}^{3} \varepsilon_{ijk} \hat{L}_k} \tag{2.1}$$

with $\varepsilon_{ijk}$ representing the antisymmetric tensor $(\varepsilon_{ijk})$ defined by

$$\varepsilon_{123} = \varepsilon_{231} = \varepsilon_{312} = 1 \ , \quad \varepsilon_{132} = \varepsilon_{321} = \varepsilon_{213} = -1 \ , \quad \text{otherwise} = 0 \ . \tag{2.2}$$

The commutator relation (2.1) follows from the basic relation

$$\left[ \hat{r}_i, \hat{p}_j \right]_- = \frac{\hbar}{i} \delta_{ij} \ , \tag{2.3}$$

and

$$\hat{L}_i = \varepsilon_{ijk} \hat{r}_j \hat{P}_k \ , \tag{2.4}$$

but is kept also for spin operators, which do not have a classical analogue. The corresponding square operator is

$$\hat{L}^2 = \sum_{j=1}^{3} \hat{L}_j^2 \ . \tag{2.5}$$

These are *Hermitian operators*. However, very often also non-Hermitian operators are useful:

$$\hat{L}_\pm = \hat{L}_x \pm i\hat{L}_y \tag{2.6}$$

with

$$\left( \hat{L}_+ \right)^\dagger = \hat{L}_- \ . \tag{2.7}$$

From (2.1), the commutator relations

$$\left[\hat{L}^2, \hat{L}_j\right]_- = 0 \tag{2.8}$$

and

$$
\begin{aligned}
\left[\hat{L}_z, \hat{L}_\pm\right]_- &= \pm\hbar\hat{L}_\pm \;, \\
\left[\hat{L}_+, \hat{L}_-\right]_- &= 2\hbar\hat{L}_z \;, \\
\left[\hat{L}^2, \hat{L}_\pm\right]_- &= 0
\end{aligned}
\tag{2.9}
$$

are easily shown to follow.

Due to (2.8), a joint set of *eigenvalue equations* exists,

$$
\begin{aligned}
\hat{L}^2 \,|l,m\rangle &= \hbar^2 l(l+1)\,|l,m\rangle \;,\; l = 0, \tfrac{1}{2}, 1, \tfrac{3}{2}, \ldots \;, \\
\hat{L}_z \,|l,m\rangle &= \hbar m\,|l,m\rangle \;,\; -l \le m \le l \;,\; |\Delta m| = 1 \;,\; 2l+1 \text{ values} \;,
\end{aligned}
\tag{2.10}
$$

which determines the eigenvalues $\hbar^2 l(l+1)$ of $\hat{L}^2$ as well as the eigenvalues $\hbar m$ of the $z$ component $\hat{L}_z$ of the angular momentum. $|l,m\rangle$ then represent the corresponding eigenvectors. These eigenvectors are assumed orthonormalized, i.e.

$$\langle l,m \mid l'm'\rangle = \delta_{ll'}\delta_{mm'} \;. \tag{2.11}$$

The angular momentum operators are traceless, i.e.

$$\mathrm{Tr}\{\hat{L}_j\} = \sum_{l,m} \left\langle l,m \left| \hat{L}_j \right| l,m \right\rangle = 0 \;. \tag{2.12}$$

The eigenvectors $|l,m\rangle$ form a *Hilbert space* $\mathbb{H}$. This Hilbert space can be decomposed into subspaces $\mathbb{H}_l$ given by the direct sum

$$\mathbb{H} = \sum_l \oplus\, \mathbb{H}_l \;. \tag{2.13}$$

Here, $\mathbb{H}_l$ defines a finite Hilbert space of dimension $n = 2l+1$. Figures 2.1 and 2.2 illustrate examples. (In the special case of an orbital angular momentum, the orbital quantum number is restricted to $l = 0, 1, 2, \ldots$, while in the case of spins the spin quantum number $l$ may be integer or half-integer.) In the following we will focus on these finite-dimensional Hilbert spaces $\mathbb{H}_l$.

Completeness guarantees

$$\hat{1} = \sum_{m=-l}^{l} |l,m\rangle\langle l,m| \tag{2.14}$$

so that any vector $|\chi\rangle$ in Hilbert space $\mathbb{H}_l$ can be represented by

$$|\chi\rangle = \hat{1}\,|\chi\rangle = \sum_m |l,m\rangle\langle l,m|\chi\rangle = \sum_m |l,m\rangle\,\chi_{lm} \;, \tag{2.15}$$

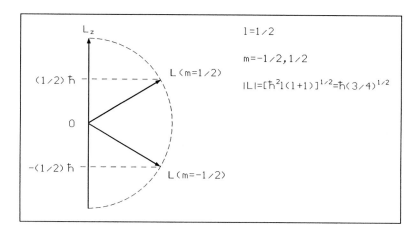

**Fig. 2.1.** Eigenvalues of the angular momentum $l = 1/2$

where

$$\langle \chi \,|\, \chi \rangle = \sum_{m,m'} \chi^*_{l,m'} \chi_{l,m} \, \langle l, m \,|\, l, m' \rangle = \sum_m \chi^*_{l,m} \chi_{l,m} = 1 \,, \qquad (2.16)$$

and * denotes conjugate complex quantities.

Correspondingly, the operators $\hat{L}_j$ can be represented by matrices restricted to obey the commutator relation (2.1): the non-zero matrix elements are

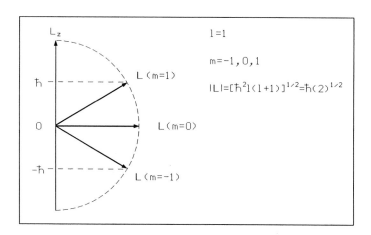

**Fig. 2.2.** Eigenvalues of the angular momentum $l = 1$

$$\left\langle l, m \pm 1 \left| \hat{L}_x \right| l, m \right\rangle = \frac{\hbar}{2} \left[(l \mp m)(l \pm m + 1)\right]^{1/2},$$
$$\left\langle l, m \pm 1 \left| \hat{L}_y \right| l, m \right\rangle = \mp \frac{i\hbar}{2} \left[(l \mp m)(l \pm m + 1)\right]^{1/2},$$
$$\left\langle l, m \left| \hat{L}_z \right| l, m \right\rangle = \hbar m. \tag{2.17}$$

For the non-Hermitian operators $\hat{L}_\pm$ one finds $(m = -l, -l+1, \ldots, l-1, l)$

$$\left\langle l, m + 1 \left| \hat{L}_+ \right| l, m \right\rangle = \hbar \left[(l - m)(l + m + 1)\right]^{1/2}$$
$$= \left\langle l, m \left| \hat{L}_- \right| l, m + 1 \right\rangle. \tag{2.18}$$

Equations (2.17) and (2.18) define a representation of the operators $\hat{L}_j$, $\hat{L}_\pm$ on the basis of the eigenvectors $|l, m\rangle$ of $\hat{L}_z$, the so-called $\hat{L}_z$ representation.

**2.2.1.2 Spin 1/2.** For $l = 1/2$ it is convenient to introduce the operators $\hat{\sigma}_j$ defined by

$$\hat{\sigma}_j^2 = \hat{1}, \quad \hat{\sigma}_j = \frac{2}{\hbar} \hat{L}_j, \tag{2.19}$$

and to rewrite the commutator relation as

$$[\hat{\sigma}_i \hat{\sigma}_j]_- = 2i \sum_{k=1}^{3} \varepsilon_{ijk} \hat{\sigma}_k. \tag{2.20}$$

With the eigenvalue equation in the form (cf. (2.10))

$$\hat{\sigma}_z |i\rangle = i |i\rangle \quad (i = \mp 1) \tag{2.21}$$

the eigenstates $|+1/2, -1/2\rangle$, $|+1/2, +1/2\rangle$ will in the following be denoted by $|-1\rangle$, $|1\rangle$. The anticommutator relation

$$[\hat{\sigma}_i \hat{\sigma}_j]_+ = 2\delta_{ij} \hat{1} \tag{2.22}$$

and the trace-relations

$$\mathrm{Tr}\{\hat{\sigma}_j\} = \sum_m \langle m | \hat{\sigma}_j | m \rangle = 0, \tag{2.23}$$

$$\mathrm{Tr}\{\hat{\sigma}_i \hat{\sigma}_j\} = \sum_m \langle m | \hat{\sigma}_i \hat{\sigma}_j | m \rangle = 2\delta_{ij}, \tag{2.24}$$

$$\mathrm{Tr}\{\hat{\sigma}_i \hat{\sigma}_j \hat{\sigma}_k\} = \sum_m \langle m | \hat{\sigma}_i \hat{\sigma}_j \hat{\sigma}_k | m \rangle = 2i\varepsilon_{ijk} \tag{2.25}$$

follow.

The $\hat{\sigma}_z$ representation of these spin operators is a special case of (2.17):

$$
\begin{aligned}
(\sigma^x) = \frac{2}{\hbar}\left\langle m\left|\hat{L}_x\right|m'\right\rangle &= \frac{2}{\hbar}\begin{pmatrix} \left\langle -1\left|\hat{L}_x\right|-1\right\rangle & \left\langle -1\left|\hat{L}_x\right|1\right\rangle \\ \left\langle 1\left|\hat{L}_x\right|-1\right\rangle & \left\langle 1\left|\hat{L}_x\right|1\right\rangle \end{pmatrix} \\
&= \begin{pmatrix} 0 & 1 \\ 1 & 0 \end{pmatrix},
\end{aligned}
\tag{2.26}
$$

$$
(\sigma^y) = \frac{2}{\hbar}\left\langle m\left|\hat{L}_y\right|m'\right\rangle = i\begin{pmatrix} 0 & +1 \\ -1 & 0 \end{pmatrix},
\tag{2.27}
$$

$$
(\sigma^z) = \frac{2}{\hbar}\left\langle m\left|\hat{L}_z\right|m'\right\rangle = \begin{pmatrix} -1 & 0 \\ 0 & +1 \end{pmatrix},
\tag{2.28}
$$

with $m = \mp 1$. (Note the ordering of the matrix elements starting with the lowest eigenvalue, here $-1$! This convention will apply to angular momenta and energy eigenstates to be discussed later. In most books restricted to a discussion of the angular momentum only, the reverse order is used.) The elements of these so-called *Pauli matrices* will be denoted by $\sigma_{ij}^x$, $\sigma_{ij}^y$, $\sigma_{ij}^z$. They form a special matrix representation ($\hat{\sigma}_z$ *representation*) of the operators $\hat{L}_j$ ($j = 1, 2, 3$ or $x, y, z$) for $l = 1/2$.

The eigenvectors of the eigenvalue equations

$$
\hat{\sigma}_x\left|e_\nu^x\right\rangle = \nu\left|e_\nu^x\right\rangle , \quad \nu = -1, +1 ,
\tag{2.29}
$$

$$
\hat{\sigma}_y\left|e_\nu^y\right\rangle = \nu\left|e_\nu^y\right\rangle , \quad \nu = -1, +1 ,
\tag{2.30}
$$

can then be represented by (modulo a phase factor)

$$
\begin{aligned}
\left|e_1^x\right\rangle &= \frac{1}{\sqrt{2}}\left(|1\rangle + |-1\rangle\right) , \\
\left|e_{-1}^x\right\rangle &= \frac{1}{\sqrt{2}}\left(|1\rangle - |-1\rangle\right)
\end{aligned}
\tag{2.31}
$$

and

$$
\begin{aligned}
\left|e_1^y\right\rangle &= \frac{1}{\sqrt{2}}\left(|1\rangle - i|-1\rangle\right) , \\
\left|e_{-1}^y\right\rangle &= \frac{1}{\sqrt{2}}\left(|1\rangle + i|-1\rangle\right) ,
\end{aligned}
\tag{2.32}
$$

respectively.

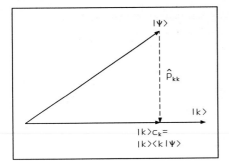

**Fig. 2.3.** The effect of the projection operator $\hat{P}_{kk}$

**2.2.1.3 Projection Operators.** Let $|k\rangle$, $|l\rangle$ be eigenvectors spanning a given Hilbert space (dimension $n$), then *projection operators* $\hat{P}_{kl}$ are defined by

$$\hat{P}_{kl} = |k\rangle \langle l| \quad (k, l = 1, 2, \ldots, n) \,.$$
(2.33)

The operators $\hat{P}_{kl}$ are termed *projection operators* because they generate "projections" onto a state vector $|k\rangle$. For example, applying the projection operator $\hat{P}_{kk}$ to a state vector $|\psi\rangle$, one obtains

$$\hat{P}_{kk} |\psi\rangle = |k\rangle c_k$$
(2.34)

with

$$c_k = \langle k | \psi \rangle \,.$$
(2.35)

If $c_k > 0$ and real, $|k\rangle c_k$ points in the same direction as $|k\rangle$, compare Fig. 2.3.

Taking the orthonormality of the eigenvectors into account, the relation

$$\hat{P}_{kl}\hat{P}_{mn} = |k\rangle \langle l| m \rangle \langle n| = \hat{P}_{kn}\delta_{lm}$$
(2.36)

implies the commutator relation

$$\left[\hat{P}_{kl}, \hat{P}_{mn}\right]_{-} = \hat{P}_{kn}\delta_{lm} - \hat{P}_{ml}\delta_{nk} \,.$$
(2.37)

In particular, it follows that

$$\left[\hat{P}_{ll}, \hat{P}_{nn}\right]_{-} = \left(\hat{P}_{ln} - \hat{P}_{nl}\right)\delta_{ln} = 0 \,.$$
(2.38)

Useful trace-relations are

$$\mathrm{Tr}\left\{\hat{P}_{kl}\right\} = \delta_{kl} \,,$$
(2.39)

$$\mathrm{Tr}\left\{\hat{P}_{kl}\hat{P}_{mn}\right\} = \delta_{lm}\delta_{kn} \,.$$
(2.40)

Any operator defined in the Hilbert space $\mathbb{H}$ of dimension $n$ can be expressed in terms of these $n^2$ projection operators. For example, the spin $1/2$ operators read (cf. (2.26)–(2.28))

$$\boxed{\begin{aligned}
\hat{\sigma}_x &= \frac{2}{\hbar}\hat{s}_x = \hat{P}_{12} + \hat{P}_{21}\ , \\
\hat{\sigma}_y &= \frac{2}{\hbar}\hat{s}_y = \mathrm{i}\left(\hat{P}_{12} - \hat{P}_{21}\right)\ , \\
\hat{\sigma}_z &= \frac{2}{\hbar}\hat{s}_z = -\hat{P}_{11} + \hat{P}_{22}\ ,
\end{aligned}}$$
(2.41)

where we have identified former $|-1\rangle$ with $|1\rangle$ and former $|1\rangle$ with $|2\rangle$. One may also introduce the non-Hermitian *raising* and *lowering operators*, respectively, defined by

$$\boxed{\begin{aligned}
\hat{\sigma}^+ &= \frac{1}{2}\left(\hat{\sigma}_x + \mathrm{i}\hat{\sigma}_y\right) = \hat{P}_{21}\ , \\
\hat{\sigma}^- &= \frac{1}{2}\left(\hat{\sigma}_x - \mathrm{i}\hat{\sigma}_y\right) = \hat{P}_{12}\ .
\end{aligned}}$$
(2.42)

In the same way, for the angular momentum components $\hat{L}_i$ we get for $l = 1$, now identifying the states $|1,-1\rangle$, $|1,0\rangle$, $|1,1\rangle$ with $|1\rangle$, $|2\rangle$, $|3\rangle$:

$$\boxed{\begin{aligned}
\frac{1}{\hbar}\hat{L}_x &= \frac{1}{\sqrt{2}}\left(\hat{P}_{12} + \hat{P}_{21}\right) + \frac{1}{\sqrt{2}}\left(\hat{P}_{23} + \hat{P}_{32}\right)\ , \\
\frac{1}{\hbar}\hat{L}_y &= \frac{\mathrm{i}}{\sqrt{2}}\left(\hat{P}_{12} - \hat{P}_{21}\right) + \frac{\mathrm{i}}{\sqrt{2}}\left(\hat{P}_{23} - \hat{P}_{32}\right)\ , \\
\frac{1}{\hbar}\hat{L}_z &= -\hat{P}_{11} + \hat{P}_{33}\ .
\end{aligned}}$$
(2.43)

Here, $\hat{\sigma}_z$, $\hat{L}_z$ are given in spectral representation, i.e. expressed in terms of their eigenvalues and eigenfunctions: in general, for $l = (n-1)/2$

$$\frac{1}{\hbar}\hat{L}_z = \sum_{j=1}^{n}(j - l - 1)\hat{P}_{jj}\ .$$
(2.44)

### 2.2.2 $SU(n)$ Algebra

In Sect. 2.2.1 the basic properties of angular momenta have been studied. In general, the eigenvectors of any complete set of operators can be taken as a representation of states and operators: in Sect. 2.2.1 the $\hat{L}_z$ representation has been examined. Another possibility is to use *operator representations*: states (*density operators*) and observables are then both defined as traces over so-called generating operators. Specific combinations of *projection operators* form the *generating operators* of the group $SU(n)$. In this case those traces make up the components of a real vector with interesting geometric properties; unitary transformations, e.g., appear as rotations, expectation values as scalar products.

**2.2.2.1 The Group of Infinitesimal Rotations.** The rotation of a vector $r = (x, y, z)$ is given by the linear system of equations

$$
\begin{pmatrix} x' \\ y' \\ z' \end{pmatrix} = \begin{pmatrix} R_{xx} & R_{xy} & R_{xz} \\ R_{yx} & R_{yy} & R_{yz} \\ R_{zx} & R_{zy} & R_{zz} \end{pmatrix} \begin{pmatrix} x \\ y \\ z \end{pmatrix} . \tag{2.45}
$$

In the case of small rotations, the equation

$$
\boxed{r' = r + \phi \times r} \tag{2.46}
$$

holds. Here, the unit vector $\phi/\phi$ specifies the rotational axis. This equation is equivalent to (2.45) if the matrix $R_\phi$ is chosen as

$$
\boxed{R_\phi = \begin{pmatrix} 1 & -\phi_z & \phi_y \\ \phi_z & 1 & -\phi_x \\ -\phi_y & \phi_x & 1 \end{pmatrix} .} \tag{2.47}
$$

The inverse of (2.46) and (2.47) then is given by

$$
r = r' - \phi \times r' \tag{2.48}
$$

and

$$
R_\phi^{-1} = \begin{pmatrix} 1 & \phi_z & -\phi_y \\ -\phi_z & 1 & \phi_x \\ \phi_y & -\phi_x & 1 \end{pmatrix} . \tag{2.49}
$$

The matrices $R_\phi$ represent a special group in 3 dimensions, the *group of infinitesimal rotations* O(3).

**2.2.2.2 Field Transformations.** (See [127].) Consider a scalar field represented by the wave function $\Psi(r)$. For an infinitesimal rotation of the vector $r$ the field will change from $\Psi$ to $\Psi'$:

$$
\Psi'(r) = \Psi(R_\phi^{-1} r) . \tag{2.50}
$$

Then, an operator $\hat{U}_R(\phi)$ with the property

$$
\hat{U}_R(\phi)\Psi(r) = \Psi'(r) \tag{2.51}
$$

can be introduced. Using (2.50), (2.51) yields

$$
\hat{U}_R(\phi)\Psi(r) = \Psi(R_\phi^{-1} r) = \Psi(r - \phi \times r) , \tag{2.52}
$$

where a Taylor expansion then leads to

$$
\begin{aligned}
\hat{U}_R(\phi)\Psi(r) &\approx \Psi(r) - (\phi \times r)\nabla\Psi(r) \\
&= \Psi(r) - \frac{i}{\hbar}(\phi \times r)\hat{p}\Psi(r) \\
&= \Psi(r) - \frac{i}{\hbar}\phi(r \times \hat{p})\Psi(r) .
\end{aligned} \tag{2.53}
$$

In this formula $\hat{\boldsymbol{p}} = (\hat{p}_x, \hat{p}_y, \hat{p}_z) = \frac{\hbar}{i}\nabla$ denotes the *momentum operator*, connected to the operator of the angular momentum, $\hat{\boldsymbol{L}} = \left(\hat{L}_x, \hat{L}_y, \hat{L}_z\right)$, by

$$\hat{\boldsymbol{L}} = \boldsymbol{r} \times \hat{\boldsymbol{p}} \ . \tag{2.54}$$

Comparing the l.h.s. and the r.h.s. of (2.53), and taking (2.54) into account, one obtains

$$\boxed{\hat{U}_R(\boldsymbol{\phi}) = \hat{1} - \frac{i}{\hbar}\boldsymbol{\phi}\hat{\boldsymbol{L}} \ .} \tag{2.55}$$

The relation (2.55) then shows that the operators $\hat{L}_x$, $\hat{L}_x$, $\hat{L}_x$ generate the infinitesimal rotation: these operators are therefore called *generating operators*.

If after a finite rotation $\phi_x$ with respect to the $x$ axis an additional infinitesimal rotation $\Delta\phi_x$ is applied, the result can be described by

$$\hat{U}_R(\phi_x + \Delta\phi_x) = \hat{U}_R(\Delta\phi_x)\hat{U}_R(\phi_x) \ . \tag{2.56}$$

Using (2.55), (2.56) reads

$$\hat{U}_R(\phi_x + \Delta\phi_x) = \left(\hat{1} - \frac{i}{\hbar}\Delta\phi_x\hat{L}_x\right)\hat{U}_R(\phi_x) \tag{2.57}$$

or

$$\frac{\hat{U}_R(\phi_x + \Delta\phi_x) - \hat{U}_R(\phi_x)}{\Delta\phi_x} = -\frac{i}{\hbar}\hat{L}_x\hat{U}_R(\phi_x) \ . \tag{2.58}$$

Equation (2.58) is equivalent to the differential equation

$$\frac{\mathrm{d}\hat{U}_R(\phi_x)}{\mathrm{d}\phi_x} = -\frac{i}{\hbar}\hat{L}_x\hat{U}_R(\phi_x) \ , \tag{2.59}$$

where the solution

$$\hat{U}_R(\phi_x) = \exp\left(-\frac{i}{\hbar}\phi_x\hat{L}_x\right) \ , \quad \hat{U}_R(\phi_x = 0) = 1 \tag{2.60}$$

is valid also for any finite rotations. In the same way, operators for rotations with respect to the $y$ axis and $z$ axis, respectively, can be derived so that the general solution finally reads

$$\boxed{\hat{U}_R(\boldsymbol{\phi}) = \exp\left(-\frac{i}{\hbar}\boldsymbol{\phi}\hat{\boldsymbol{L}}\right) \ .} \tag{2.61}$$

Any operator $\exp(i\hat{A})$ is a unitary operator if $\hat{A}$ is Hermitian. Thus, $\hat{U}_R(\boldsymbol{\phi})$ represents a unitary operator because $\hat{\boldsymbol{L}}$ is a Hermitian operator. Due to the fact that in such a case $\det\left[\exp\left(\hat{A}\right)\right] = \exp\left(\text{Tr}\{\hat{A}\}\right)$ holds, the relation

$$\det\left[\hat{U}_R(\boldsymbol{\phi})\right] = 1 \tag{2.62}$$

follows because (cf. (2.12))

$$\text{Tr}\left\{\phi_j \hat{L}_j\right\} = 0 . \tag{2.63}$$

If we assume that

$$\left[\hat{U}_{\text{R}}(\phi), \hat{H}\right]_- = 0 \tag{2.64}$$

and thus

$$\left[\hat{L}_j, \hat{H}\right]_- = 0 \tag{2.65}$$

holds, the operators $\hat{L}_j$ and $\hat{U}_{\text{R}}(\phi)$ specify constants of motion (cf. the Heisenberg picture in Sect. 3.2.4).

**2.2.2.3 Generating Operators of $SU(n)$.** (See [69, 39, 68, 60].) All unitary matrices with $n$ rows and $n$ columns form the group $U(n)$. If, in addition, the constraint

$$\det(\mathsf{U}) = 1 \tag{2.66}$$

is imposed, one obtains the subgroup $SU(n)$. Due to the fact that elements of a unitary matrix can be represented by

$$U_{nm} = \exp(\mathrm{i}A_{nm}) , \tag{2.67}$$

$$A_{nm} = A^*_{mn} \tag{2.68}$$

(*Hermitian matrix*), the corresponding matrix A and thus U contain

$$
\begin{aligned}
s &= \text{real and imaginary parts of all elements } A_{nm} - \\
&\quad \text{constraint (2.66)} - \text{constraints (2.68)} \\
&= 2n^2 - 1 - n^2 \\
&= n^2 - 1 \tag{2.69}
\end{aligned}
$$

parameters so that there are also $s$ independent generating operators.

The generating operators of the group $SU(n)$ will be denoted by $\hat{\lambda}_j$, $(j = 1, 2, \ldots, s,$ Hermitian); they are defined by the commutator relations

$$\boxed{\left[\hat{\lambda}_i, \hat{\lambda}_j\right]_- = 2\mathrm{i} \sum_{k=1}^{s} f_{ijk} \hat{\lambda}_k} \tag{2.70}$$

and the trace-relations

$$\boxed{\text{Tr}\left\{\hat{\lambda}_j\right\} = 0 , \quad \text{Tr}\left\{\hat{\lambda}_j \hat{\lambda}_k\right\} = 2\delta_{jk} .} \tag{2.71}$$

(The trace-relation (2.25) does not hold in the general case $n > 2$.) From (2.70) and (2.71) one readily concludes that

$$4\mathrm{i}f_{jkl} = \text{Tr}\left\{\left[\hat{\lambda}_j, \hat{\lambda}_k\right]_- \hat{\lambda}_l\right\} . \tag{2.72}$$

The *structure constants* of $SU(n)$, $f_{ijk}$, form a totally antisymmetric matrix, and the rank of the group $SU(n)$ is given by $r = n - 1$. An infinitesimal element of the group $SU(n)$ can be represented by

$$\hat{U}(\phi) = \hat{1} - \frac{\mathrm{i}}{\hbar} \sum_{j=1}^{s} \phi_j \hat{\lambda}_j \ . \tag{2.73}$$

**Theorem 2.2.1 (Racah).** *The number of linear independent operators in a Hilbert space of dimension n, which commute with all generators $\hat{\lambda}_j$ of $SU(n)$, is equal to the rank of the group $r = n - 1$. Such operators are usually called Casimir operators.*

*Example 2.2.1.* The group $SU(2)$ is characterized by $n = 2$, $s = n^2 - 1 = 3$, and $r = 1$. In this case, the generating operators $\hat{\lambda}_j$ can be identified with the operators $\hat{\sigma}_j$ ($i = 1, 2, 3$) according to (2.41), where $\hat{\sigma}_x := \hat{u}_{12}$, $\hat{\sigma}_y := \hat{v}_{12}$, $\hat{\sigma}_z := \hat{w}_1$ with

$$\hat{u}_{12} = \hat{P}_{12} + \hat{P}_{21} \ ,$$
$$\hat{v}_{12} = \mathrm{i} \left( \hat{P}_{12} - \hat{P}_{21} \right) \ , \tag{2.74}$$
$$\hat{w}_1 = -\hat{P}_{11} + \hat{P}_{22} \ ,$$

and

$$\hat{C} = \sum_{i=1}^{3} \hat{\lambda}_i^2 = \frac{4}{\hbar^2} \hat{\boldsymbol{L}}^2 \tag{2.75}$$

is a Casimir operator. Equation (2.70) then coincides with (2.20). In $SU(2)$ the structure constants $f_{ijk}$ are given by the $\varepsilon$ tensor (cf. (2.2)).

*Example 2.2.2.* In the case of $SU(3)$ ($n = 3$, $s = 8$, $r = 2$), the three operators given by (2.74) have to be supplemented by

$$\hat{u}_{13} = \hat{P}_{13} + \hat{P}_{31} \ ,$$
$$\hat{u}_{23} = \hat{P}_{23} + \hat{P}_{32} \ ,$$
$$\hat{v}_{13} = \mathrm{i} \left( \hat{P}_{13} - \hat{P}_{31} \right) \ ,$$
$$\hat{v}_{23} = \mathrm{i} \left( \hat{P}_{23} - \hat{P}_{32} \right) \ ,$$
$$\hat{w}_2 = -\frac{1}{\sqrt{3}} \left( \hat{P}_{11} + \hat{P}_{22} - 2\hat{P}_{33} \right) \ . \tag{2.76}$$

Using the notation $\hat{\boldsymbol{\lambda}} = \left\{ \hat{\lambda}_1, \hat{\lambda}_2, \dots, \hat{\lambda}_i, \dots, \hat{\lambda}_s \right\}$, the set of generating operators now reads

$$\hat{\boldsymbol{\lambda}} = \{ \hat{u}_{12}, \hat{u}_{13}, \hat{u}_{23}, \hat{v}_{12}, \hat{v}_{13}, \hat{v}_{23}, \hat{w}_1, \hat{w}_2 \} \ . \tag{2.77}$$

In $SU(3)$ the structure constants are given by

$$f_{ijk} = \begin{cases} +1 & \text{for} \quad (i,j,k) = (1,4,7),(4,7,1),(7,1,4) \\ -1 & \text{for} \quad (i,j,k) = (4,1,7),(1,7,4),(7,4,1) \\ +\dfrac{1}{2} & \text{for} \quad (i,j,k) = (2,1,6),(1,6,2),(6,2,1), \\ & \qquad\qquad\qquad (3,1,5),(1,5,3),(5,3,1), \\ & \qquad\qquad\qquad (3,2,4),(2,4,3),(4,3,2), \\ & \qquad\qquad\qquad (2,5,7),(5,7,2),(7,2,5), \\ & \qquad\qquad\qquad (3,7,6),(7,6,3),(6,3,7), \\ & \qquad\qquad\qquad (5,4,6),(4,6,5),(6,5,4) \\ -\dfrac{1}{2} & \text{for} \quad (i,j,k) = (1,2,6),(2,6,1),(6,1,2), \\ & \qquad\qquad\qquad (1,3,5),(3,5,1),(5,1,3), \\ & \qquad\qquad\qquad (2,3,4),(3,4,2),(4,2,3), \\ & \qquad\qquad\qquad (5,2,7),(2,7,5),(7,5,2), \\ & \qquad\qquad\qquad (7,3,6),(3,6,7),(6,7,3), \\ & \qquad\qquad\qquad (4,5,6),(5,6,4),(6,4,5) \\ +\dfrac{1}{2}\sqrt{3} & \text{for} \quad (i,j,k) = (3,6,8),(6,8,3),(8,3,6), \\ & \qquad\qquad\qquad (2,5,8),(5,8,2),(8,2,5) \\ -\dfrac{1}{2}\sqrt{3} & \text{for} \quad (i,j,k) = (6,3,8),(3,8,6),(8,6,3), \\ & \qquad\qquad\qquad (5,2,8),(2,8,5),(8,5,2) \\ 0 & \text{for} \quad (i,j,k) = \text{otherwise} \end{cases} \qquad (2.78)$$

In the general case,

$$\hat{\boldsymbol{\lambda}} = \{\hat{u}_{12}, \hat{u}_{13}, \hat{u}_{23}, \ldots, \hat{v}_{12}, \hat{v}_{13}, \hat{v}_{23}, \ldots, \hat{w}_1, \hat{w}_2, \ldots, \hat{w}_{n-1}\} \qquad (2.79)$$

with

$$\boxed{\begin{aligned} \hat{u}_{jk} &= \hat{P}_{jk} + \hat{P}_{kj}\,, \\ \hat{v}_{jk} &= \mathrm{i}\left(\hat{P}_{jk} - \hat{P}_{kj}\right)\,, \\ \hat{w}_l &= -\sqrt{\frac{2}{l(l+1)}}\left(\hat{P}_{11} + \ldots + \hat{P}_{ll} - l\hat{P}_{l+1,l+1}\right) \end{aligned}} \qquad (2.80)$$

applies, where

$$\boxed{1 \le j < k \le n\,, \quad 1 \le l \le n-1\,, \quad n^2 - 1 \text{ operators}\,.} \qquad (2.81)$$

One may note in passing that the operator relations defining $\hat{w}_l$ can directly be compared with coordinate relations defining the so-called *Jacobian coor-*

*dinates* (in one dimension) of classical mechanics:

$$\xi_1 = r_1 - r_2 \ ,$$

$$\xi_2 = \tfrac{1}{2}\left(r_1 + r_2 - 2r_3\right) \ ,$$

$$\xi_3 = \tfrac{1}{3}\left(r_1 + r_2 + r_3 - 3r_4\right) \ , \tag{2.82}$$

$$\cdot \quad \cdot \qquad \cdot \qquad \cdot \qquad \cdot$$

$$\xi_l = \tfrac{1}{l}\left(r_1 + r_2 + r_3 + \ldots + r_l - l r_{l+1}\right) \ .$$

In general, these terms are vectors. The "relative" operators $\hat{w}_l$ share with these relative coordinates the disadvantage that they hardly correspond to a direct and intuitively clear observable: they are mathematical constructs.

### 2.2.2.4 Operator Representations in $\mathbb{H}$.

Consider $n^2$ operators $\hat{G}_j$ defined in Hilbert space $\mathbb{H}$ of dimension $n$ which fulfil the orthogonality relation

$$\mathrm{Tr}\left\{\hat{G}_j \hat{G}_k\right\} = \delta_{jk} \ , \tag{2.83}$$

then any operator $\hat{A}$ can be represented as

$$\hat{A} = \sum_{j=1}^{n^2} \hat{G}_j \,\mathrm{Tr}\left\{\hat{G}_j \hat{A}\right\} \tag{2.84}$$

(compare the state representation (2.15)).

With $s = n^2 - 1$ generating operators $\hat{\lambda}_j$ according to (2.70)–(2.71), the operator representation of the Hermitian operator $\hat{A}$ reads

$$\boxed{\hat{A} = \frac{1}{n} \mathcal{A}_0 \hat{1} + \frac{1}{2} \sum_{j=1}^{s} \mathcal{A}_j \hat{\lambda}_j \ ,} \tag{2.85}$$

wherein

$$\mathcal{A}_0 = \mathrm{Tr}\left\{\hat{A}\right\} \ , \quad \mathcal{A}_j = \mathrm{Tr}\left\{\hat{A}\hat{\lambda}_j\right\} = \text{real} \ . \tag{2.86}$$

$\mathcal{A}_0$ and the $s$-dimensional vector $(\mathcal{A}_j)$ uniquely specify the operator $\hat{A}$ with respect to a given basis $|m\rangle$. These vectors are elements of an Euclidean vector space $\mathbb{V}$ of dimension $s$.

*Example 2.2.3.* Consider in $SU(2)$ the operator $\hat{A} = \hat{\boldsymbol{\sigma}}\boldsymbol{a}$ with $\hat{\boldsymbol{\sigma}} = (\hat{\sigma}_1, \hat{\sigma}_2, \hat{\sigma}_3)$, and $\boldsymbol{a}$ any 3-dimensional vector. In this case we get $\mathcal{A}_0 = 0$ and

$$\mathcal{A}_k = \mathrm{Tr}\left\{(\hat{\boldsymbol{\sigma}}\boldsymbol{a})\,\hat{\lambda}_k\right\} \ . \tag{2.87}$$

Taking, e.g.,

$$\boldsymbol{a} = (\sin\Theta, 0, \cos\Theta) \ , \tag{2.88}$$

and observing (2.24),

$$\text{Tr}\{\hat{\sigma}_i\hat{\sigma}_j\} = 2\delta_{ij} \ , \tag{2.89}$$

(2.87) implies the vector components

$$
\begin{aligned}
\mathcal{A}_1 &= \text{Tr}\left\{(\hat{\sigma}_1 \sin\Theta + \hat{\sigma}_3 \cos\Theta)\,\hat{\sigma}_1\right\} = 2\sin\Theta \ , \\
\mathcal{A}_2 &= 0 \ , \\
\mathcal{A}_3 &= \text{Tr}\left\{(\hat{\sigma}_1 \sin\Theta + \hat{\sigma}_3 \cos\Theta)\,\hat{\sigma}_3\right\} = 2\cos\Theta \ .
\end{aligned}
\tag{2.90}
$$

*Example 2.2.4.* The $SU(2)$ representation of the operator product $\hat{A} = \hat{\sigma}_i\hat{\sigma}_j$ leads to the vector components

$$\mathcal{A}_k = 2\mathrm{i}\varepsilon_{ijk} \ , \tag{2.91}$$

where we have used (2.25).

*Example 2.2.5.* The $x$ and $y$ components of the angular momentum for $l = 1$ can be written in terms of the generating operators of $SU(3)$ as (cf. (2.43))

$$\frac{1}{\hbar}\hat{L}_x = \frac{1}{\sqrt{2}}\left(\hat{u}_{12} + \hat{u}_{23}\right) \ , \quad \frac{1}{\hbar}\hat{L}_y = \frac{1}{\sqrt{2}}\left(\hat{v}_{12} + \hat{v}_{23}\right) \ . \tag{2.92}$$

Correspondingly, for $l = 3/2$ we find the representation in $SU(4)$:

$$\frac{1}{\hbar}\hat{L}_x = \frac{\sqrt{3}}{2}\left(\hat{u}_{12} + \hat{u}_{34}\right) + \hat{u}_{23} \ , \quad \frac{1}{\hbar}\hat{L}_y = \frac{\sqrt{3}}{2}\left(\hat{v}_{12} + \hat{v}_{34}\right) + \hat{v}_{23} \ . \tag{2.93}$$

Comparison with (2.85) directly allows to identify $\mathcal{A}_0$ and $\mathcal{A}_k$ for $\hat{A} = \hat{L}_x/\hbar$, $\hat{L}_y/\hbar$.

**2.2.2.5 Anticommutator of $\hat{\lambda}_i$.** For later reference we also consider the anticommutator

$$\left[\hat{\lambda}_i, \hat{\lambda}_k\right]_+ = \begin{cases} 2\hat{\lambda}_i^2 \\ \hat{\lambda}_i\hat{\lambda}_k + \hat{\lambda}_k\hat{\lambda}_i \end{cases} \text{for} \begin{array}{c} i = k \\ i \neq k \end{array} \ . \tag{2.94}$$

We apply the operator representation (2.85) to

$$\hat{\lambda}_i^2 = \frac{2}{n}\hat{1} + \frac{1}{2}\sum_{p=1}^{s}\text{Tr}\left\{\hat{\lambda}_i^2\hat{\lambda}_p\right\}\hat{\lambda}_p \tag{2.95}$$

and to

$$\hat{\lambda}_i\hat{\lambda}_k = \frac{1}{2}\sum_{p=1}^{s}\text{Tr}\left\{\hat{\lambda}_i\hat{\lambda}_k\hat{\lambda}_p\right\}\hat{\lambda}_p \ , \tag{2.96}$$

where

$$\text{Tr}\left\{\hat{\lambda}_i\hat{\lambda}_k\right\} = 2\delta_{ik} \tag{2.97}$$

(cf. (2.71)) has been used. We thus obtain

$$\boxed{\left[\hat{\lambda}_i, \hat{\lambda}_k\right]_+ = \frac{4}{n}\hat{1}\delta_{ik} + 2\sum_{p=1}^{s}d_{ikp}\hat{\lambda}_p \ ,} \tag{2.98}$$

$$4d_{ikp} = \mathrm{Tr}\left\{\left[\hat{\lambda}_i, \hat{\lambda}_k\right]_+ \hat{\lambda}_p\right\} \tag{2.99}$$

(cf. Sect. 2.2.2.3). The tensor coefficients $d_{ikp}$ (like the structure constants $f_{ijk}$ defined by (2.70)) can be found tabulated in the literature. In $SU(2)$ those coefficients are identically zero (cf. (2.25)); in $SU(3)$ the following relations hold:

$$d_{ikp} = \begin{cases} +\dfrac{1}{2} & \text{for} \quad (i,k,p) = (1,2,3),(2,3,1),(3,1,2), \\ & \qquad\qquad\quad (2,1,3),(1,3,2),(3,2,1), \\ & \qquad\qquad\quad (1,5,6),(5,6,1),(6,1,5), \\ & \qquad\qquad\quad (5,1,6),(1,6,5),(6,5,1), \\ & \qquad\qquad\quad (3,4,5),(4,5,3),(5,3,4), \\ & \qquad\qquad\quad (4,3,5),(3,5,4),(5,4,3), \\ & \qquad\qquad\quad (3,3,7),(3,7,3),(7,3,3), \\ & \qquad\qquad\quad (6,6,7),(6,7,6),(7,6,6) \\[4pt] -\dfrac{1}{2} & \text{for} \quad (i,k,p) = (2,4,6),(4,6,2),(6,2,4), \\ & \qquad\qquad\quad (4,2,6),(2,6,4),(6,4,2), \\ & \qquad\qquad\quad (2,2,7),(2,7,2),(7,2,2), \\ & \qquad\qquad\quad (5,5,7),(5,7,5),(7,5,5) \\[4pt] -\dfrac{1}{\sqrt{3}} & \text{for} \quad (i,k,p) = (1,1,8),(1,8,1),(8,1,1), \\ & \qquad\qquad\quad (4,4,8),(4,8,4),(8,4,4), \\ & \qquad\qquad\quad (7,7,8),(7,8,7),(8,7,7) \\[4pt] +\dfrac{1}{\sqrt{3}} & \text{for} \quad (i,k,p) = (8,8,8) \\[4pt] +\dfrac{1}{2\sqrt{3}} & \text{for} \quad (i,k,p) = (2,2,8),(2,8,2),(8,2,2), \\ & \qquad\qquad\quad (3,3,8),(3,8,3),(8,3,3), \\ & \qquad\qquad\quad (5,5,8),(5,8,5),(8,5,5), \\ & \qquad\qquad\quad (6,6,8),(6,8,6),(8,6,6) \\[4pt] 0 & \text{for} \quad (i,k,p) = \text{otherwise} \end{cases} \tag{2.100}$$

**2.2.2.6 Complete Set of Operators in $SU(n)$.** One easily convinces oneself that

$$[\hat{w}_l, \hat{w}_{l'}]_- = 0 \quad (l, l' = 1, 2, \ldots, n-1) . \tag{2.101}$$

These $n-1$ operators together with the unit operator form a complete set in the sense that any other operator is either "incompatible" (i.e. does not commute with all these $\hat{w}_l$) or is a linear combination. Then let $|\nu\rangle$ denote the joint eigenstates obeying the respective eigenvalue equations

**Table 2.1.** Eigenvalues $w_l^{(\nu)}$ of the operators $\hat{w}_l$ in $SU(n)$

| $n = 2$ | $n = 3$ | $n = 3$ |
|---|---|---|
| $l = 1$ | $l = 1$ | $l = 2$ |
| $w_1^{(1)} = -1$ | $w_1^{(1)} = -1$ | $w_2^{(1)} = -\frac{1}{\sqrt{3}}$ |
| $w_1^{(2)} = 1$ | $w_1^{(2)} = 1$ | $w_2^{(2)} = -\frac{1}{\sqrt{3}}$ |
| | $w_1^{(3)} = 0$ | $w_2^{(3)} = \frac{2}{\sqrt{3}}$ |

| $n = 4$ | $n = 4$ | $n = 4$ |
|---|---|---|
| $l = 1$ | $l = 2$ | $l = 3$ |
| $w_1^{(1)} = -1$ | $w_2^{(1)} = -\frac{1}{\sqrt{3}}$ | $w_3^{(1)} = -\frac{1}{\sqrt{6}}$ |
| $w_1^{(2)} = 1$ | $w_2^{(2)} = -\frac{1}{\sqrt{3}}$ | $w_3^{(2)} = -\frac{1}{\sqrt{6}}$ |
| $w_1^{(3)} = 0$ | $w_2^{(3)} = \frac{2}{\sqrt{3}}$ | $w_3^{(3)} = -\frac{1}{\sqrt{6}}$ |
| $w_1^{(4)} = 0$ | $w_2^{(4)} = 0$ | $w_3^{(4)} = \frac{3}{\sqrt{6}}$ |

$$\hat{w}_l \, |\nu\rangle = w_l^{(\nu)} \, |\nu\rangle \ , \quad 1 \le \nu \le n \ , \quad 1 \le l \le n - 1 \ . \tag{2.102}$$

The eigenvalues are from (2.80):

$$
\begin{aligned}
w_l^{(\nu)} &= -\sqrt{\frac{2}{l(l+1)}} & &: 1 \le \nu \le l \ , \\
w_l^{(l+1)} &= l\sqrt{\frac{2}{l(l+1)}} & &: \nu = l + 1 \ , \\
w_l^{(\nu)} &= 0 & &: l + 1 < \nu \le n \ .
\end{aligned}
\tag{2.103}
$$

Explicit examples are shown in Table 2.1.

**2.2.2.7 Operator in $w_l$ Subspace.** For an operator $\hat{A}$ in an $n$-dimensional Hilbert space the eigenvalue equation reads

$$\hat{A} \, |\nu\rangle = A^{(\nu)} \, |\nu\rangle \ , \quad \nu = 1, 2, \ldots, n \ . \tag{2.104}$$

For

$$\left[ \hat{A}, \hat{w}_l \right]_- = 0 \ , \quad l = 1, 2, \ldots, n - 1 \tag{2.105}$$

any eigenvector of $\hat{A}$ is also eigenvector of the operators $\hat{w}_l$,

$$\hat{w}_l \, |\nu\rangle = w_l^{(\nu)} \, |\nu\rangle \ , \tag{2.106}$$

and can thus be specified by the $n - 1$ values $w_l^{(\nu)}$ as introduced above. In the following we study the problem of finding the eigenvalues of an operator $\hat{A}$ given its $SU(n)$ representation $\mathcal{A}_l$ and $\mathcal{A}_0$.

Let $\hat{A}$ have the operator representation

$$\boxed{\hat{A} = \frac{1}{n}\mathcal{A}_0\hat{1} + \frac{1}{2}\sum_{l=1}^{n-1}\mathcal{A}_l\hat{w}_l ,}$$
(2.107)

$$\mathcal{A}_l = \text{Tr}\left\{\hat{A}\hat{w}_l\right\} , \quad \mathcal{A}_0 = \text{Tr}\left\{\hat{A}\right\} .$$
(2.108)

Inserting (2.107) into (2.104), one obtains

$$\hat{A}\,|\nu\rangle = \left(\frac{1}{n}\mathcal{A}_0\hat{1} + \frac{1}{2}\sum_{l=1}^{n-1}\mathcal{A}_l\hat{w}_l\right)|\nu\rangle = A^{(\nu)}\,|\nu\rangle ,$$
(2.109)

and, with (2.106), one obtains

$$\boxed{\hat{A}\,|\nu\rangle = \left(\frac{1}{n}\mathcal{A}_0 + \frac{1}{2}\sum_{l=1}^{n-1}\mathcal{A}_l w_l^{(\nu)}\right)|\nu\rangle = A^{(\nu)}\,|\nu\rangle ,}$$
(2.110)

from which one can read off the eigenvalues $A^{(\nu)}$ in terms of $\mathcal{A}_l$ and the eigenvalues of $\hat{w}_l$. If $A^{(\nu)}$ are non-degenerate, $\hat{A}$ is said to be "complete".

*Example 2.2.6.* In $SU(n)$ the projectors $\hat{P}_{mm}$ are operators in $\hat{w}_l$ subspace, i.e. they have the representation

$$\hat{P}_{mm} = \frac{1}{n}\hat{1} + \frac{1}{2}\sum_{l=1}^{n-1}w_l^{(m)}\hat{w}_l \quad (m = 1, 2, \ldots, n) .$$
(2.111)

This form follows from the fact that $\text{Tr}\left\{\hat{P}_{mm}\right\} = 1$ and

$$\text{Tr}\left\{\hat{P}_{mm}\hat{\lambda}_j\right\} = \begin{cases} w_l^{(m)} & : \quad j = s' + l \\ 0 & : \quad \text{otherwise} \end{cases} ,$$
(2.112)

where $s' = n^2 - n$ and the eigenvalues $w_l^{(m)}$ are given by (2.103).

If we apply (2.110) to $\hat{A} = \hat{P}_{mm}$ and observe that the eigenvalues of $\hat{P}_{mm}$ are simply

$$A^{(\nu)} = \delta_{\nu m} ,$$
(2.113)

we end up with

$$\boxed{\frac{1}{n} + \frac{1}{2}\sum_{l=1}^{n-1}w_l^{(m)}w_l^{(\nu)} = \delta_{\nu m} .}$$
(2.114)

This sum rule is easily confirmed based on Table 2.1 or (2.103).

Equation (2.111) together with Table 2.1 is useful for transforming operators in terms of projectors to the $SU(n)$ form:

*Example 2.2.7.*

$$\frac{1}{\hbar}\hat{L}_z(l=1) = -\hat{P}_{11} + \hat{P}_{33} = \frac{1}{2}\left(\hat{w}_1 + \sqrt{3}\hat{w}_2\right) , \tag{2.115}$$

$$\frac{1}{\hbar}\hat{L}_z(l=3/2) = -\frac{3}{2}\hat{P}_{11} - \frac{1}{2}\hat{P}_{22} + \frac{1}{2}\hat{P}_{33} + \frac{3}{2}\hat{P}_{44}$$

$$= \frac{1}{2}\left(\hat{w}_1 + \sqrt{3}\hat{w}_2 + \frac{\sqrt{6}}{4}\hat{w}_3\right) . \tag{2.116}$$

$\hat{L}_z$ is "complete" for any $l$.

### 2.2.3 Unitary Transformations

Unitary transformations $\hat{U}$ with

$$\hat{U}\hat{U}^\dagger = \hat{U}^\dagger\hat{U} = \hat{1} \tag{2.117}$$

play a very specific role in quantum mechanics: leaving the basic quantum structure in terms of commutator relations, expectation values, etc. invariant, they allow for flexible mathematical manipulations. Basis changes and unitary time evolution are important applications.

**2.2.3.1 Basis Transformations.** Any operator in Hilbert space $\mathbb{H}$ of dimension $n$, e.g. a wave function $|\psi\rangle$, can be represented with respect to various basis states. The transformation from one basis to another is described by a unitary operator $\hat{U}$: let $|\alpha\rangle$ be a new basis, where

$$\boxed{|\alpha\rangle = \sum_j |j\rangle\langle j|\alpha\rangle = \sum_j |j\rangle U_{j\alpha}} \tag{2.118}$$

expresses $|\alpha\rangle$ in terms of the original eigenvectors $|j\rangle$ and the unitary matrix $U_{j\alpha}$. Then the new matrix elements are

$$\boxed{A_{\alpha\beta} = \left\langle\alpha\left|\hat{A}\right|\beta\right\rangle = \sum_{j,k} A_{kj} U_{\alpha k} U_{j\beta} ,} \tag{2.119}$$

wherein $A_{kj}$ denotes the matrix elements in $|j\rangle$ basis.

Also the corresponding vector $\mathcal{A}_j$ defined in an Euclidean vector space $\mathbb{V}$ of dimension $s = n^2 - 1$ is defined with respect to a specific basis $|i\rangle$. A unitary transformation to a new basis implies an orthogonal transformation in $O(s)$,

$$\left(\mathcal{A}'_j\right) = \mathsf{X}\left(\mathcal{A}_j\right) \tag{2.120}$$

with

$$\det\left(\mathsf{X}\right) = 1 . \tag{2.121}$$

*Example 2.2.8.* The expression

$$A = \frac{1}{2} \left( \begin{array}{cc} \mathcal{A}_0 - \mathcal{A}_3 & \mathcal{A}_1 + \mathrm{i}\mathcal{A}_2 \\ \mathcal{A}_1 - \mathrm{i}\mathcal{A}_2 & \mathcal{A}_0 + \mathcal{A}_3 \end{array} \right) \qquad (2.122)$$

defines the matrix representation of operator $\hat{A}$ in terms of $\mathcal{A}_0$ and the $SU(2)$ vector components $\mathcal{A}_1$, $\mathcal{A}_2$, $\mathcal{A}_3$. As the Pauli matrices have been used in the $\hat{\sigma}_z$ representation, (2.122) defines the $\hat{\sigma}_z$ representation of operator $\hat{A}$. The eigenvectors $|e_1^x\rangle$, $|e_{-1}^x\rangle$ of the operator $\hat{\sigma}_x$ can be written in the form (2.31) with

$$\langle e_1^x | 1 \rangle = \langle 1 | e_1^x \rangle = \frac{1}{\sqrt{2}}, \quad \langle e_{-1}^x | 1 \rangle = \langle 1 | e_{-1}^x \rangle = \frac{1}{\sqrt{2}},$$

$$\langle e_1^x | {-1} \rangle = \langle {-1} | e_1^x \rangle = \frac{1}{\sqrt{2}}, \quad \langle e_{-1}^x | {-1} \rangle = \langle {-1} | e_{-1}^x \rangle = -\frac{1}{\sqrt{2}} \qquad (2.123)$$

so that

$$U = \frac{1}{\sqrt{2}} \left( \begin{array}{cc} -1 & 1 \\ 1 & 1 \end{array} \right), \quad U_{j\alpha} = U_{\alpha j} \qquad (2.124)$$

and

$$\left\langle e_{-1}^x \left| \hat{A} \right| e_{-1}^x \right\rangle = \frac{1}{2} \left( A_{11} - A_{1-1} - A_{-11} + A_{-1-1} \right), \qquad (2.125)$$

$$\left\langle e_1^x \left| \hat{A} \right| e_{-1}^x \right\rangle = \frac{1}{2} \left( A_{11} - A_{1-1} + A_{-11} - A_{-1-1} \right), \qquad (2.126)$$

$$\left\langle e_{-1}^x \left| \hat{A} \right| e_1^x \right\rangle = \frac{1}{2} \left( A_{11} + A_{1-1} - A_{-11} - A_{-1-1} \right), \qquad (2.127)$$

$$\left\langle e_1^x \left| \hat{A} \right| e_1^x \right\rangle = \frac{1}{2} \left( A_{11} + A_{1-1} + A_{-11} + A_{-1-1} \right). \qquad (2.128)$$

Then, using (2.122), i.e.

$$A_{-1-1} = \frac{1}{2}(\mathcal{A}_0 - \mathcal{A}_3), \quad A_{11} = \frac{1}{2}(\mathcal{A}_0 + \mathcal{A}_3),$$

$$A_{-11} = \frac{1}{2}(\mathcal{A}_1 + \mathrm{i}\mathcal{A}_2), \quad A_{1-1} = \frac{1}{2}(\mathcal{A}_1 - \mathrm{i}\mathcal{A}_2), \qquad (2.129)$$

one obtains the result

$$A = \frac{1}{2} \left( \begin{array}{cc} \mathcal{A}_0 - \mathcal{A}_1^{(z)} & \mathcal{A}_3^{(z)} - \mathrm{i}\mathcal{A}_2^{(z)} \\ \mathcal{A}_3^{(z)} + \mathrm{i}\mathcal{A}_2^{(z)} & \mathcal{A}_0 + \mathcal{A}_1^{(z)} \end{array} \right), \qquad (2.130)$$

which represents the matrix of $\hat{A}$ in the $\hat{\sigma}_x$ basis. The upper index $z$ indicates that the vector components $\mathcal{A}_i^{(z)}$ refer to the $\hat{\sigma}_z$ representation. Comparing with the standard from (2.122) this result can be written as

$$
\begin{pmatrix} A_1^{(x)} \\ A_2^{(x)} \\ A_3^{(x)} \end{pmatrix} = \begin{pmatrix} 0 & 0 & 1 \\ 0 & -1 & 0 \\ 1 & 0 & 0 \end{pmatrix} \begin{pmatrix} A_1^{(z)} \\ A_2^{(z)} \\ A_3^{(z)} \end{pmatrix} . \tag{2.131}
$$

This is an example of an orthogonal transformation X in three dimensions.

*Example 2.2.9.* In the same way, a matrix representation of $\hat{A}$ on the basis of eigenvectors $|e_1^y\rangle$, $|e_{-1}^y\rangle$ (see (2.32)) of the operator $\hat{\sigma}_y$ can be calculated:

$$
\boxed{A = \frac{1}{2} \begin{pmatrix} A_0 - A_2^{(z)} & A_3^{(z)} + i A_1^{(z)} \\ A_3^{(z)} - i A_1^{(z)} & A_0 + A_2^{(z)} \end{pmatrix} .} \tag{2.132}
$$

This result defines the orthogonal transformation

$$
\begin{pmatrix} A_1^{(y)} \\ A_2^{(y)} \\ A_3^{(y)} \end{pmatrix} = \begin{pmatrix} 0 & 0 & 1 \\ 1 & 0 & 0 \\ 0 & 1 & 0 \end{pmatrix} \begin{pmatrix} A_1^{(z)} \\ A_2^{(z)} \\ A_3^{(z)} \end{pmatrix} . \tag{2.133}
$$

The possibility of such transformations underlies complete (ensemble) measurements of the density matrix (cf. Sect. 2.3.6.4).

### 2.2.3.2 Factorization of Unitary Transformations: Quantum Gates.
It turns out that a general rotation in the $(2^2 - 1)$-dimensional vector space of $SU(2)$ specified by the Euler angles $\alpha$, $\beta$, $\gamma$, i.e.

$$
\begin{pmatrix} A_x' \\ A_y' \\ A_z' \end{pmatrix} = \begin{pmatrix} X_{xx} & X_{xy} & X_{xz} \\ X_{yx} & X_{yy} & X_{yz} \\ X_{zx} & X_{zy} & X_{zz} \end{pmatrix} \begin{pmatrix} A_x \\ A_y \\ A_z \end{pmatrix} , \tag{2.134}
$$

$$
\begin{aligned}
X_{xx} &= \cos\gamma \cos\beta \cos\alpha - \sin\gamma \sin\alpha , \\
X_{xy} &= -\sin\gamma \cos\beta \cos\alpha - \cos\gamma \sin\alpha , \\
X_{xz} &= \sin\beta \cos\alpha , \\
X_{yx} &= \cos\gamma \cos\beta \sin\alpha + \sin\gamma \cos\alpha , \\
X_{yy} &= -\sin\gamma \cos\beta \sin\alpha + \cos\gamma \cos\alpha , \\
X_{yz} &= \sin\beta \sin\alpha , \\
X_{zx} &= -\cos\gamma \sin\beta , \\
X_{zy} &= \sin\gamma \cos\beta , \\
X_{zz} &= \cos\beta ,
\end{aligned} \tag{2.135}
$$

is connected with a product of unitary operators in terms of just two different generators, i.e.

$$
\boxed{\hat{R}(\alpha, \beta, \gamma) = e^{-i\alpha\hat{\lambda}_z/2} e^{-i\beta\hat{\lambda}_y/2} e^{-i\gamma\hat{\lambda}_z/2} ,} \tag{2.136}
$$

acting in the 2-dimensional Hilbert space (cf. [94]).
    Let us consider

$$\hat{U}^j(A_j) = e^{iA_j\hat{\lambda}_j/2}$$

$$= 1 + \frac{i}{1!}\frac{A_j}{2}\hat{\lambda}_j - \frac{1}{2!}\left(\frac{A_j}{2}\right)^2\hat{\lambda}_j^2 - \frac{i}{3!}\left(\frac{A_j}{2}\right)^3\hat{\lambda}_j^3 +$$

$$\frac{1}{4!}\left(\frac{A_j}{2}\right)^4\hat{\lambda}_j^4 + \dots. \tag{2.137}$$

With $\hat{\sigma}_j^2 = \hat{1}$ we find

$$\hat{U}^j(A_j) = 1 - \frac{1}{2!}\left(\frac{A_j}{2}\right)^2 + \frac{1}{4!}\left(\frac{A_j}{2}\right)^4 - \dots +$$

$$i\hat{\lambda}_j\left[\left(\frac{A_j}{2}\right) - \frac{1}{3!}\left(\frac{A_j}{2}\right)^3 + \dots\right] \tag{2.138}$$

and thus

$$\hat{U}^j(A_j) = \hat{1}\cos\left(\frac{A_j}{2}\right) + i\hat{\lambda}_j\sin\left(\frac{A_j}{2}\right) \quad (j = 1, 2, 3). \tag{2.139}$$

Using the matrix representation for $\hat{1}$ and $\hat{\lambda}_j$, one finds

$$U_{ik}^x(A_x) = \begin{pmatrix} \cos(A_x/2) & i\sin(A_x/2) \\ i\sin(A_x/2) & \cos(A_x/2) \end{pmatrix}, \tag{2.140}$$

$$U_{ik}^y(A_y) = \begin{pmatrix} \cos(A_y/2) & \sin(A_y/2) \\ -\sin(A_y/2) & \cos(A_y/2) \end{pmatrix}, \tag{2.141}$$

$$U_{ik}^z(A_z) = \begin{pmatrix} e^{-iA_z/2} & 0 \\ 0 & e^{iA_z/2} \end{pmatrix}. \tag{2.142}$$

The matrix representation of $\hat{R}(\alpha, \beta, \gamma)$ is thus given by

$$R_{ik}(\alpha, \beta, \gamma)$$
$$= \begin{pmatrix} e^{i\alpha/2}\cos(\beta/2)e^{i\gamma/2} & e^{i\alpha/2}\sin(\beta/2)e^{-i\gamma/2} \\ -e^{-i\alpha/2}\sin(\beta/2)e^{i\gamma/2} & e^{-i\alpha/2}\cos(\beta/2)e^{-i\gamma/2} \end{pmatrix}. \tag{2.143}$$

It is clear that in this 3-dimensional space the two vectors $(A_j)$ and $(A_j')$ define a plane: if we choose a reference frame in which the unit vector $e_z$, say, coincides with the normal vector of this plane, the general rotation described by three Euler angles reduces to a single rotation around the $z$ axis, by an angle $\varphi$. In a vector space $\mathbb{V}$ of higher dimension $s > 3$ one shows in linear algebra that there exists an orthonormal basis such that any orthogonal transformation X can be decomposed into a block-diagonal form, where each block is either a number $\pm 1$ or a $2 \times 2$-matrix of the form

$$Y_i = \begin{pmatrix} \cos(\varphi_i) & -\sin(\varphi_i) \\ \sin(\varphi_i) & \cos(\varphi_i) \end{pmatrix}, \tag{2.144}$$

i.e. a rotation in a 2-dimensional subspace. X thus factorizes into a finite set of actions in dimension $\leq 2$ ("local"). This also applies to (2.120) interpreted as an "active" transformation for a fixed basis. If $(A_j)$ is interpreted to stand for an input state $\lambda$, $(A_j')$ for the output state $\lambda'$, the output can be thought to be generated by a sequence of local manipulations, sometimes called *quantum gates*.

*Factorization* means that one and the same type of local unitary transformation (with parameter $\varphi_i$) can be combined in a gate array to produce any desired unitary transformation in the total state space. Such gates are consequently called *universal*. However, the factorization only holds for this very specific basis. With respect to a general basis, the action of those gates will be non-local, and this non-locality is just typical for quantum networks.

### 2.2.3.3 Unitary Transformations Generated by $\hat{P}_{kk}$.

We consider the parameter-dependent transformation

$$\hat{U}_{kk}(\alpha_k) := e^{i\alpha_k \hat{P}_{kk}} = \hat{1} + i\alpha_k \hat{P}_{kk} + \frac{(i\alpha_k)^2}{2} \hat{P}_{kk}^2 + \cdots . \tag{2.145}$$

Observing that

$$\hat{P}_{kk}\hat{P}_{kk} = \hat{P}_{kk} , \tag{2.146}$$

$\hat{U}_{kk}(\alpha_k)$ can be condensed into

$$\hat{U}_{kk}(\alpha_k) = \hat{1} + \hat{P}_{kk}\left[i\alpha_k + \frac{(i\alpha_k)^2}{2} + \cdots\right] = \hat{1} + \hat{P}_{kk}\left(e^{i\alpha_k} - 1\right) . \tag{2.147}$$

It thus follows that

$$\left[\hat{P}_{nn}, \hat{U}_{kk}(\alpha_k)\right]_- = 0 ,$$

$$\left[\hat{P}_{nk}, \hat{U}_{kk}(\alpha_k)\right]_- = \left[\hat{P}_{nk}, \hat{P}_{kk}\right]_- \left(e^{i\alpha_k} - 1\right) = -\hat{P}_{nk}\left(e^{i\alpha_k} - 1\right) , \tag{2.148}$$

$$\left[\hat{P}_{kn}, \hat{U}_{kk}(\alpha_k)\right]_- = \left[\hat{P}_{kn}, \hat{P}_{kk}\right]_- \left(e^{i\alpha_k} - 1\right) = \hat{P}_{kn}\left(e^{i\alpha_k} - 1\right)$$

$(n, k = 1, 2, \ldots)$ and

$$\left[\hat{P}_{nm}, \hat{U}_{kk}(\alpha_k)\right]_- = 0 \tag{2.149}$$

$(n \neq m, \; n, m \neq k)$, and the transformed projection operators read

$$\hat{U}_{kk}^{\dagger}(\alpha_k)\hat{P}_{nn}\hat{U}_{kk}(\alpha_k) = \hat{P}_{nn} + \hat{U}_{kk}^{\dagger}(\alpha_k)\left[\hat{P}_{nn}, \hat{U}_{kk}(\alpha_k)\right]_{-}$$
$$= \hat{P}_{nn} ,$$
$$\hat{U}_{kk}^{\dagger}(\alpha_k)\hat{P}_{kn}\hat{U}_{kk}(\alpha_k) = \hat{P}_{kn} + \hat{U}_{kk}^{\dagger}(\alpha_k)\left[\hat{P}_{kn}, \hat{U}_{kk}(\alpha_k)\right]_{-}$$
$$= \hat{P}_{kn} -$$
$$\left[\hat{1} + \hat{P}_{kk}\left(e^{-i\alpha_k} - 1\right)\right]\hat{P}_{kn}\left(e^{i\alpha_k} - 1\right) , \qquad (2.150)$$
$$\hat{U}_{kk}^{\dagger}(\alpha_k)\hat{P}_{nk}\hat{U}_{kk}(\alpha_k) = \hat{P}_{nk} + \hat{U}_{kk}^{\dagger}(\alpha_k)\left[\hat{P}_{nk}, \hat{U}_{kk}(\alpha_k)\right]_{-}$$
$$= \hat{P}_{nk} +$$
$$\left[\hat{1} + \hat{P}_{kk}\left(e^{-i\alpha_k} - 1\right)\right]\hat{P}_{nk}\left(e^{i\alpha_k} - 1\right) .$$

Taking

$$\hat{P}_{kk}\hat{P}_{kn} = \hat{P}_{kn} ,$$
$$\hat{P}_{kk}\hat{P}_{nk} = 0 \qquad\qquad (2.151)$$

into account, the transformed projection operators can be written as

$$\boxed{\begin{aligned}
\hat{U}_{kk}^{\dagger}(\alpha_k)\hat{P}_{mn}\hat{U}_{kk}(\alpha_k) &= \hat{P}_{mn} \quad (m, n \neq k) , \\
\hat{U}_{kk}^{\dagger}(\alpha_k)\hat{P}_{kn}\hat{U}_{kk}(\alpha_k) &= \hat{P}_{kn}e^{-i\alpha_k} \quad (n \neq k) , \\
\hat{U}_{kk}^{\dagger}(\alpha_k)\hat{P}_{nk}\hat{U}_{kk}(\alpha_k) &= \hat{P}_{nk}e^{i\alpha_k} \quad (n \neq k) , \\
\hat{U}_{kk}^{\dagger}(\alpha_k)\hat{P}_{kk}\hat{U}_{kk}(\alpha_k) &= \hat{P}_{kk} .
\end{aligned}} \qquad (2.152)$$

Similar relations are known for the *creation* and *annihilation operators* in second quantization. In terms of the generating operators $\hat{\lambda}_i$ of $SU(n)$, these transformations $\hat{U}_{kk}$ leave the $\hat{w}_l$ subspace invariant and only affect the coherence terms $\hat{u}_{ij}$, $\hat{v}_{ij}$ for $i$ or $j = k$.

### 2.2.4 Raising and Lowering Operators

Besides the $n^2$ projection operators, and the $n^2 - 1$ generating operators of $SU(n)$ (plus the operator $\hat{1}$), there is another set which is often used: the raising and lowering operators which are related to the creation operators and destruction operators.

**2.2.4.1 Basic Definition.** Generalizing (2.42) we introduce the $n(n-1)$ non-Hermitian operators

$$\boxed{\begin{aligned}
\hat{\sigma}_{ij}^{+} &= \frac{1}{2}\left(\hat{u}_{ij} + i\hat{v}_{ij}\right) = \hat{P}_{ji} , \\
\hat{\sigma}_{ij}^{-} &= \frac{1}{2}\left(\hat{u}_{ij} - i\hat{v}_{ij}\right) = \hat{P}_{ij}
\end{aligned}} \qquad (2.153)$$

with

$$i < j \tag{2.154}$$

and

$$\left[\hat{\sigma}_{ij}^{-}, \hat{\sigma}_{kl}^{+}\right]_{-} = \left[\hat{P}_{ij}, \hat{P}_{lk}\right]_{-} = \hat{P}_{ik}\delta_{jl} - \hat{P}_{lj}\delta_{ik} . \tag{2.155}$$

Representing the diagonal projectors by products

$$\begin{aligned}
\hat{\sigma}_{ij}^{+}\hat{\sigma}_{ij}^{-} &= \hat{P}_{jj} , \quad j = 2, 3, \ldots, n , \\
\hat{\sigma}_{ij}^{-}\hat{\sigma}_{ij}^{+} &= \hat{P}_{ii} , \quad i = 1, 2, \ldots, n-1 ,
\end{aligned} \tag{2.156}$$

the operators $\hat{\sigma}_{ij}^{\pm}$ also suffice to represent any operator $\hat{A}$ in that $n$-dimensional Hilbert space. Important trace-relations are

$$\begin{aligned}
\mathrm{Tr}\left\{\hat{\sigma}_{ij}^{\pm}\right\} &= 0 , \\
\mathrm{Tr}\left\{\hat{\sigma}_{ij}^{+}\hat{\sigma}_{kl}^{-}\right\} &= \delta_{ik}\delta_{jl} .
\end{aligned} \tag{2.157}$$

*Example 2.2.10.* In a 2-dimensional Hilbert space there is only one operator $\hat{\sigma}^{+} = \hat{P}_{12}$ and one operator $\hat{\sigma}^{-} = \hat{P}_{21}$. For

$$\begin{aligned}
\hat{\sigma}^{+}\hat{\sigma}^{-} &= \hat{P}_{22} , \\
\hat{\sigma}^{-}\hat{\sigma}^{+} &= \hat{P}_{11} ,
\end{aligned} \tag{2.158}$$

completeness of the basis implies

$$\hat{1} = \sum_{j} \hat{P}_{jj} = \hat{P}_{11} + \hat{P}_{22} = \left[\hat{\sigma}^{-}, \hat{\sigma}^{+}\right]_{+} , \tag{2.159}$$

while

$$\left[\hat{\sigma}^{-}, \hat{\sigma}^{-}\right]_{+} = \left[\hat{\sigma}^{+}, \hat{\sigma}^{+}\right]_{+} = 0 . \tag{2.160}$$

The matrix representation is

$$\left(\sigma_{ij}^{+}\right) = \begin{pmatrix} 0 & 1 \\ 0 & 0 \end{pmatrix} , \quad \left(\sigma_{ij}^{-}\right) = \begin{pmatrix} 0 & 0 \\ 1 & 0 \end{pmatrix} . \tag{2.161}$$

Furthermore, the "raising" and "lowering" character is demonstrated by

$$\begin{aligned}
\hat{\sigma}^{+}|1\rangle &= |2\rangle , \\
\hat{\sigma}^{-}|2\rangle &= |1\rangle .
\end{aligned} \tag{2.162}$$

**2.2.4.2 Fermion Number States.** If state $|1\rangle$ denotes the vacuum state $|0\rangle$ of a fermion system, state $|2\rangle$ the 1-particle state of a particular mode $\nu$, the operator $\hat{\sigma}^{+}$ is identical with the so-called *creation operator* $\hat{a}^{+}$ and the operator $\hat{\sigma}^{-}$ with the so-called *destruction operator* $\hat{a}$ (also called *annihilation operator*):

$$\begin{aligned}
\hat{a}_{\nu}^{+}|0\rangle &= |1_{\nu}\rangle , \\
\hat{a}_{\nu}|1_{\nu}\rangle &= |0\rangle
\end{aligned} \tag{2.163}$$

$(\nu = 1, 2, 3, \ldots)$ with

$$\boxed{\begin{aligned}
\left[\hat{a}_\nu, \hat{a}_\mu^+\right]_+ &= \hat{1}\delta_{\nu\mu} \,, \\
\left[\hat{a}_\nu, \hat{a}_\mu\right]_+ &= \left[\hat{a}_\nu^+, \hat{a}_\mu^+\right]_+ = 0 \,.
\end{aligned}} \tag{2.164}$$

The so-called *particle number operator* is

$$\hat{N}_\nu = \hat{a}_\nu^+ \hat{a}_\nu = |1_\nu\rangle \langle 1_\nu| \,. \tag{2.165}$$

These operators describe 2-dimensional (number) spaces associated with each mode $\nu$. This 2-dimensional space is a consequence of the *Pauli principle*, according to which each mode can be occupied at most by 1 particle.

**2.2.4.3 Boson Number States.** For boson particles the Pauli principle does not apply: the occupation number space is discrete but infinite. We restrict ourselves here to a truncated subspace of finite dimension $n$ with the basis

$$\hat{1} = \sum_{j=1}^{n} |j\rangle \langle j| = \sum_{j=1}^{n} \hat{P}_{jj} \tag{2.166}$$

and introduce the operator sums

$$\hat{b}^+ = \sum_{j=1}^{n-1} \sqrt{j}\,\hat{\sigma}_{j,j+1}^+ \,, \tag{2.167}$$

$$\hat{b} = \sum_{j=1}^{n-1} \sqrt{j}\,\hat{\sigma}_{j,j+1}^- \,. \tag{2.168}$$

These operators are "non-selective" raising and lowering operators, i.e. they act on any state $|j\rangle$:

$$\hat{b}^+ |j-1\rangle = \sqrt{j-1}\,|j\rangle \quad (j \geq 2) \,, \tag{2.169}$$

$$\hat{b}|j\rangle = \sqrt{j-1}\,|j-1\rangle \quad (j \geq 2) \tag{2.170}$$

and

$$\hat{N}|j\rangle = \hat{b}^+\hat{b}|j\rangle = (j-1)|j\rangle \,. \tag{2.171}$$

Any operator $\hat{A}$ can be represented by products of $\left(\hat{b}^+\right)^m$ and $\left(\hat{b}\right)^n$. One easily shows that

$$\hat{N} = \hat{b}^+\hat{b} = \sum_{j=1}^{n}(j-1)\hat{P}_{jj} = \frac{1}{\hbar}\hat{L}_z(l) + l\hat{1} \tag{2.172}$$

with $l = (n-1)/2$: here we have applied (2.44). Using (2.37), one finds from

$$\left[\hat{b}, \hat{b}^+\right]_- = \sum_{j,j'=1}^{n-1} \sqrt{jj'} \left[\hat{\sigma}_{j,j+1}^-, \hat{\sigma}_{j',j'+1}^+\right]_- : \tag{2.173}$$

$$\boxed{\begin{aligned} \left[\hat{b}, \hat{b}^+\right]_- &= \hat{1} - n\hat{P}_{nn} , \\ \left[\hat{b}, \hat{b}\right]_- &= 0 , \\ \left[\hat{b}^+, \hat{b}^+\right]_- &= 0 \end{aligned}} \tag{2.174}$$

so that

$$\text{Tr}\left\{\left[\hat{b}, \hat{b}^+\right]_-\right\} = 0 . \tag{2.175}$$

For $n \to \infty$ the traceless commutator becomes that of the conventional boson creation and destruction operator $(\hat{a}^+, \hat{a})$:

$$[\hat{a}, \hat{a}^+]_- = \hat{1} . \tag{2.176}$$

One should note, however, that certain quantum mechanical rules cannot be represented within any finite basis.

### 2.2.5 Discrete Hamilton Models

We now apply the results of the preceding sections to specific operators other than angular momenta. Important examples are (non-degenerate) $n$-level Hamilton operators, the eigenstates of which may be taken to define the complete basis $|n)$ underlying the $SU(n)$ representation. These operators represent relevant finite subspaces within a complete (more realistic) model. A most convenient starting point is the *spectral representation*, in which the respective operator is expressed in terms of its eigenvalues and eigenstates. Contrary to fundamental spins, these few-level models typically contain phenomenological parameters: they have to be specified by a concrete implementation.

**2.2.5.1 2-Level Systems.** The Hamiltonian $\hat{H}$ with the spectrum of Fig. 2.4 can be represented by

$$\hat{H} = \sum_{j=1}^{2} E_j |j\rangle \langle j| = \sum_{j=1}^{2} E_j \hat{P}_{jj} , \tag{2.177}$$

wherein $E_j$ denotes the eigenenergies, and $\hat{P}_{jj}$ the projection operators of type (2.33). In the operator representation (2.85) this Hamiltonian reads

$$\begin{aligned} \hat{H} &= \frac{1}{2}\mathcal{H}_0\hat{1} + \frac{1}{2}\sum_{\nu=1}^{3}\mathcal{H}_\nu\hat{\lambda}_\nu \\ &= \frac{1}{2}\mathcal{H}_0\hat{1} + \frac{1}{2}\mathcal{H}_3\hat{\lambda}_3 \end{aligned} \tag{2.178}$$

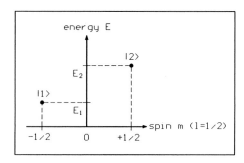

Fig. 2.4. Energy versus pseudospin eigenvalues of a 2-level system

with

$$\mathcal{H}_0 = \mathrm{Tr}\left\{\hat{H}\right\} = E_1 + E_2 \tag{2.179}$$

and

$$\mathcal{H}_\nu = \mathrm{Tr}\left\{\hat{H}\hat{\lambda}_\nu\right\} \,, \quad \nu = 1, 2, 3 \,, \quad \mathcal{H}_1 = \mathcal{H}_2 = 0 \,,$$

$$\mathcal{H}_3 = \mathrm{Tr}\left\{\hat{H}\hat{w}_1\right\} = \mathrm{Tr}\left\{\left(E_1\hat{P}_{11} + E_2\hat{P}_{22}\right)\left(-\hat{P}_{11} + \hat{P}_{22}\right)\right\} \,. \tag{2.180}$$

Observing

$$\mathrm{Tr}\left\{\hat{P}_{ll}\hat{P}_{mm}\right\} = \delta_{lm} \tag{2.181}$$

(see (2.40)), (2.180) then yields

$$\mathcal{H}_3 = E_2 - E_1 = \hbar\omega_{21} \,. \tag{2.182}$$

Inserting (2.182) and (2.179) into (2.178), one finally obtains

$$\boxed{\hat{H} = \frac{1}{2}\left[(E_1 + E_2)\,\hat{1} + (E_2 - E_1)\,\hat{w}_1\right]} \,. \tag{2.183}$$

$\hat{H}$ is thus specified by its trace $\mathcal{H}_0$ and a 3-dimensional vector of the form $(0, 0, E_2 - E_1)$. The eigenstates are eigenstates of $\hat{w}_1 = \hat{\sigma}_z$, i.e. spin states, see Fig. 2.4. $\mathcal{H}_0$ and $E_2 - E_1$ are phenomenological parameters.

**2.2.5.2 3-Level Systems.** The spectral representation of the Hamiltonian is

$$\hat{H} = \sum_{j=1}^{3} E_j \hat{P}_{jj} \,. \tag{2.184}$$

In $SU(3)$ (n=3, s=8, r=2) with generating operators (2.74) and (2.76) this Hamiltonian reads

$$\hat{H} = \frac{1}{3}\mathcal{H}_0\hat{1} + \frac{1}{2}\mathcal{H}_7\hat{w}_1 + \frac{1}{2}\mathcal{H}_8\hat{w}_2 \,. \tag{2.185}$$

With $\mathcal{H}_0 = E_1 + E_2 + E_3$, and observing

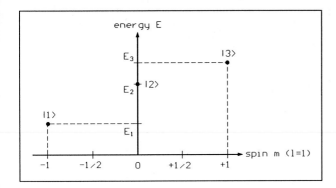

**Fig. 2.5.** Energy versus pseudospin eigenvalues of a 3-level system

$$\mathcal{H}_7 = \text{Tr}\left\{\hat{H}\hat{w}_1\right\} = E_2 - E_1 = \hbar\omega_{21}\ , \tag{2.186}$$

$$\mathcal{H}_8 = \text{Tr}\left\{\hat{H}\hat{w}_2\right\} = -\frac{1}{\sqrt{3}}\left(E_1 + E_2 - 2E_3\right) = \frac{\hbar}{\sqrt{3}}\left(\omega_{31} + \omega_{32}\right)\ , \tag{2.187}$$

one obtains the representation

$$
\begin{aligned}
\hat{H} &= \frac{1}{3}\left(E_1 + E_2 + E_3\right)\hat{1} + \frac{1}{2}\left(E_2 - E_1\right)\hat{w}_1 - \\
&\quad \frac{1}{2\sqrt{3}}\left(E_1 + E_2 - 2E_3\right)\hat{w}_2 \\
&= \frac{1}{3}\left(E_1 + E_2 + E_3\right)\hat{1} + \frac{1}{2}\hbar\omega_{21}\hat{w}_1 - \\
&\quad \frac{1}{2\sqrt{3}}\left(\hbar\omega_{31} + \hbar\omega_{32}\right)\hat{w}_2\ .
\end{aligned}
\tag{2.188}
$$

$\hat{H}$ is specified by its trace $\mathcal{H}_0$ and the 8-dimensional vector $(\mathcal{H}_l)$ which contains the two non-zero vector components $\mathcal{H}_7$ and $\mathcal{H}_8$.

Figure 2.5 illustrates the corresponding energy spectrum; note that the eigenvectors are also eigenvectors of $\hat{L}_z(|m| \le l)$ for spin $l = 1$ (see (2.92)),

$$\hat{L}_z = \frac{\hbar}{\sqrt{2}}\left(\hat{w}_1 + \sqrt{3}\hat{w}_2\right)\ , \tag{2.189}$$

$$
\begin{aligned}
|1\rangle &= |m = -1\rangle\ , \\
|2\rangle &= |m = 0\rangle\ , \\
|3\rangle &= |m = +1\rangle\ .
\end{aligned}
\tag{2.190}
$$

In the special case of an equidistant energy spectrum

$$\Delta E = E_{i+1} - E_i \tag{2.191}$$

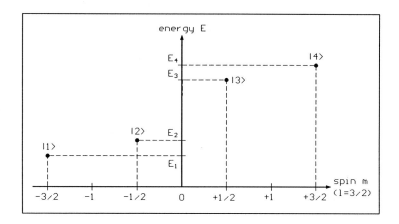

**Fig. 2.6.** Energy versus pseudospin eigenvalues of a 4-level system

the identification

$$E_2 - E_1 = \Delta E , \quad E_3 - E_2 = \Delta E , \quad E_3 - E_1 = 2\Delta E \qquad (2.192)$$

reduces (2.188) to

$$\hat{H} = \frac{1}{3}\mathcal{H}_0\hat{1} + \frac{1}{2}\Delta E\left(\hat{w}_1 + \sqrt{3}\hat{w}_2\right) . \qquad (2.193)$$

Only in this case can one express $\hat{H}$ in terms of a specific spin momentum $\hat{L}_z$: Inserting $\hat{w}_1$, $\hat{w}_2$ defined by (2.74) and (2.76), (2.193) yields

$$\hat{H} = \frac{1}{3}\mathcal{H}_0\hat{1} + \frac{1}{2}\Delta E\left(-\hat{P}_{11} + \hat{P}_{22} - \hat{P}_{11} - \hat{P}_{22} + 2\hat{P}_{33}\right)$$
$$= \frac{1}{3}\mathcal{H}_0\hat{1} + \Delta E\left(-\hat{P}_{11} + \hat{P}_{33}\right) \qquad (2.194)$$

so that with

$$\left(-\hat{P}_{11} + \hat{P}_{33}\right) = \frac{1}{\hbar}\hat{L}_z(l = 1) \qquad (2.195)$$

(see (2.43)), one obtains the representation

$$\hat{H} = \frac{1}{3}\mathcal{H}_0\hat{1} + \frac{1}{\hbar}\Delta E\hat{L}_z(l = 1) . \qquad (2.196)$$

**2.2.5.3 4-Level Systems.** The representation of the Hamiltonian $\hat{H}$ in $SU(4)$ (n=4, s=15, r=3) – compare Fig. 2.6 – with generating operators defined by (2.80) can be calculated as described above. The result is

$$\hat{H} = \frac{1}{4}\left(E_1 + E_2 + E_3 + E_4\right)\hat{1} + \frac{1}{2}\left(E_2 - E_1\right)\hat{w}_1 -$$
$$\frac{1}{2\sqrt{3}}\left(E_1 + E_2 - 2E_3\right)\hat{w}_2 - \tag{2.197}$$
$$\frac{1}{2\sqrt{6}}\left(E_1 + E_2 + E_3 - 3E_4\right)\hat{w}_3 \; .$$

Here, $\hat{H}$ is specified by its trace $\mathcal{H}_0$ and a 15-dimensional vector $\mathcal{H}_l$ ($l = 1, 2, \ldots, 15$). In the case of an equidistant spectrum, (2.197) can be reduced to

$$\hat{H} = \frac{1}{4}\mathcal{H}_0\hat{1} + \frac{1}{\hbar}\Delta E \hat{L}_z (l = 3/2) \; . \tag{2.198}$$

Generalizing to any finite $n$ we get

$$\hat{H} = \frac{1}{n}\mathcal{H}_0\hat{1} + \Delta E \frac{1}{\hbar}\hat{L}_z(l) \tag{2.199}$$

with

$$l = (n - 1)/2 \tag{2.200}$$

and

$$\mathcal{H}_0 = nE_1 + (n - 1)\Delta E \; . \tag{2.201}$$

One should note that all these simple models are exclusively defined in the $\hat{w}_l$ subspace.

**2.2.5.4 Harmonic Oscillator and Charge States.** The harmonic oscillator is a special model with an equidistant spectrum for $n \to \infty$. The corresponding Hamiltonian is usually defined in terms of the creation operator and destruction operator $\hat{a}^+$, $\hat{a}$ (see Sect. 2.2.4):

$$\hat{H} = \Delta E \left(\frac{1}{2}\hat{1} + \hat{N}\right) \tag{2.202}$$

(operator in second quantization), where

$$\hat{N} = \hat{a}^+\hat{a} \; , \tag{2.203}$$

$$\hat{N}|j\rangle = (j - 1)|j\rangle \; . \tag{2.204}$$

$j - 1$ is interpreted as the number of particles (*phonons*) in the oscillation mode. From the normalized ground state $|1\rangle$ with no phonons (*vacuum*) any higher excited state is obtained by

$$|j\rangle = \frac{1}{\sqrt{(j - 1)!}}\left(\hat{a}^+\right)^{j-1}|1\rangle \; . \tag{2.205}$$

As the number of phonons is not restricted, these phonons obey *Bose statistics*. Fermions would be constrained to occupation numbers $0, 1$ (*Fermi statistics*).

Truncated to finite dimension $n$ we get, using

$$\hat{P}_{mm} = \frac{1}{n}\hat{1} + \frac{1}{2}\sum_{l=1}^{n-1} w_l^{(m)}\hat{w}_l \quad (m = 1, 2, \ldots, n) \tag{2.206}$$

(see (2.111)), the Hamiltonian

$$\boxed{\hat{H} = \frac{\Delta E}{n}\left(\frac{n}{2} + \sum_{j=1}^{n-1} j\right)\hat{1} + \frac{1}{2}\Delta E \sum_{l=1}^{n-1}\hat{w}_l \sum_{j=2}^{n}(j-1)w_l^{(j)}.} \tag{2.207}$$

$\hat{H}$ generalizes the result of (2.183) and (2.193) for

$$\mathcal{H}_0 = \frac{n}{2}\Delta E + \Delta E \sum_{j=1}^{n-1} j. \tag{2.208}$$

In $SU(n)$ the Hamilton operator is thus specified by the vector component $(s' = n^2 - n)$:

$$\mathcal{H}_{s'+l} = \Delta E \sum_{j=2}^{n}(j-1)w_l^{(j)} \quad (l = 1, 2, \ldots, n-1). \tag{2.209}$$

The other components are zero. One easily convinces oneself that

$$\hat{H}|m\rangle = (m - 1/2)\Delta E |m\rangle \quad (m = 1, 2, \ldots, n), \tag{2.210}$$

where we have applied (2.114).

**2.2.5.5 2-Level Systems with Transfer Coupling.** In the case of transfer coupling, the Hamiltonian reads

$$\hat{H} = \hat{H}_0 + \hat{H}', \tag{2.211}$$

where $\hat{H}_0$ is given by (2.177) and

$$\hat{H}' = \frac{1}{2}\sum_{j,j',j\neq j'} V_{jj'}\hat{P}_{jj'}, \quad V_{jj'} = V_{j'j}^*. \tag{2.212}$$

Observing that

$$\begin{aligned}\text{Re}\{V_{jj'}\} &= \text{Re}\{V_{j'j}\}, \\ \text{Im}\{V_{jj'}\} &= -\text{Im}\{V_{j'j}\}\end{aligned} \tag{2.213}$$

(Re denotes the real, Im the imaginary part), and inserting (2.213) into the coupling term $\hat{H}'$, leads to

$$\begin{aligned}\hat{H}' &= \frac{1}{2}\left(\text{Re}\{V_{12}\} + i\text{Im}\{V_{12}\}\right)\hat{P}_{12} + \frac{1}{2}\left(\text{Re}\{V_{12}\} - i\text{Im}\{V_{12}\}\right)\hat{P}_{21} \\ &= \frac{1}{2}\text{Re}\{V_{12}\}\left(\hat{P}_{12} + \hat{P}_{21}\right) + i\frac{1}{2}\text{Im}\{V_{12}\}\left(\hat{P}_{12} - \hat{P}_{21}\right),\end{aligned} \tag{2.214}$$

and with

$$\hat{u}_{12} = \left(\hat{P}_{12} + \hat{P}_{21}\right) \ , \quad \hat{v}_{12} = i\left(\hat{P}_{12} - \hat{P}_{21}\right) \tag{2.215}$$

(compare (2.74)) finally to

$$\hat{H}' = \frac{1}{2}\mathrm{Re}\left\{V_{12}\right\}\hat{u}_{12} + \frac{1}{2}\mathrm{Im}\left\{V_{12}\right\}\hat{v}_{12} \ . \tag{2.216}$$

The total Hamiltonian $\hat{H}$ can then be represented by

$$\begin{aligned}
\hat{H} &= \hat{H}^0 + \hat{H}' \\
&= \frac{1}{2}\left[(E_1 + E_2)\,\hat{1} + (E_2 - E_1)\,\hat{w}_1\right] + \frac{1}{2}\mathrm{Re}\left\{V_{12}\right\}\hat{u}_{12} + \\
&\quad \frac{1}{2}\mathrm{Im}\left\{V_{12}\right\}\hat{v}_{12} \ .
\end{aligned} \tag{2.217}$$

$\hat{H}$ is specified by its trace $\mathcal{H}_0 = E_1 + E_2$ and the vector

$$(\mathcal{H}_l) = (\mathrm{Re}\left\{V_{12}\right\}, \mathrm{Im}\left\{V_{12}\right\}, E_2 - E_1) \ . \tag{2.218}$$

*Example 2.2.11.* Special versions of (2.217) occur, e.g., in the context of tunneling between degenerate states:

$$\hat{H} = \frac{1}{2}\mathrm{Re}\left\{V_{12}\right\}\hat{u}_{12} \ . \tag{2.219}$$

Then, eigenfunctions of the Hamiltonian $\hat{H}$ are eigenfunctions of the operator $\hat{u}_{12}$.

## 2.2.5.6 Few-electron Basis and Charge States.
The basis states $|i(\nu)\rangle$, $i = 1, 2, \ldots, n$ of subsystem $\nu$ have been introduced as single-particle states: the operators, angular momentum (spin) and Hamiltonians considered so far refer to the same level: only then do those states serve as a complete basis. On the other hand, the detailed nature of those states is not required at all

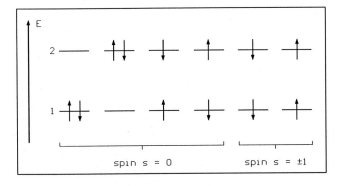

**Fig. 2.7.** A spin-degenerate 2-level system

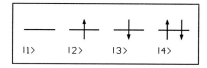

$|1\rangle$    $|2\rangle$    $|3\rangle$    $|4\rangle$

**Fig. 2.8.** Local charge states

for the application of the $SU(n)$ algebra: any definite collection of states will do. A finite subset of few-electron states in atoms, molecules or quantum dots is a pertinent example. An example for fixed local charge (electron number $N = 2$) is shown in Fig. 2.7. Here the *Pauli exclusion principle* constrains the state space. It is characterized by various energy states and total spin states.

A typical situation is encountered for local charge states, where each spin state $i =\uparrow, \downarrow$ can be occupied by 0 or 1 electron (cf. Fig. 2.8). The corresponding Hamiltonian model requires the use of second quantization: the number of electrons is not fixed. Contrary to the case studied in Sect. 2.2.5.4, electrons are fermions: for each spin state $i$ the associated space of number states is 2-dimensional, i.e. $|n_i\rangle$, $n_i = 0$ (vacuum), $n_i = 1$ (1-electron number state), with

$$\langle n_i | n_j \rangle = \delta_{ij} . \tag{2.220}$$

The 4 charge states are

$$|1\rangle = |0_\downarrow 0_\uparrow\rangle , \quad |2\rangle = |0_\downarrow 1_\uparrow\rangle , \quad |3\rangle = |1_\downarrow 0_\uparrow\rangle , \quad |4\rangle = |1_\downarrow 1_\uparrow\rangle . \tag{2.221}$$

In second quantization one writes

$$|1_i\rangle = \hat{a}_i^+ |0> \tag{2.222}$$

with (cf. Sect. 2.2.4.2)

$$\hat{a}_i^+ = \hat{P}_{10}^i = |1_i\rangle \langle 0_i| , \tag{2.223}$$

$$\hat{a}_i = \hat{P}_{01}^i = |0_i\rangle \langle 1_i| , \tag{2.224}$$

$$\left[\hat{a}_i, \hat{a}_j^+\right]_+ = \hat{1}\delta_{ij} , \tag{2.225}$$

$$\left[\hat{a}_i^+, \hat{a}_j^+\right]_+ = [\hat{a}_i, \hat{a}_j]_+ = 0 , \tag{2.226}$$

$$\hat{N}_i = \hat{a}_i^+ \hat{a}_i = |1_i\rangle \langle 1_i| = \hat{P}_{11}^i . \tag{2.227}$$

The Hamiltonian for the 4-state node can thus be written in the occupation number space as

$$\hat{H}_0 = E_0 \left(\hat{P}_{11}^\downarrow + \hat{P}_{11}^\uparrow\right) = E_0 \left(\hat{N}_\downarrow + \hat{N}_\uparrow\right) . \tag{2.228}$$

Charge transfer is not only limited by the Pauli principle, but also by Coulomb interaction: including an on-site Coulomb energy $E_Q$ (*Hubbard-type interaction*), we obtain

$$\hat{H} = \hat{H}_0 + E_Q \left(\hat{P}_{11}^\downarrow \otimes \hat{P}_{11}^\uparrow\right) = \hat{H}_0 + E_Q \left(\hat{N}_\downarrow \otimes \hat{N}_\uparrow\right) . \tag{2.229}$$

This is nothing but a specific interpretation of the general 4-level model discussed already in Sect. 2.2.5.3:

$$\hat{H} = \sum_{j=1}^{4} E_j \hat{P}_{jj} \tag{2.230}$$

with

$$E_1 = 0 \ ,$$
$$E_2 = E_3 = E_0 \ , \quad \text{(degenerate)} \tag{2.231}$$
$$E_4 = 2E_0 + E_Q$$

and

$$|1\rangle = |0\rangle \ ,$$
$$|2\rangle = \hat{a}_{\downarrow}^{+} |0\rangle \ , \quad |3\rangle = \hat{a}_{\uparrow}^{+} |0\rangle \ , \tag{2.232}$$
$$|4\rangle = \hat{a}_{\uparrow}^{+} \hat{a}_{\downarrow}^{+} |0\rangle \ .$$

Mesoscopic metallic islands between external voltage gates show incremental charging effects (cf. [10]). Depending on the voltage, 0, 1, 2, ... additional electrons can overcome the charging energy expressed in terms of a phenomenological capacity (Coulomb blockade). However, these islands do not represent quantum nodes in the sense of this book: they are not described by a $SU(n)$ algebra. The conductivity experiments usually probe ensembles of such islands.

## 2.3 Density Operator

Contents: expectation values, general state as operator, density matrix, restrictions, constants (measures), independent parameters, coherence vectors, partial polarization, measure of uncertainty, Jaynes' Principle, canonical statistical operator, measurements, projection postulate.

### 2.3.1 Fundamental Properties

Expectation values of operators representing observables are introduced with respect to state vectors. However, general states require a state description by means of an operator called *density operator*. Basic definitions and properties are studied (cf. [20, 28, 44]).

**2.3.1.1 Special Expectation Values.** *Expectation values* are of direct experimental interest. Pertinent examples are the magnetic moment,

$$M_\alpha = g_{\mathrm{s}} \frac{e}{m_0} \langle \chi | \hat{L}_\alpha | \chi \rangle \ , \tag{2.233}$$

and the so-called polarization,

$$P_\alpha = \frac{1}{\hbar l} \langle \chi | \hat{L}_\alpha | \chi \rangle \ , \qquad (2.234)$$

with $\langle \chi | \hat{L}_\alpha | \chi \rangle$ defined by (cf. (2.15))

$$\langle \chi | \hat{L}_\alpha | \chi \rangle = \sum_{m,m'=-l}^{l} \chi_{m'}^* \chi_m \, \langle m | \hat{L}_\alpha | m' \rangle \ , \quad \alpha = x, y, z \ . \qquad (2.235)$$

($g_s$ denotes the Landé $g$ factor of a spin, $e$ the electron charge, $m_0$ the mass of the particle, and $\hbar$ represents Planck's constant.)

In the case of $l = 1/2$, the spin polarization reads

$$P_\alpha = \frac{2}{\hbar} \langle \chi | \hat{L}_\alpha | \chi \rangle = \langle \chi | \hat{\sigma}_\alpha | \chi \rangle \ . \qquad (2.236)$$

*Example 2.3.1.* For the superposition state

$$|\chi\rangle = \frac{1}{\sqrt{2}} \left( |1\rangle + e^{i\varphi} |-1\rangle \right) \ , \qquad (2.237)$$

one obtains

$$\begin{aligned} P_\alpha = {} & \frac{1}{2} \langle 1 | \hat{\sigma}_\alpha | 1 \rangle + \frac{1}{2} e^{i\varphi} \langle 1 | \hat{\sigma}_\alpha | -1 \rangle + \frac{1}{2} e^{-i\varphi} \langle -1 | \hat{\sigma}_\alpha | 1 \rangle + \\ & \frac{1}{2} \langle -1 | \hat{\sigma}_\alpha | -1 \rangle \ , \end{aligned} \qquad (2.238)$$

and with the Pauli matrices (2.26)–(2.28),

$$\begin{aligned} P_x &= \frac{1}{2} \left( e^{i\varphi} + e^{-i\varphi} \right) = \cos\varphi \ , \\ P_y &= \frac{i}{2} \left( e^{i\varphi} - e^{-i\varphi} \right) = \sin\varphi \ , \\ P_z &= 0 \ . \end{aligned} \qquad (2.239)$$

*Example 2.3.2.* For the eigenvector

$$|\chi\rangle = |\pm 1\rangle \ , \qquad (2.240)$$

the polarization is in the $z$ direction:

$$\begin{aligned} P_x &= \frac{2}{\hbar} \langle \pm 1 | \hat{L}_x | \pm 1 \rangle = 0 \ , \\ P_y &= 0 \ , \\ P_z &= \pm 1 \ . \end{aligned} \qquad (2.241)$$

In both cases the relation

$$|\boldsymbol{P}|^2 = P_x^2 + P_y^2 + P_z^2 = \frac{4}{\hbar^2} \left( \left\langle \hat{L}_x \right\rangle^2 + \left\langle \hat{L}_y \right\rangle^2 + \left\langle \hat{L}_z \right\rangle^2 \right) = 1 \qquad (2.242)$$

holds which defines *complete polarization*.

*Example 2.3.3.* For $l = 1$ (e.g. atomic polarization), the expectation value of the polarization reads

$$P_\alpha = \frac{1}{\hbar} \langle \chi | \hat{L}_\alpha | \chi \rangle \ . \tag{2.243}$$

Inserting the eigenvectors $|-1\rangle$, $|0\rangle$, $|1\rangle$, one obtains polarization vectors represented by

$$|\chi\rangle = |-1\rangle : \ P_x = P_y = 0 \ , \ P_z = -1 \ , \tag{2.244}$$

$$|\chi\rangle = |0\rangle : \ P_x = P_y = P_z = 0 \ , \ |\boldsymbol{P}|^2 = 0 \ , \tag{2.245}$$

$$|\chi\rangle = |1\rangle : \ P_x = P_y = 0 \ , \ P_z = 1 \ . \tag{2.246}$$

The states $|1\rangle$, $|-1\rangle$ show atomic polarization in the direction $\pm \boldsymbol{e}_z$. The second equation demonstrates that a pure state now no longer implies complete polarization.

**2.3.1.2 The Density Operator of a Pure State.** Let $|\mu\rangle$ denote a complete orthonormal set of functions spanning a Hilbert space of dimension $n$. Introducing the *density operator*

$$\boxed{\hat{\rho} = |\mu\rangle \langle \mu| = \hat{P}_{\mu\mu} \ ,} \tag{2.247}$$

we find

$$\sum_\nu \langle \nu| \hat{\rho} \hat{A} |\nu\rangle = \sum_{\nu,\nu'} \langle \nu| \hat{\rho} |\nu'\rangle \langle \nu'| \hat{A} |\nu\rangle = \left\langle \mu \middle| \hat{A} \middle| \mu \right\rangle \tag{2.248}$$

so that

$$\left\langle \hat{A} \right\rangle = \mathrm{Tr} \left\{ \hat{\rho} \hat{A} \right\} \ . \tag{2.249}$$

The density operator (2.247) is an alternative definition of the pure state $|\mu\rangle$. According to (2.40), we have

$$\mathrm{Tr} \left\{ \hat{P}_{\mu\mu} \hat{P}_{\nu\nu} \right\} = \delta_{\mu\nu} \ . \tag{2.250}$$

**2.3.1.3 Density Operator in Eigenrepresentation.** More general expectation values result if (2.247) is replaced by the density operator

$$\boxed{\hat{\rho} = \sum_{\nu=1}^n \rho_\nu |\nu\rangle \langle \nu| = \sum_\nu \rho_\nu \hat{P}_{\nu\nu} \ .} \tag{2.251}$$

Equation (2.251) specifies the eigenrepresentation of the general density operator with trace

$$\boxed{\mathrm{Tr} \left\{ \hat{\rho} \right\} := C(n,1) = \sum_{\nu=1}^n \rho_\nu = 1 \ .} \tag{2.252}$$

Here, $\rho_\nu$ (the eigenvalues of the density operator) are defined by

$$\boxed{\rho_\nu = \mathrm{Tr}\left\{\hat{\rho}\hat{P}_{\nu\nu}\right\} \geq 0 \ , \quad \text{real} \ .}$$
(2.253)

Due to the fact that these eigenvalues are real and non-negative, $\hat{\rho}$ must be a Hermitian, positive definite operator. Then, the expectation value of $\hat{A}$ can be expressed as

$$\left\langle \hat{A} \right\rangle = \mathrm{Tr}\left\{\hat{\rho}\hat{A}\right\} = \sum_{\nu=1}^{n} \rho_\nu A_{\nu\nu} \ ,$$
(2.254)

where the set $\rho_\nu$ functions as a "classical" probability distribution. $\hat{\rho}$ is not a conventional observable: the eigenvalues of the operator $\hat{\rho}$ have the meaning of a probability (relative frequency). Therefore, $\hat{\rho}$ is often called *statistical operator*.

**Theorem 2.3.1 (v. Neumann 1927, Wigner 1932).** *Any quantum mechanical state can be defined by a density operator $\hat{\rho}$ on a corresponding Hilbert space $\mathbb{H}$.*

Equation (2.251) may be interpreted as an orthogonal decomposition of $\hat{\rho}$ into pure states. Such decompositions are, in general, ambiguous, unless specific conditions are fulfilled:

**Theorem 2.3.2 (Fano).** *A unique orthogonal decomposition of $\hat{\rho}$ into pure states exists if the eigenvalues $\rho_\nu \neq 0$ of a density operator are non-degenerate.*

**2.3.1.4 General Representation.** Using the unitary transformation of the kind (2.15), i.e.

$$|\nu\rangle = \sum_i |i\rangle \langle i|\, \nu \rangle$$
(2.255)

with

$$\langle i|\,\nu \rangle = \chi_{i\nu} = \chi_{\nu i}^* \ ,$$
(2.256)

the density operator (2.251) can be transformed:

$$\hat{\rho} = \sum_\nu \rho_\nu |\nu\rangle \langle \nu|$$

$$= \sum_\nu \rho_\nu \sum_{i,j} |i\rangle \langle i|\nu\rangle \langle \nu|j\rangle \langle j|$$

$$= \sum_{i,j} \sum_\nu \rho_\nu \chi_{i\nu} \chi_{j\nu}^* |i\rangle \langle j| \ .$$
(2.257)

Introducing the matrix elements

$$\boxed{\rho_{ij} = \sum_\nu \rho_\nu \chi_{i\nu} \chi_{j\nu}^* = \rho_{ji}^* \ , \quad \rho_{ij} = \mathrm{Tr}\left\{\hat{\rho}\hat{P}_{ji}\right\} = \left\langle \hat{P}_{ji} \right\rangle = \overline{P}_{ji} \ ,}$$
(2.258)

where the first equation of (2.258) defines the Hermitian matrix $(\rho_{ij})$, (2.257) yields

$$\hat{\rho} = \sum_{i,j} \rho_{ij} |i\rangle \langle j| = \sum_{ij} \rho_{ij} \hat{P}_{ij} \; . \tag{2.259}$$

On the basis of this representation, expectation values are expressed as

$$\left\langle \hat{A} \right\rangle = \sum_{i,j} \rho_{ij} A_{ji} \; . \tag{2.260}$$

The matrix elements $B_{ij}$ of any Hermitian, positive definite operator obey the *Cauchy-Schwarz inequality*,

$$|B_{ji}| \leq \sqrt{B_{ii} B_{jj}} \; , \tag{2.261}$$

which implies

$$|\rho_{ji}| \leq \sqrt{\rho_{ii} \rho_{jj}} \; . \tag{2.262}$$

**2.3.1.5 Trace-relations.** Let us define for $q = 1, 2, 3, \ldots$

$$C(n, q) := \mathrm{Tr} \left\{ \hat{\rho}^q \right\} \; . \tag{2.263}$$

In general, the matrix representations of $C(n, q)$ read:

$$C(n, 2) = \sum_{j,k=1}^{n} \rho_{jk} \rho_{kj} \; , \quad C(n, 3) = \sum_{j,k,l=1}^{n} \rho_{jk} \rho_{kl} \rho_{lj} \; , \quad \text{etc.} \tag{2.264}$$

Then the inequalities

$$C(n, q) \leq C(n, q - 1) \leq 1 \tag{2.265}$$

hold in any Hilbert space of dimension $n$.

*Proof.* For a pure state $|\mu\rangle$ we have

$$\hat{\rho}^q = \hat{P}_{\mu\mu}^q = \hat{P}_{\mu\mu} = \hat{\rho} \tag{2.266}$$

so that

$$C(n, q) = C(n, 1) = 1 \; . \tag{2.267}$$

Now, turning to the general density operator (2.251) in eigenrepresentation, one obtains

$$\hat{\rho}^q = \sum_{\nu} \rho_{\nu}^q \hat{P}_{\nu\nu} \tag{2.268}$$

so that

$$\mathrm{Tr} \left\{ \hat{\rho}^q \right\} = \sum_{\nu} \rho_{\nu}^q \tag{2.269}$$

and

$$\sum_\nu \rho_\nu^q \le \left( \sum_\nu \rho_\nu^{q-1} \right) \left( \sum_\nu \rho_\nu \right) = \sum_\nu \rho_\nu^{q-1} . \tag{2.270}$$

Combining (2.270) and (2.259), one obtains the result (2.265), observing that trace-properties are independent of the representation chosen.

### 2.3.2 Coherence Vector

The density operator in Hilbert space of dimension $n$, like any operator, can be represented by the generating operators of $SU(n)$. The expectation values of those generating operators form a vector called *coherence vector*. The properties of this coherence vector are considered in this section.

**2.3.2.1 Definitions.** Using the operator representation (2.85), identifying the operator $\hat{A}$ with the density operator $\hat{\rho}$, and taking (2.252) into account, one obtains the representation ($s = n^2 - 1$)

$$\boxed{\hat{\rho} = \frac{1}{n}\hat{1} + \frac{1}{2}\sum_{j=1}^{s} \lambda_j \hat{\lambda}_j} \tag{2.271}$$

with the elements $\lambda_j$ being defined by

$$\boxed{\lambda_j = \left\langle \hat{\lambda}_j \right\rangle = \mathrm{Tr}\left\{ \hat{\rho}\hat{\lambda}_j \right\} = \text{real (because } \hat{\rho} \text{ is Hermitian)} .} \tag{2.272}$$

According to (2.249), $\lambda_j$ is the expectation value of $\hat{\lambda}_j$.

The $s$-dimensional vector

$$\boxed{\boldsymbol{\lambda} = \{\lambda_j , \; j = 1, 2, \ldots, s\}} \tag{2.273}$$

denotes the so-called *coherence vector*, which may be considered as a *generalized Bloch vector* (see Chap. 3). $\boldsymbol{\lambda}$ has the same mathematical structure as the vector $(\mathcal{H}_l)$ defining Hamiltonian models. In the following sections, we will discuss some state models; however, it will be more important to understand how states are prepared, how they change with time, and how they are measured.

Using the representation (2.271), the trace $\mathrm{Tr}\left\{\hat{\rho}^2\right\}$ is given by

$$\mathrm{Tr}\left\{\hat{\rho}^2\right\} = \frac{1}{n^2}\mathrm{Tr}\left\{\hat{1}\hat{1}\right\} + \frac{1}{4}\sum_{i,j=1}^{s} \lambda_i \lambda_j \mathrm{Tr}\left\{\hat{\lambda}_i\hat{\lambda}_j\right\}$$

$$= \frac{1}{n} + \frac{1}{4}\sum_{i,j=1}^{s} \lambda_i \lambda_j \mathrm{Tr}\left\{\hat{\lambda}_i\hat{\lambda}_j\right\} \tag{2.274}$$

so that with

$$\mathrm{Tr}\left\{\hat{\lambda}_i\hat{\lambda}_j\right\} = 2\delta_{ij} , \tag{2.275}$$

one obtains

$$\boxed{\text{Tr}\left\{\hat{\rho}^2\right\} = \frac{1}{n} + \frac{1}{2}|\lambda|^2 = C(n,2) \ .}$$  (2.276)

Observing (2.265) and $|\lambda|^2 \geq 0$ one thus ends up with the inequality

$$\boxed{\frac{1}{n} \leq C(n,2) \leq 1 \ .}$$  (2.277)

This means that the length of the coherence vector is limited by

$$|\lambda|^2 \leq 2\left(1 - \frac{1}{n}\right) \ .$$  (2.278)

The next higher trace-relation

$$\begin{aligned} C(n,3) &= \text{Tr}\left\{\hat{\rho}^3\right\} \\ &= \frac{1}{n^3}\text{Tr}\left\{\hat{1}\right\} + \frac{1}{2n^2}\sum_{i,j}\lambda_i\lambda_j\text{Tr}\left\{\hat{\lambda}_i\hat{\lambda}_j\right\} + \\ &\quad \frac{1}{8}\sum_{i,j,k}\lambda_i\lambda_j\lambda_k\text{Tr}\left\{\hat{\lambda}_i\hat{\lambda}_j\hat{\lambda}_k\right\} \end{aligned}$$  (2.279)

reduces to

$$\boxed{C(n,3) = \frac{1}{n^2}\left(1 + |\lambda|^2\right) + \sum_{i,j,k}d_{ijk}\lambda_i\lambda_j\lambda_k \ .}$$  (2.280)

The last term containing the structure constants $d_{ijk}$ vanishes for $n = 2$: $C(2,3)$ does not constitute an additional constraint.

**2.3.2.2 Relation to Density Matrix.** For a density operator $\hat{\rho}$ defined by (2.271), one obtains the density matrix $(\rho_{ij})$ by using a matrix representation of the generating operators.

*Example 2.3.4.* For $SU(2)$ one finds

$$\begin{aligned} (\rho_{ij}) = \frac{1}{2}\Bigg[ &\begin{pmatrix} 1 & 0 \\ 0 & 1 \end{pmatrix} + u_{12}\begin{pmatrix} 0 & 1 \\ 1 & 0 \end{pmatrix} + iv_{12}\begin{pmatrix} 0 & 1 \\ -1 & 0 \end{pmatrix} + \\ &w_1\begin{pmatrix} -1 & 0 \\ 0 & 1 \end{pmatrix}\Bigg] \ , \end{aligned}$$  (2.281)

$$u_{12} = \text{Tr}\left\{\hat{\rho}\hat{u}_{12}\right\} \ , \quad v_{12} = \text{Tr}\left\{\hat{\rho}\hat{v}_{12}\right\} \ , \quad w_1 = \text{Tr}\left\{\hat{\rho}\hat{w}_1\right\} \ ,$$  (2.282)

which can be condensed into

$$\boxed{(\rho_{ij}) = \frac{1}{2}\begin{pmatrix} 1 - w_1 & u_{12} + iv_{12} \\ u_{12} - iv_{12} & 1 + w_1 \end{pmatrix} = \begin{pmatrix} \rho_{11} & \rho_{12} \\ \rho_{21} & \rho_{22} \end{pmatrix} \ .}$$  (2.283)

*Example 2.3.5.* For $SU(3)$ we get (using matrices corresponding to the generating operators defined by (2.74)–(2.76), and inserting these matrices into (2.271))

$$
(\rho_{ij}) = \frac{1}{2} \begin{pmatrix} \frac{2}{3} - w_1 - \frac{1}{\sqrt{3}}w_2 & u_{12} + iv_{12} & u_{13} + iv_{13} \\ u_{12} - iv_{12} & \frac{2}{3} + w_1 - \frac{1}{\sqrt{3}}w_2 & u_{23} + iv_{23} \\ u_{13} - iv_{13} & u_{23} - iv_{23} & \frac{2}{3} + \frac{2}{\sqrt{3}}w_2 \end{pmatrix} . \qquad (2.284)
$$

These two examples illustrate the relation between the coherence vector $\boldsymbol{\lambda}$ and the density matrix defined by (2.258). In eigenrepresentation we have in general:

$$
\hat{\rho} = \frac{1}{n}\hat{1} + \frac{1}{2}\sum_{l=1}^{n-1} w_l \hat{w}_l \qquad (2.285)
$$

with

$$
w_l = \mathrm{Tr}\{\hat{\rho}\hat{w}_l\} . \qquad (2.286)
$$

Now, according to (2.111),

$$
\hat{P}_{ii} = \frac{1}{n}\hat{1} + \frac{1}{2}\sum_{l=1}^{n-1} w_l^{(i)} \hat{w}_l \qquad (2.287)
$$

so that

$$
\rho_{ii} = \mathrm{Tr}\left\{\hat{P}_{ii}\hat{\rho}\right\} = \frac{1}{n} + \frac{1}{2}\sum_{l=1}^{n-1} w_l^{(i)} w_l , \quad \rho_{ij} = 0 \ (i \neq j) . \qquad (2.288)
$$

The corresponding coherence vector is of the form

$$
\boldsymbol{\lambda} = (0, 0, \ldots, 0, w_1, w_2, \ldots, w_{n-1}) . \qquad (2.289)
$$

In general, (2.288) is supplemented by

$$
\rho_{ij} = \begin{cases} \dfrac{1}{2}(u_{ij} + iv_{ij}) & (i < j) \\[2mm] \dfrac{1}{2}(u_{ji} - iv_{ji}) & (i > j) \end{cases} . \qquad (2.290)
$$

**2.3.2.3 Independent Parameters.** The coherence vector and the corresponding density matrix contain $s = n^2 - 1$ independent parameters. (A density matrix consists of $2n^2$ complex numbers. Due to the fact that there are $n^2$ constraints $\rho_{ij} = \rho_{ji}^*$ and 1 constraint $\mathrm{Tr}\{\hat{\rho}\} = 1$ in the case of $SU(n)$, the total number of independent parameters is $n^2 - 1$. Compare with (2.69).)

If the density matrix of an operator $\hat{\rho}$ is diagonal within the reference frame considered, the coherence vector and the density matrix contain $n - 1$ independent parameters. However, in general, the additional transformation matrix $U_{\mu i} = U_{i\mu}^*$ to this diagonal representation with $n^2$ independent parameters has to be taken into account, with the $n$ constraints

$$\sum_j \rho_{ij} \langle j | \nu \rangle = \rho_{\nu\nu} \langle i | \nu \rangle \ . \tag{2.291}$$

Thus, in all, the number of independent parameters is again given by $(n-1) + n^2 - n = n^2 - 1$.

In the case of a pure state represented by

$$|\psi\rangle = \sum_j a_j |j\rangle \ , \quad |\psi\rangle \langle\psi| = \sum_{j,k} a_j a_k^* |j\rangle \langle k| \ , \quad \rho_{jk} = a_j a_k^* \tag{2.292}$$

(see (2.255) and (2.259)), the density matrix consists of $2(n-1)$ independent parameters. ($n$ complex amplitudes $a_j$ consist of $2n$ independent parameters. Taking the necessary normalization condition as well as one ambiguous overall phase into account, one obtains $2n-2 = 2(n-1)$ independent parameters.)

**2.3.2.4 Expectation Values in $SU(n)$.** The expectation values have been defined by (2.249)

$$\langle \hat{A} \rangle = \mathrm{Tr} \left\{ \hat{\rho} \hat{A} \right\} := \overline{A} \ . \tag{2.293}$$

Inserting the operator representation (2.85), i.e.

$$\hat{A} = \frac{1}{n} \mathrm{Tr} \left\{ \hat{A} \right\} \hat{1} + \frac{1}{2} \sum_{j=1}^s \mathcal{A}_j \hat{\lambda}_j \tag{2.294}$$

with

$$\mathcal{A}_j = \mathrm{Tr} \left\{ \hat{A} \hat{\lambda}_j \right\} \ , \tag{2.295}$$

and using (2.252), one obtains the expression

$$\boxed{\overline{A} = \overline{A}(\boldsymbol{\lambda}) = \frac{1}{n} \mathrm{Tr} \left\{ \hat{A} \right\} + \frac{1}{2} \sum_{j=1}^s \mathcal{A}_j \lambda_j \ ,} \tag{2.296}$$

where $\lambda_j$ is defined by (2.272). The second term of the r.h.s. of (2.296) is proportional to the scalar product of vectors $(\mathcal{A}_j)$ and $\boldsymbol{\lambda}$.

*Example 2.3.6.* Consider the expectation value in $SU(3)$,

$$P_\alpha = \mathrm{Tr} \left\{ \hat{\rho} \frac{\hat{L}_\alpha}{\hbar} \right\} \ , \tag{2.297}$$

i.e. the polarization vector defined in (2.243) with $\hat{L}_\alpha/\hbar$ given by (2.92). The result is

$$P_x = \frac{1}{\sqrt{2}} \left( u_{12} + u_{23} \right) ,$$

$$P_y = \frac{1}{\sqrt{2}} \left( v_{12} + v_{23} \right) , \tag{2.298}$$

$$P_z = \frac{1}{2} \left( w_1 + \sqrt{3} w_2 \right) .$$

Contrary to $SU(2)$, the polarization vector in $SU(3)$ is only a function of the coherence vector components, not proportional to $\boldsymbol{\lambda}$. $\boldsymbol{P}$ can thus be zero even if $|\boldsymbol{\lambda}|^2 = 4/3$ (pure state): an example is given by (2.245).

The expectation value $\left\langle \hat{A}\hat{B} \right\rangle = \overline{AB}$ has, using the operator representation (2.294), the following representation:

$$\boxed{\overline{AB} = \frac{1}{n^2} \mathrm{Tr}\left\{ \hat{A} \right\} \mathrm{Tr}\left\{ \hat{B} \right\} + \frac{1}{2n} \mathrm{Tr}\left\{ \hat{A} \right\} \sum_{j=1}^{s} \mathcal{A}_j \lambda_j + \\ \frac{1}{2n} \mathrm{Tr}\left\{ \hat{B} \right\} \sum_{k=1}^{s} \mathcal{B}_k \lambda_k + \frac{1}{4} \sum_{j,k=1}^{s} \mathcal{A}_j \mathcal{B}_k \mathrm{Tr}\left\{ \hat{\rho} \hat{\lambda}_j \hat{\lambda}_k \right\} .} \tag{2.299}$$

With the help of (2.296) and (2.272), (2.299) can be recast into the form

$$\overline{AB} = \overline{A}\,\overline{B} + \frac{1}{4} \sum_{j,k=1}^{s} \mathcal{A}_j \mathcal{B}_k \left[ \mathrm{Tr}\left\{ \hat{\rho} \hat{\lambda}_j \hat{\lambda}_k \right\} - \mathrm{Tr}\left\{ \hat{\rho} \hat{\lambda}_j \right\} \mathrm{Tr}\left\{ \hat{\rho} \hat{\lambda}_k \right\} \right] . \tag{2.300}$$

Restricting ourselves to

$$\mathrm{Tr}\left\{ \hat{A} \right\} = \mathrm{Tr}\left\{ \hat{B} \right\} = 0 , \tag{2.301}$$

(2.299) yields

$$\boxed{\overline{AB} = \frac{1}{4} \sum_{j,k=1}^{s} \mathcal{A}_j \mathcal{B}_k \mathrm{Tr}\left\{ \hat{\rho} \hat{\lambda}_j \hat{\lambda}_k \right\} .} \tag{2.302}$$

*Example 2.3.7.* Using (2.271) in the case of $SU(2)$,

$$\hat{\rho} = \frac{1}{2}\hat{1} + \frac{1}{2} \sum_{l=1}^{3} \lambda_l \hat{\lambda}_l , \tag{2.303}$$

the expectation value (2.302) results in

$$\overline{AB} = \frac{1}{8} \sum_{j,k=1}^{3} \mathcal{A}_j \mathcal{B}_k \mathrm{Tr}\left\{ \hat{\lambda}_j \hat{\lambda}_k \right\} + \frac{1}{8} \sum_{j,k,l=1}^{3} \mathcal{A}_j \mathcal{B}_k \lambda_l \mathrm{Tr}\left\{ \hat{\lambda}_j \hat{\lambda}_k \hat{\lambda}_l \right\} . \tag{2.304}$$

Identifying the generating operators with the operators $\hat{\sigma}_j$, the relations (2.24) and (2.25) apply, and (2.304) reduces to

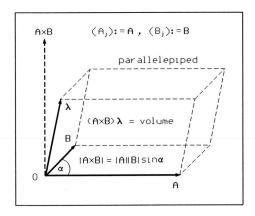

**Fig. 2.9.** The coherence vector and the vectors $(\mathcal{A}_j)$, $(\mathcal{B}_j)$ form a parallelepiped, the volume of which can be identified <u>with</u> part of the expectation value $\overline{AB}$

$$\overline{AB} = \frac{1}{4}\sum_{j=1}^{3} \mathcal{A}_j \mathcal{B}_j + \frac{i}{4}\sum_{j,k,l=1}^{3} \mathcal{A}_j \mathcal{B}_k \lambda_l \varepsilon_{jkl} \; . \tag{2.305}$$

The first term of this expression does not depend on the coherence vector.

For the second term a geometrical interpretation is available: observing that

$$(\boldsymbol{a} \times \boldsymbol{b})\,\boldsymbol{c} = \sum_{j,k,l=1}^{3} a_j b_k c_l \varepsilon_{jkl} \; , \tag{2.306}$$

it is obvious that this second term is proportional to the volume of the parallelepiped formed by vectors $(\mathcal{A}_j)$, $(\mathcal{B}_j)$ and $\boldsymbol{\lambda}$. Figure 2.9 illustrates this geometrical situation.

In the same way, the expectation value $\overline{BA}$ can be calculated:

$$\overline{BA} = \frac{1}{4}\sum_{j=1}^{3} \mathcal{A}_j \mathcal{B}_j - \frac{i}{4}\sum_{j,k,l=1}^{3} \mathcal{A}_j \mathcal{B}_k \lambda_l \varepsilon_{jkl} \tag{2.307}$$

so that the symmetrized expectation value is independent of $\boldsymbol{\lambda}$ and given by

$$\frac{1}{2}\left(\overline{AB} + \overline{BA}\right) := \left\langle \hat{A}\hat{B} \right\rangle_{\mathrm{s}}$$

$$= \frac{1}{4}\sum_{j=1}^{3} \mathcal{A}_j \mathcal{B}_j \; . \tag{2.308}$$

## 2.3.3 State Models in $SU(n)$

Contrary to Hamilton models, states, in general, are not fixed. Nevertheless, they can be classified in terms of some basic models.

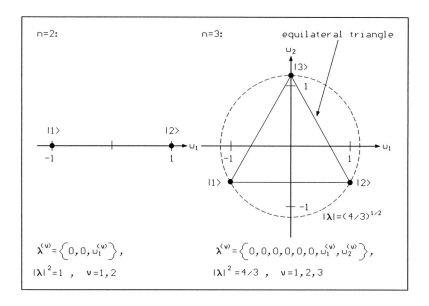

**Fig. 2.10.** Pure states in $w_l$ subspace (black dots)

**2.3.3.1 Pure States.** Special pure states are the eigenstates of $\hat{w}_l$, $l = 1, 2, \ldots, n - 1$. Figure 2.10 shows a graphic representation of the coherence vector $\boldsymbol{\lambda}^{(\nu)} = \left(0, 0, \ldots, 0, w_1^{(\nu)}, \ldots, w_{n-1}^{(\nu)}\right)$ for $n = 2, 3$. The case $n = 4$ can be represented by the corners of a regular tetrahedron. In the general case, equilateral polygons with the dimension $n - 1$ result. These states are eigenstates of any operator $\hat{A}$ which can be written as a linear combination of $\hat{1}$ and the operators $\hat{w}_l$. These operators could thus be represented by vectors in the same subspace,

$$\mathcal{A} = (0, 0, 0, \ldots, \mathcal{A}_1, \mathcal{A}_2, \ldots, \mathcal{A}_{n-1}) \ . \tag{2.309}$$

Examples are the Hamilton models (2.183), (2.188), (2.197), (2.207).

**2.3.3.2 Mixed States.** The density operator in $\hat{w}_l$ subspace reads (cf. (2.285))

$$\hat{\rho} = \frac{1}{n}\hat{1} + \frac{1}{2}\sum_{l=1}^{n-1} w_l \hat{w}_l \tag{2.310}$$

with

$$w_l = \text{Tr}\left\{\hat{\rho}\hat{w}_l\right\} \ . \tag{2.311}$$

Special examples for $SU(3)$ are listed in Table 2.2.

All states in $SU(n)$ are restricted by (2.278),

$$\boldsymbol{\lambda}^2 \leq \lambda_{\text{max}}^2 = 2\left(1 - \frac{1}{n}\right) \ . \tag{2.312}$$

**Table 2.2.** Expectation values of special mixed states (1)–(4) in $SU(3)$

| (1) | (2) | (3) | (4) |
|---|---|---|---|
| $\hat{\rho} =$ | $\hat{\rho} =$ | $\hat{\rho} =$ | $\hat{\rho} =$ |
| $\frac{1}{2}\left(\hat{P}_{11} + \hat{P}_{22}\right)$ | $\frac{1}{2}\left(\hat{P}_{11} + \hat{P}_{33}\right)$ | $\frac{1}{2}\left(\hat{P}_{22} + \hat{P}_{33}\right)$ | $\frac{1}{3}\left(\hat{P}_{11} + \hat{P}_{22} + \hat{P}_{33}\right)$ |
| $w_1 = 0$ | $w_1 = -\frac{1}{2}$ | $w_1 = \frac{1}{2}$ | $w_1 = 0$ |
| $w_2 = -\frac{1}{\sqrt{3}}$ | $w_2 = \frac{1}{2\sqrt{3}}$ | $w_2 = \frac{1}{2\sqrt{3}}$ | $w_2 = 0$ |

This, however, does not mean that the possible states form a full $\left(n^2 - 1\right)$ sphere. Using the matrix elements $\rho_{ii}$ according to (2.288), one obtains, e.g., for $SU(3)$

$$2\rho_{11} = \frac{2}{3} - w_1 - \frac{1}{\sqrt{3}}w_2 \geq 0 \, ,$$

$$2\rho_{22} = \frac{2}{3} + w_1 - \frac{1}{\sqrt{3}}w_2 \geq 0 \, , \qquad (2.313)$$

$$2\rho_{33} = \frac{2}{3} + \frac{2}{\sqrt{3}}w_2 \geq 0 \, .$$

Rewriting (2.313), one obtains the inequalities

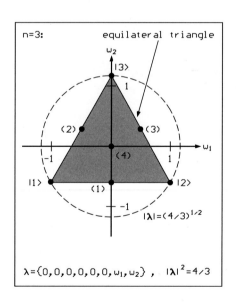

**Fig. 2.11.** Mixed states in $w_l$ subspace (shaded area). The states of Table 2.2 are marked

$$w_1 \leq \frac{2}{3} - \frac{1}{\sqrt{3}} w_2 \ ,$$

$$w_1 \geq -\frac{2}{3} + \frac{1}{\sqrt{3}} w_2 \ , \tag{2.314}$$

$$w_2 \geq -\frac{1}{\sqrt{3}} \ ,$$

which determine the domain of possible expectation values $w_1$, $w_2$, i.e. the mixed states $\boldsymbol{\lambda} = (0,0,0,0,0,0,w_1,w_2)$. This domain appears in Fig. 2.11 as the shaded equilateral triangle. The specific mixed states (1)–(4) of Table 2.2 are represented as black dots. There are no allowed states in this plane outside the shaded area.

**2.3.3.3 Coherent States.** Any pure state can be expressed as a superposition of an appropriate set of basis states. If a fixed phase relation between the basis states exists, this superposition defines a (one-node) *coherent state*. Using the $|\nu\rangle$ states as a basis, a coherent state can be defined as a state $\boldsymbol{\lambda}$ not contained in the $w_l$ subspace: the coherence is specified by the $u, v$ components (see (2.80)).

*Example 2.3.8.* In the case $n = 2$, consider the special state

$$|\psi\rangle = \frac{1}{\sqrt{2}} \left( |1\rangle + e^{i\alpha} |2\rangle \right) \tag{2.315}$$

corresponding to the density matrix

$$\hat{\rho} = |\psi\rangle \langle \psi| = \frac{1}{2} \left( \hat{P}_{11} + \hat{P}_{22} + \hat{P}_{12} e^{i\alpha} + \hat{P}_{21} e^{-i\alpha} \right) \ . \tag{2.316}$$

The coherence vector then is

$$w_1 = \mathrm{Tr} \left\{ \hat{\rho} \left( -\hat{P}_{11} + \hat{P}_{22} \right) \right\} = 0 \ , \tag{2.317}$$

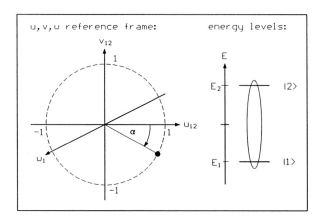

**Fig. 2.12.** A coherent state in the $u, v$ plane

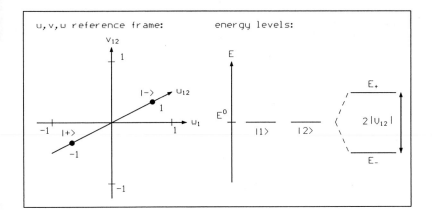

**Fig. 2.13.** Eigenvectors as coherent states

$$u_{12} = \mathrm{Tr}\left\{ \hat{\rho} \left( \hat{P}_{12} + \hat{P}_{21} \right) \right\} = \frac{1}{2} \left( e^{i\alpha} + e^{-i\alpha} \right) = \cos\alpha \; , \qquad (2.318)$$

$$v_{12} = i\mathrm{Tr}\left\{ \hat{\rho} \left( \hat{P}_{12} - \hat{P}_{21} \right) \right\} = \frac{i}{2} \left( e^{i\alpha} - e^{-i\alpha} \right) = -\sin\alpha \; , \qquad (2.319)$$

$$\lambda = (u_{1,2}, v_{12}, w_1) = (\cos\alpha, -\sin\alpha, 0) \; . \qquad (2.320)$$

This coherent state defined in the $u, v$ plane is illustrated in Fig. 2.12.

Such coherent states, for example, result as eigenstates of the Hamiltonian

$$\hat{H} = \hat{H}^0 + V_{12} \left| 1 \right\rangle \left\langle 2 \right| + V_{21} \left| 2 \right\rangle \left\langle 1 \right| \; , \quad |V_{12}| = |V_{21}| \; , \qquad (2.321)$$

$$\hat{H}^0 = \left( \hat{P}_{11} + \hat{P}_{22} \right) E^0 \; . \qquad (2.322)$$

The corresponding eigenvectors and eigenvalues are in terms of the eigenstates of $\hat{H}^0$:

$$\left| + \right\rangle = \frac{1}{\sqrt{2}} \left( - \left| 1 \right\rangle + \left| 2 \right\rangle \right) \; , \quad E_+ = E^0 + |V_{12}| \; , \qquad (2.323)$$

$$\left| - \right\rangle = \frac{1}{\sqrt{2}} \left( \left| 1 \right\rangle + \left| 2 \right\rangle \right) \; , \quad E_- = E^0 - |V_{12}| \; . \qquad (2.324)$$

These states coincide with (2.315) for $\alpha = \pi$ (0). In Fig. 2.13 the energy spectrum and the states $\left| \pm \right\rangle$ in the $u, v, w$ reference frame are illustrated.

Other coherent states located, for example, in the $v, w$ plane, are possible, too.

*Example 2.3.9.* Consider the state vector

$$\left| \psi \right\rangle = \cos\alpha \left| 1 \right\rangle + i \sin\alpha \left| 2 \right\rangle \qquad (2.325)$$

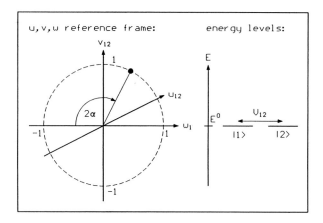

u, v, u reference frame:          energy levels:

**Fig. 2.14.** A coherent state in the $v, w$ plane

implying the density matrix

$$\hat{\rho} = |\psi\rangle \langle\psi| = \cos^2 \alpha \hat{P}_{11} + \sin^2 \alpha \hat{P}_{22} + \mathrm{i} \sin \alpha \cos \alpha \hat{P}_{12} - \mathrm{i} \sin \alpha \cos \alpha \hat{P}_{21} \ . \tag{2.326}$$

Using (2.74) and (2.326), one obtains for the components of the coherence vector

$$w_1 = \cos^2 \alpha - \sin^2 \alpha = -2 \cos 2\alpha \ ,$$
$$u_{12} = 0 \ , \tag{2.327}$$
$$v_{12} = 2 \sin \alpha \cos \alpha = \sin 2\alpha \ .$$

Figure 2.14 illustrates a possible position of this coherence vector. A realization could be obtained by the state vector

$$|\psi\rangle = \frac{1}{\sqrt{2}} \left( \mathrm{e}^{-\mathrm{i}\alpha} |-\rangle - \mathrm{e}^{\mathrm{i}\alpha} |+\rangle \right) \ , \tag{2.328}$$

which can be interpreted as a superposition of eigenvectors $|+\rangle$, $|-\rangle$ of the Hamiltonian (2.321). Using (2.323) and (2.324), (2.328) results in

$$|\psi\rangle = \cos \alpha |1\rangle - \mathrm{i} \sin \alpha |2\rangle \ . \tag{2.329}$$

**2.3.3.4 Phase States and Glauber States.** Special superposition states for the linear harmonic oscillator in $SU(n)$ are the *phase states*

$$|\Theta\rangle = \frac{1}{\sqrt{n}} \sum_{j=1}^{n} \mathrm{e}^{\mathrm{i}(j-1)\Theta} |j\rangle \tag{2.330}$$

and the *Glauber states*

$$|\alpha\rangle = N \sum_{j=1}^{n} \frac{\alpha^{j-1}}{\sqrt{(j-1)!}} |j\rangle \tag{2.331}$$

with

$$\langle \alpha | \alpha \rangle = 1 \tag{2.332}$$

and

$$N^{-2} = \sum_{j=1}^{n} \frac{|\alpha|^{2(j-1)}}{(j-1)!} \ . \tag{2.333}$$

Both sets of states are non-orthogonal and overcomplete. For $n \to \infty$, the relation

$$N^{-2} = e^{|\alpha|^2} \tag{2.334}$$

holds and the probability to find a specific state $|j\rangle$ is Poisson-distributed:

$$w_j = |\langle j | \alpha \rangle|^2 = e^{-|\alpha|^2} \frac{|\alpha|^{2(j-1)}}{(j-1)!} \ . \tag{2.335}$$

With

$$\hat{a} |j\rangle = \sqrt{j-1} |j-1\rangle \tag{2.336}$$

one shows that

$$\hat{a} |\alpha\rangle = \alpha |\alpha\rangle \ , \tag{2.337}$$

$$\langle \alpha | \hat{a}^+ \hat{a} | \alpha \rangle = \langle \hat{a}\alpha | \hat{a}\alpha \rangle = |\alpha|^2 \tag{2.338}$$

and

$$\overline{H} = \Delta E \left( |\alpha|^2 + \frac{1}{2} \right) \ . \tag{2.339}$$

The $|j\rangle$ states are often interpreted as photon (or phonon) number states (*Fock states*). $|\alpha|^2$ then defines the average particle number in state $|\alpha\rangle$. Nevertheless, these *Glauber states* are *single-node coherent states*, where *node* refers here to the oscillator under consideration (not to the individual quanta making up the state). As in second quantization the number of particles is not conserved, a node can no longer refer to a particle. A node (in the way we use this term) refers to a certain state subspace of fixed dimension within a given model space.

### 2.3.4 Entropy

Entropy measures the lack of information about a physical state. It does not tell us anything about the origin of this indeterminacy.

A typical situation, which one encounters in classical as well as in quantum models, concerns ensembles of $N \gg 1$ identical objects, the behaviour of which we cannot (or do not want to) follow in detail. *Ensemble averaging* (for $N \to \infty$) then leads (for quantum systems) to a density matrix with a finite entropy, the *von Neumann entropy*. The resulting *lack of knowledge* (lack of control) implies a certain indeterminacy of any experiment performed on this ensemble. This uncertainty, of course, should not be confused with that underlying the *Heisenberg uncertainty relation*: the latter limits the *predictability of incompatible observables* with respect to a single quantum object even if its initial state has no entropy at all (pure state).

We know from experience that the extremely large amount of information necessary to completely specify the state of a macroscopic system (*macro state*) is, in typical cases, not required to predict its behaviour reliably. Few pertinent macroscopic properties define, for example, the thermodynamic equilibrium states. The corresponding density operator can be derived from an extremum principle (*Jaynes' principle*).

Non-local correlations between quantum objects give rise to a strange indeterminacy of local (i.e. subsystem) properties. This situation is entirely non-classical. One can even reverse the argument, i.e. if a quantum object is in a zero-entropy state, it cannot be correlated to any other quantum object. An immediate consequence here is that entropy is no longer an additive property. This will be discussed in Sect. 2.4.

**2.3.4.1 von Neumann Entropy.** The quantity

$$S(\hat{\rho}) = -k \text{Tr} \{\hat{\rho} \ln \hat{\rho}\} \ , \tag{2.340}$$

wherein $k$ denotes a proportional constant, is a measure for the uncertainty of a quantum mechanical state. This measure is usually called *von Neumann entropy*. It has the following basic properties (cf. [20]):

*Property 1.* For a density operator in eigenrepresentation (cf. (2.251))

$$\hat{\rho} = \sum_{\nu=1}^{n} \rho_\nu \hat{P}_{\nu\nu} \ (0 \le \rho_\nu \le 1) \ , \tag{2.341}$$

the measure (2.340) reads

$$S(\hat{\rho}) = -k \sum_{\nu=1}^{n} \rho_\nu \ln \rho_\nu \ , \tag{2.342}$$

and the inequality

$$S(\hat{\rho}) \ge 0 \tag{2.343}$$

holds.

*Property 2.* A pure state

$$\hat{\rho} = |m\rangle \langle m| \tag{2.344}$$

is a special case of (2.341) with $\rho_\nu = \delta_{\nu m}$; its entropy is

$$S(\hat{\rho}) = 0 \tag{2.345}$$

observing $(\nu \ne m)$

$$\lim_{\rho_\nu \to 0} \rho_\nu \ln \rho_\nu \to 0 \ . \tag{2.346}$$

In the case of mixed states $S(\hat{\rho}) > 0$ holds. Due to the fact that a mixed state is characterized by a distribution of quantum mechanical state vectors (compare Sect. 1.6.3.7), the entropy $S$ is a measure for the uncertainty of the occurrence of a particular pure state.

*Property 3.* The maximum uncertainty for a state in an $n$-dimensional Hilbert space is given by

$$S(\hat{\rho}) = k \ln n \ . \tag{2.347}$$

*Proof.* This relation can be proved as follows: the expression (2.342) can be rewritten as

$$S(\hat{\rho}) = k \sum_{\nu=1}^{n} \rho_\nu \ln \left( \frac{1}{\rho_\nu} \right) = k \sum_{\nu=1}^{n} \rho_\nu \ln \left( \frac{n}{n\rho_\nu} \right)$$

$$= k \sum_{\nu=1}^{n} \rho_\nu \ln n + k \sum_{\nu=1}^{n} \rho_\nu \ln \left( \frac{1}{n\rho_\nu} \right) \ . \tag{2.348}$$

With

$$\ln x \le x - 1 \ , \quad \sum_{\nu=1}^{n} \rho_\nu = 1 \ , \tag{2.349}$$

the inequality

$$S(\hat{\rho}) = k \sum_{\nu=1}^{n} \rho_\nu \ln n + k \sum_{\nu=1}^{n} \rho_\nu \ln \left( \frac{1}{n\rho_\nu} \right)$$

$$\le k \ln n + k \sum_{\nu=1}^{n} \rho_\nu \left( \frac{1}{n\rho_\nu} - 1 \right) \tag{2.350}$$

results. Finally, as

$$\sum_{\nu=1}^{n} \frac{1}{n} = 1 \ , \tag{2.351}$$

the last sum vanishes, and we are left with

$$S(\hat{\rho}) \le k \ln n \ . \tag{2.352}$$

Therefore, the maximum uncertainty is, indeed, given by (2.347). This maximum is obtained for the broadest distribution possible, i.e.

$$\rho_\nu = \frac{1}{n} \ , \quad \nu = 1, 2, \ldots, n \ . \tag{2.353}$$

*Example 2.3.10 (Entropy in $SU(2)$).* We show that the entropy can be expressed in terms of $|\lambda|$. For this purpose we consider the general density matrix

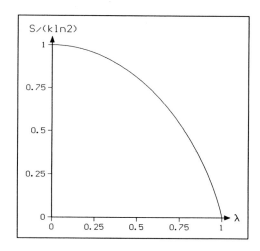

Fig. 2.15. Entropy in $SU(2)$ as a function of the length of the coherence vector

$$(\rho_{ij}) = \frac{1}{2} \begin{pmatrix} 1 - w_1 & u_{12} + iv_{12} \\ u_{12} - iv_{12} & 1 + w_1 \end{pmatrix} . \tag{2.354}$$

Its eigenvalues $\rho_i$ are obtained from the determinant

$$\begin{vmatrix} 1 - w_1 - \rho_i & u_{12} + iv_{12} \\ u_{12} - iv_{12} & 1 + w_1 - \rho_i \end{vmatrix} = \rho_i^2 - 2\rho_i - \left( u_{12}^2 + v_{12}^2 + w_1^2 \right) + 1$$

$$= 0 \tag{2.355}$$

implying

$$\rho_{1,2} = \frac{1}{2} \left( 1 \mp \sqrt{u_{12}^2 + v_{12}^2 + w_1^2} \right) = \frac{1}{2} \left( 1 \mp |\boldsymbol{\lambda}| \right) := \frac{1}{2} \left( 1 \mp \lambda \right) , \tag{2.356}$$

where $\boldsymbol{\lambda}$ denotes the corresponding coherence vector with absolute value $\lambda$. In the case of a diagonal density matrix, the diagonal matrix

$$(\rho_{ij}) = \frac{1}{2} \begin{pmatrix} 1 - |\boldsymbol{\lambda}| & 0 \\ 0 & 1 + |\boldsymbol{\lambda}| \end{pmatrix} , \tag{2.357}$$

$S$ is defined by

$$S = -k \left( \rho_1 \ln \rho_1 + \rho_2 \ln \rho_2 \right) \tag{2.358}$$

(compare with (2.342)) so that with (2.356)

$$S = \frac{k}{2} \ln \frac{4 \left( 1 - \lambda \right)^{\lambda - 1}}{\left( 1 + \lambda \right)^{\lambda + 1}} \tag{2.359}$$

(see Fig. 2.15). The limiting cases are, as expected,

$$S \left( \lambda = +1 \right) = 0 , \tag{2.360}$$

$$S \left( \lambda = 0 \right) = k \ln 2 . \tag{2.361}$$

**2.3.4.2 Indeterminacy of Experimental Results.** Let a single measurement be described by the operator $\hat{F}$ with $F_\alpha$, $\alpha = 1, 2, \ldots, n$ being the possible measurement results, and $p_\alpha$,

$$0 \le p_\alpha \le 1, \quad \sum_\alpha p_\alpha = 1, \tag{2.362}$$

denoting the probability to observe the result $F_\alpha$. Then

$$\boxed{\eta_F\,(p_1, p_2, \ldots, p_n) = -k \sum_\alpha p_\alpha \ln p_\alpha} \tag{2.363}$$

defines a measure for the indeterminacy of the experiment $F$ (cf. [48]).

If $\hat{F}$ is identified with a complete operator $\hat{M}$ which satisfies the commutator relation

$$\left[\hat{M}, \hat{\rho}\right]_- = 0, \tag{2.364}$$

the eigenvectors $|\nu\rangle$ of the density operator $\hat{\rho}$ are also eigenvectors of the operator $\hat{M}$. Then,

$$p_\nu = \langle\nu|\,\hat{\rho}\,|\nu\rangle = \rho_\nu \tag{2.365}$$

holds so that the measure

$$\eta_M = -k \sum_\nu w_\nu \ln w_\nu \tag{2.366}$$

is identical with

$$S\,(\hat{\rho}) = -k \sum_\nu \rho_\nu \ln \rho_\nu, \tag{2.367}$$

i.e. the relation

$$\eta_M = S \tag{2.368}$$

shows that the introduced measure for the uncertainty of a state, the measure $S$, can be identified with the measure for the indeterminacy of a complete observable, the measure $\eta_M$.

If $\hat{F}$ is complete but does not satisfy (2.364), the relation

$$p_\alpha = \langle u_\alpha|\,\hat{\rho}\,|u_\alpha\rangle = \sum_\nu \rho_\nu\,|\langle u_\alpha|\,\nu\rangle|^2 \tag{2.369}$$

holds, where $|u_\alpha\rangle$ denotes the eigenvectors of the operator $\hat{F}$, and $|\nu\rangle$ the eigenvectors of the density operator $\hat{\rho}$. The difference $\eta_F - \eta_M$ is then

$$\eta_F - \eta_M = k\left(\sum_\nu \rho_\nu \ln \rho_\nu - \sum_\alpha p_\alpha \ln p_\alpha\right). \tag{2.370}$$

Using (2.369), and taking $\sum_\alpha |\langle u_\alpha|\,\nu\rangle|^2 = |\langle\nu|\,\nu\rangle|^2 = 1$ into account, (2.370) results in

$$\eta_F - \eta_M = k \sum_{\nu,\alpha} |\langle u_\alpha | \nu \rangle|^2 \rho_\nu (\ln \rho_\nu - \ln p_\alpha)$$

$$= k \sum_{\nu,\alpha} |\langle u_\alpha | \nu \rangle|^2 \rho_\nu \ln \left( \frac{\rho_\nu}{p_\alpha} \right) . \tag{2.371}$$

With

$$\ln x \geq 1 - \frac{1}{x} , \quad x = \frac{\rho_\nu}{p_\alpha} \rightarrow \rho_\nu \ln \left( \frac{\rho_\nu}{p_\alpha} \right) \geq \rho_\nu - p_\alpha , \tag{2.372}$$

the difference $\eta_F - \eta_M$ can be rewritten as

$$\eta_F - \eta_M \geq k \sum_{\nu,\alpha} |\langle u_\alpha | \nu \rangle|^2 (\rho_\nu - p_\alpha)$$

$$= k \left( \sum_\nu \rho_\nu - \sum_\alpha p_\alpha \right)$$

$$= 0 . \tag{2.373}$$

Together with (2.368), one obtains *Klein's inequality* (cf. [48])

$$\boxed{\eta_F \geq \eta_M = S .} \tag{2.374}$$

Measurements in the eigenbasis of $\hat{\rho}$ are the least undetermined.

### 2.3.5 Canonical Statistical Operator

In order to calculate expectation values of a quantum system, the density operator should be known. Complete measurements of $\hat{\rho}$ are possible but practically limited to small state spaces (cf. Sect. 2.3.6). In high-dimensional Hilbert spaces so-called *macro states* will often suffice to characterize the system. These macro states are defined by a finite set $b$, $b \ll s$, of expectation values. Then, taking a special variational principle, namely *Jaynes' principle*, as a basis, an unbiased characterization of the density operator is possible. This principle and the corresponding density operator are studied in this section (cf. [48]).

**2.3.5.1 Jaynes' Principle.** A level of observation can be defined by a set of operators $\hat{G}_\nu$,

$$\left\{ \hat{G}_1, \hat{G}_2, \ldots \hat{G}_b \right\} := \left\{ \hat{G} \right\} , \tag{2.375}$$

where, typically,

$$b \ll n^2 - 1 . \tag{2.376}$$

The set of expectation values

$$\mathrm{Tr} \left\{ \hat{\rho} \hat{G}_\nu \right\} = \overline{G}_\nu \quad (\nu = 1, 2, \ldots, b) \tag{2.377}$$

are then not sufficient to determine the corresponding density operator $\hat{\rho}$ unambiguously. However, a unique determination is possible by requiring

$$S\left(\hat{\rho}\right) = -k\mathrm{Tr}\left\{\hat{\rho}\ln\hat{\rho}\right\} := \text{maximum}\ \ \text{for}\ \ \hat{\rho} = \hat{K} \tag{2.378}$$

under the constraints (2.377) (*Jaynes' principle*). The density operator $\hat{K}$ is called *canonical statistical operator*. One obtains

$$\hat{K} = \frac{1}{Z}\exp\left(-\sum_{\nu=1}^{b}\beta_{\nu}\hat{G}_{\nu}\right) \tag{2.379}$$

with the partition function $Z = Z\left(\beta_1, \beta_2, \ldots, \beta_b\right)$ being defined by

$$Z\left(\beta_1, \beta_2, \ldots, \beta_b\right) = \mathrm{Tr}\left\{\exp\left(-\sum_{\nu=1}^{b}\beta_{\nu}\hat{G}_{\nu}\right)\right\}. \tag{2.380}$$

The parameters $\beta_{\nu}$ are functions of the expectation values $\overline{G}_{\nu}$, i.e.

$$\beta_{\nu} = \beta_{\nu}\left(\overline{G}_1, \overline{G}_2, \ldots, \overline{G}_b\right). \tag{2.381}$$

Relations of the kind (2.381) occur in many fields of modern theoretical physics. For example, in the context of *thermodynamics, laser theory* and *brain research*. In particular, such relations allow a macroscopic determination of statistical distribution functions occurring in statistical theories. In many cases such functions can be calculated explicitly. For example, in the context of *Landau's theory of phase transitions* and in laser theory the parameters $\beta_{\nu}$ can be expressed by analytical power series functions which show the property *self-similarity* (see [150]).

*Proof.* In order to prove that (2.379) indeed corresponds to the maximum of $S$, a trial density operator $\hat{\rho}$ is introduced on the same level of observation:

$$\mathrm{Tr}\left\{\hat{K}\hat{G}_{\nu}\right\} = \mathrm{Tr}\left\{\hat{\rho}\hat{G}_{\nu}\right\} = \overline{G}_{\nu}\ \ (\nu = 1, 2, \ldots, b),$$
$$\mathrm{Tr}\left\{\hat{K}\right\} = \mathrm{Tr}\left\{\hat{\rho}\right\} = 1. \tag{2.382}$$

Then using (2.379) in the form

$$\ln\hat{K} = -\sum_{\nu}\beta_{\nu}\hat{G}_{\nu} - \ln Z\left(\beta_1, \beta_2, \ldots, \beta_b\right)\hat{1}, \tag{2.383}$$

the trace $\mathrm{Tr}\left\{\hat{K}\ln\hat{K}\right\}$ can be rewritten as

$$\mathrm{Tr}\left\{\hat{K}\ln\hat{K}\right\} = -\sum_{\nu}\beta_{\nu}\mathrm{Tr}\left\{\hat{K}\hat{G}_{\nu}\right\} - \ln Z\left(\beta_1, \beta_2, \ldots, \beta_b\right). \tag{2.384}$$

Due to (2.382), the expression $\mathrm{Tr}\left\{\hat{K}\hat{G}_{\nu}\right\}$ can be replaced by $\mathrm{Tr}\left\{\hat{\rho}\hat{G}_{\nu}\right\}$. Thus, the equation

$$\operatorname{Tr}\left\{\hat{K}\ln\hat{K}\right\} = -\sum_{\nu}\beta_{\nu}\operatorname{Tr}\left\{\hat{\rho}\hat{G}_{\nu}\right\} - \ln Z\left(\beta_{1},\beta_{2},\ldots,\beta_{b}\right)$$

$$= \operatorname{Tr}\left\{\hat{\rho}\ln\hat{K}\right\} \qquad (2.385)$$

follows.

Now, let $|\rho_m\rangle$ with

$$\hat{\rho}\,|\rho_m\rangle = \rho_m\,|\rho_m\rangle \qquad (2.386)$$

denote the eigenfunctions of $\hat{\rho}$, and $|k_n\rangle$ with

$$\hat{K}\,|k_n\rangle = k_n\,|k_n\rangle \qquad (2.387)$$

the eigenfunctions of $\hat{K}$. Then one obtains

$$\operatorname{Tr}\left\{\hat{\rho}\ln\hat{\rho}\right\} - \operatorname{Tr}\left\{\hat{\rho}\ln\hat{K}\right\} = \sum_{m}\rho_m\left(\ln\rho_m - \langle\rho_m|\ln\hat{K}\,|\rho_m\rangle\right). \qquad (2.388)$$

Completeness of the functions $|k_n\rangle$,

$$\hat{1} = \sum_{n}|k_n\rangle\,\langle k_n|\ , \qquad (2.389)$$

allows us to rewrite the matrix element as

$$\left\langle\rho_m\left|\ln\hat{K}\right|\rho_m\right\rangle = \sum_{n}\langle\rho_m|\,k_n\rangle\left\langle k_n\left|\ln\hat{K}\right|k_n\right\rangle\langle k_n|\,\rho_m\rangle$$

$$= \sum_{n}|\langle\rho_m|\,k_n\rangle|^{2}\ln k_n\ . \qquad (2.390)$$

Due to

$$\sum_{n}|\langle\rho_m|\,k_n\rangle|^{2} = 1\ , \qquad (2.391)$$

for any $m$, we get

$$\operatorname{Tr}\left\{\hat{\rho}\ln\hat{\rho}\right\} - \operatorname{Tr}\left\{\hat{\rho}\ln\hat{K}\right\} = \sum_{m,n}|\langle\rho_m|\,k_n\rangle|^{2}\left[\rho_m\left(\ln\rho_m - \ln k_n\right)\right]\ . \qquad (2.392)$$

Using the inequality (2.372) with $x = \rho_m/k_n$, equation (2.392) is replaced by

$$\operatorname{Tr}\left\{\hat{\rho}\ln\hat{\rho}\right\} - \operatorname{Tr}\left\{\hat{\rho}\ln\hat{K}\right\} \geq \sum_{m,n}|\langle\rho_m|\,k_n\rangle|^{2}\left[\rho_m - k_n\right] \qquad (2.393)$$

so that, observing, again, (2.391),

$$\operatorname{Tr}\left\{\hat{\rho}\ln\hat{\rho}\right\} - \operatorname{Tr}\left\{\hat{\rho}\ln\hat{K}\right\} \geq \sum_{m}\rho_m - \sum_{n}k_n = 1 - 1 = 0\ . \qquad (2.394)$$

Comparing (2.394) with (2.385), one finally obtains

$$\operatorname{Tr}\left\{\hat{\rho}\ln\hat{\rho}\right\} \geq \operatorname{Tr}\left\{\hat{K}\ln\hat{K}\right\} \qquad (2.395)$$

and therefore

$$S\left(\hat{\rho}\right) \le S\left(\hat{K}\right) , \tag{2.396}$$

as asserted.

For the canonical statistical operator (2.379), the measure of uncertainty (entropy) is given by

$$\begin{aligned} S_{\{\hat{G}\}} &= -k \text{Tr}\left\{\hat{K}\ln\hat{K}\right\} \\ &= k\sum_{\nu}\beta_{\nu}\overline{G}_{\nu} + k\ln Z , \end{aligned} \tag{2.397}$$

where the lower case symbol $\left\{\hat{G}\right\}$ indicates the corresponding level of observation.

**2.3.5.2 Equilibrium Thermodynamics.** The principle discussed above (Jaynes' principle) has found broadest application in the statistical foundation of thermodynamics. Here one studies so-called statistical (or Gibbs) ensembles, where each ensemble member is a many-particle system described by the same Hamiltonian. If the level of observation is defined exclusively by the total energy, i.e. if $\hat{G}_1 = \hat{H}$, the canonical statistical operator reads

$$\hat{K}_{\{\hat{H}\}} = \frac{\exp\left(-\beta_1\hat{H}\right)}{Z_{\{\hat{H}\}}} \tag{2.398}$$

with

$$Z_{\{\hat{H}\}} = \text{Tr}\left\{\exp\left(-\beta_1\hat{H}\right)\right\} . \tag{2.399}$$

Identifying the parameter $\beta_1$ with

$$\beta_1 = \frac{1}{k_B T} , \tag{2.400}$$

wherein $k_B$ denotes Boltzmann's constant and $T$ the temperature, the statistical operator (2.398) represents the *canonical ensemble* of equilibrium thermodynamics, where

$$\text{Tr}\left\{\hat{K}\hat{H}\right\} = \frac{\text{Tr}\left\{\hat{H}\exp\left(-\beta_1\hat{H}\right)\right\}}{\text{Tr}\left\{\exp\left(-\beta_1\hat{H}\right)\right\}} = \overline{H} := U \tag{2.401}$$

defines the *internal energy* $U$ of a thermodynamic system. If, as an additional operator $\hat{G}_2$, the *particle number operator* $\hat{N}$ is introduced, the *grand canonical ensemble* results. Equation (2.342) thus includes the statistical formulation of the *thermodynamic entropy* if the constant $k$ is identified with

*Boltzmann's constant* $k_B$ and if the element $\rho_\nu$ is identified with the probability of finding the thermodynamic system in a particular microstate $\nu$ within the *phase space* considered.

From (2.397) we obtain the well-known relation for the thermodynamic entropy

$$S_{\{\hat{H}\}} = k_B \left(\beta_1 U + \ln Z_{\{\hat{H}\}}\right) . \tag{2.402}$$

Multiplying this equation with the absolute temperature $T$, one obtains

$$TS_{\{\hat{H}\}} = U + k_B T \ln Z_{\{\hat{H}\}} \tag{2.403}$$

so that the *free energy* $F$ defined by

$$F = U - TS_{\{\hat{H}\}} \tag{2.404}$$

can be calculated from the partition function $Z$:

$$F = -k_B T \ln Z_{\{\hat{H}\}} . \tag{2.405}$$

Such equilibrium states will be assumed to apply to systems acting as a bath (see Sect. 3.3.2).

### 2.3.6 Direct Ensemble Measurements

In this section, the notion *quantum measurement* is introduced based on *axiomatic measurement theory*. Here, we restrict ourselves to *direct measurements*. *Indirect measurements* are generic for composite systems to be discussed in Sect. 2.4. Detailed measurement models require dynamical investigations to be postponed to Chap. 3 and Chap. 4.

**2.3.6.1 Direct and Indirect Measurements.** *Direct measurements* require the free transport of the particle to be measured to the macroscopic measurement device. We thus are typically concerned with "beams". The *measurement environment* consists of two essential parts: a *filter* (like a polarizing filter, an energy filter or a momentum filter) and a *detector* (like a photon counter):

$$\boxed{\text{measurement} = \text{filter} + \text{detector} .} \tag{2.406}$$

The idealized filter functions as a beam splitter, mapping the eigenstates $|a_m\rangle$ of the measured observable $\hat{A}$ into corresponding beam directions. Thus counting in a specific direction means counting particles with that specific property. This case of a *direct measurement* is illustrated in Fig. 2.16. Here a binary decision based on a 2-dimensional Hilbert space is shown.

In contrast, in an *indirect measurement* a correlated system 2 is measured in order to gain information about the system 1. If the correlation is known, conclusions can be drawn about system 1. In other words, theoretical *prior*

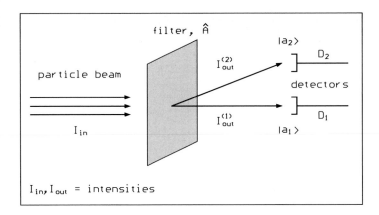

**Fig. 2.16.** A direct ensemble measurement: particles of a beam are split (states $|a_1\rangle$, $|a_2\rangle$) and counted

*knowledge* is necessary. This type is illustrated in Fig. 2.17. We will return to this issue in Sect. 2.4.5.

The probabilities for a specific measurement outcome can be inferred from the relative intensities of the outgoing beams. These probabilities are controlled by the density operator of the incoming beam. In this way, information about the density matrix becomes available. If $\hat{A}$ is an observable defined in the $\hat{w}_l$ subspace (cf. Sect. 2.2.2.7), the measurement states are joint eigenstates of the operators $\hat{w}_l$ with

$$\hat{w}_l \, |m\rangle = w_l^{(m)} \, |m\rangle \ , \ \ 1 \le m \le n \ , \ \ 1 \le l \le n-1 \tag{2.407}$$

(compare (2.102)). The measurement states of an observable $\hat{B}$ with

$$\left[\hat{B}, \hat{A}\right]_- \ne 0 \tag{2.408}$$

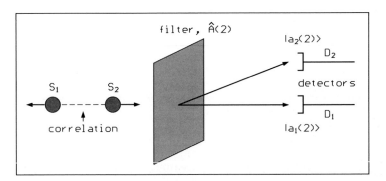

**Fig. 2.17.** An indirect ensemble measurement: particles $S_1$ are measured by observing correlated particles $S_2$ reaching the detector in state $|a_m(2)\rangle$, $m = 1, 2$

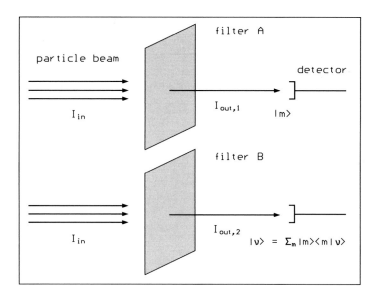

**Fig. 2.18.** "Incompatible" ensemble measurements; only one channel is shown

will be defined by

$$|\nu\rangle = \sum_m |m\rangle \langle m\,|\nu\rangle \ . \tag{2.409}$$

With respect to a homogeneous ensemble (beam) such "incompatible" measurements can easily be performed. Figure 2.18 illustrates these cases. A set of such "incompatible" measurements is required to determine the total density matrix of a homogeneous ensemble.

In this section we will restrict ourselves to direct measurements. For *ensemble measurements*, *intensities* $I_{\text{in}}$ and $I_{\text{out}}$ as shown in Figs. 2.16 and 2.18, or relative intensities $I_{\text{out}}/I_{\text{in}}$ are the pertinent measurement quantities so that relative frequencies for the occurrence of particular particle states can be inferred.

### 2.3.6.2 Axiomatic Measurement Theory and Projection Operators.
Let $\hat{A}$ represent a measurement observable with eigenvalue equation

$$\hat{A}_i |a_i\rangle = a_i |a_i\rangle \quad (i = 1, 2, 3, \ldots) \tag{2.410}$$

and projection operators

$$\hat{P}^a_{mm} = |a_m\rangle \langle a_m| \ . \tag{2.411}$$

If a measurement represented by $\hat{A}$ is carried out on a system in state $\hat{\rho}$, the probability to find result $a_m$ is defined by

$$p_m = \text{Tr}\left\{ \hat{P}^a_{mm} \hat{\rho} \hat{P}^a_{mm} \right\} \ . \tag{2.412}$$

Observing

$$\text{Tr}\left\{\hat{A}\hat{B}\right\} = \text{Tr}\left\{\hat{B}\hat{A}\right\} \;,\quad \hat{P}^a_{mm}\hat{P}^a_{mm} = \hat{P}^a_{mm} \;, \tag{2.413}$$

the probability $p_m$ can be written as

$$p_m = \text{Tr}\left\{\hat{P}^a_{mm}\hat{\rho}\right\} \;. \tag{2.414}$$

Then after the measurement the system is in the state

$$\hat{\rho}' = \frac{1}{p_m}\hat{P}^a_{mm}\hat{\rho}\hat{P}^a_{mm} \;. \tag{2.415}$$

This is the fundamental *projection postulate* (von Neumann measurement, cf. [147]).

*Example 2.3.11.* Consider the case $n = 2$ with

$$\hat{\rho} = \sum_{ij} \rho_{ij} \,|i\rangle\,\langle j| \;. \tag{2.416}$$

The probability $p_1$ to find the result $\hat{P}^z_{11} = |1\rangle\,\langle 1|$ can then be calculated by using (2.414):

$$p_1 = \text{Tr}\left\{|1\rangle\,\langle 1|\,\hat{\rho}\,|1\rangle\,\langle 1|\right\} = \rho_{11} \;, \tag{2.417}$$

in which case after measurement the final state is

$$\hat{\rho}' = \frac{1}{\rho_{11}}\,|1\rangle\,\langle 1|\sum_{ij}\rho_{ij}\,|i\rangle\,\langle j|\,1\rangle\,\langle 1| = |1\rangle\,\langle 1| \;. \tag{2.418}$$

Alternative outcomes require consideration of a set of different *projectors*. In axiomatic measurement theory one usually restricts oneself to commuting projection operators. The projection operators defined by (2.411) obey

$$\left[\hat{P}_{mm},\hat{P}_{nn}\right]_- = 0 \;. \tag{2.419}$$

Based on these, further commuting operators can be introduced.

*Example 2.3.12.* The pairs

$$\hat{P}_{11},\hat{P} = \hat{1} - \hat{P}_{11} \tag{2.420}$$

obey

$$\left[\hat{P}_{11},\hat{P}\right]_- = \left[\hat{P}_{11},\hat{1}\right]_- - \left[\hat{P}_{11},\hat{P}_{11}\right]_- = 0 \;. \tag{2.421}$$

*Example 2.3.13.* Similarly, for

$$\hat{P}_{11},\hat{P}_a = \hat{P}_{11} + \hat{P}_{22} \tag{2.422}$$

one easily shows

$$\left[\hat{P}_{11},\hat{P}_a\right]_- = 0 \;. \tag{2.423}$$

*Example 2.3.14.* On the other hand, the pair of projection operators $\hat{P}_{11}$, $\hat{P}_{11}^x$, with

$$
\begin{aligned}
\hat{P}_{11}^x &= \frac{1}{2} \left( |1\rangle + |-1\rangle \right) \left( \langle 1| + \langle -1| \right) \\
&= \frac{1}{2} \left( \hat{P}_{11} + \hat{P}_{-1-1} + \hat{P}_{1-1} + \hat{P}_{-11} \right) ,
\end{aligned}
\tag{2.424}
$$

does not commute. Such non-commuting projection operators do not occur in direct measurements of the "either/or" type.

Projection operators are called orthogonal if they obey the relations

$$
\sum_{\alpha=1}^{M} \hat{P}_\alpha = 1 , \quad \hat{P}_\alpha \hat{P}_\beta = \delta_{\alpha\beta} \hat{P}_\alpha .
\tag{2.425}
$$

Orthogonality implies commutativity,

$$
\begin{aligned}
\left[ \hat{P}_\alpha, \hat{P}_\beta \right]_- &= \hat{P}_\alpha \hat{P}_\beta - \hat{P}_\beta \hat{P}_\alpha \\
&= \delta_{\alpha\beta} \hat{P}_\alpha - \delta_{\beta\alpha} \hat{P}_\beta \\
&= 0 ,
\end{aligned}
\tag{2.426}
$$

while the reverse is not true, i.e. commuting operators are not necessarily orthogonal: an example is given by (2.422).

The projection postulate can alternatively be expressed in terms of the $SU(n)$ algebra. Restricting ourselves to the projections $\hat{P}_{mm}$, we recall that (cf. (2.111))

$$
\hat{P}_{mm} = \frac{1}{n} \hat{1} + \frac{1}{2} \sum_{l=1}^{n-1} w_l^{(m)} \hat{w}_l \quad (m = 1, 2, \ldots, n) .
\tag{2.427}
$$

It follows that

$$
\begin{aligned}
\hat{P}_{mm} \hat{w}_l \hat{P}_{mm} &= w_l^{(m)} \hat{P}_{mm} , \\
\hat{P}_{mm} \hat{1} \hat{P}_{mm} &= \hat{P}_{mm} , \\
\hat{P}_{mm} \hat{u}_{ij} \hat{P}_{mm} &= \hat{P}_{mm} \hat{v}_{ij} \hat{P}_{mm} = 0
\end{aligned}
\tag{2.428}
$$

and

$$
\hat{P}_{mm} \hat{\rho} \hat{P}_{mm} = \left( \frac{1}{n} + \frac{1}{2} \sum_{l=1}^{n-1} w_l^{(m)} w_l \right) \hat{P}_{mm} .
\tag{2.429}
$$

The probability for outcome $m$ thus is

$$
\boxed{p_m = \mathrm{Tr} \left\{ \hat{P}_{mm} \hat{\rho} \hat{P}_{mm} \right\} = \frac{1}{n} + \frac{1}{2} \sum_{l=1}^{n-1} w_l^{(m)} w_l ,}
\tag{2.430}
$$

and after measurement the state is reduced to

$$\frac{1}{p_m}\hat{P}_{mm}\hat{\rho}\hat{P}_{mm} = \hat{P}_{mm} = \frac{1}{n}\hat{1} + \frac{1}{2}\sum_{l=1}^{n-1} w_l^{(m)}\hat{w}_l \; . \tag{2.431}$$

In terms of the coherence vector the effect of the projection can thus be summarized as $(m = 1, 2, \ldots, n)$

$$\boxed{\lambda \stackrel{\hat{P}_{mm}}{\longrightarrow} \lambda' = \begin{cases} w_l' = w_l^{(m)} & (l = 1, 2, \ldots, n-1) \\ u_{ij}' = v_{ij}' = 0 & (i < j = 1, 2, \ldots, n) \end{cases}} \tag{2.432}$$

We note that $\lambda'$ contains no information whatsoever about the previous state $\lambda$: the projection constitutes a "fundamental" Markov process. The state $\lambda$ only enters the probability $p_m$ for the respective outcome $m$. $p_m$ can be measured for a (homogeneous) ensemble. In $SU(2)$, (2.432) reads

$$\lambda \left\langle \begin{array}{l} \lambda'_- = (0, 0, -1) \quad \text{with} \quad p_{-1} = \frac{1}{2}(1 - \lambda_3) \\[2mm] \lambda'_+ = (0, 0, +1) \quad \text{with} \quad p_{+1} = \frac{1}{2}(1 + \lambda_3) \end{array} \right. \tag{2.433}$$

The ensemble average is

$$\lambda' = p_{-1}(0, 0, -1) + p_{+1}(0, 0, 1) = (0, 0, \lambda_3) \; . \tag{2.434}$$

### 2.3.6.3 Implementation of Unitary Transformations in $SU(2)$.

Typical measurement scenarios can be decomposed into a part generating the projection proper (as a decision between a given set of discrete alternatives, cf. the preceding section), and the specification of those alternative measurement states. There has been a long and controversial discussion whether any orthogonal pair of states can be made the measurement basis of a real physical apparatus (cf. [83]). Actually, this problem can be cast into a more operational form: if there is one apparatus to measure some basis states $|1\rangle$, $|2\rangle$, any other set of states can also be measured provided there is a general method available to generate unitary transformations. As discussed in Sect. 2.2.3.1, unitary transformations in Hilbert space are equivalent to orthogonal transformations in coherence vector space. For particle beams elementary devices like *beam splitters*, mirrors, and phase shifters can be combined to do the job. Let us consider a simple example.

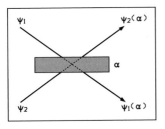

**Fig. 2.19.** Input–output modes of an ideal beam splitter

For pure states the unitary transformation may be seen as the input–output relation of a so-called *4-port* (cf. Sect. 2.2.3.2):

$$|\psi(\alpha)\rangle = \hat{U}(\alpha)\,|\psi\rangle \tag{2.435}$$

resulting in

$$\left\langle i\left|\hat{U}(\alpha)\right|\psi\right\rangle = \sum_{j=1}^{2}\left\langle i\left|\hat{U}(\alpha)\right|j\right\rangle\langle j\,|\psi(\alpha)\rangle \quad (i=1,2)\,. \tag{2.436}$$

This equation can either be interpreted as a transformation of the basis vectors $|i(\alpha)\rangle = \hat{U}^{\dagger}(\alpha)\,|i\rangle$ with respect to a fixed vector $|\psi\rangle$, or as a transformation of the vector $|\psi(\alpha)\rangle$ with respect to a fixed basis $|i\rangle$:

$$\psi_i(\alpha) = \sum_{j=1}^{2} U_{ij}(\alpha)\psi_j \quad (i=1,2) \tag{2.437}$$

(compare Fig. 2.19). The two modes $|1\rangle$, $|2\rangle$ could represent the two basic polarization states of a spin $1/2$, the polarization states of a light beam, the vacuum and 1-photon state of a cavity, or the two beam directions crossing at a beam splitter. Such a beam splitter with transmission index

$$T(i) = |\tau(i)|^2 \tag{2.438}$$

and reflection index

$$R(i) = |\varrho(i)|^2 \tag{2.439}$$

for mode $i = 1, 2$ could be characterized by

$$U_{ij}(\alpha) = \begin{pmatrix} \tau(1) & \varrho(1) \\ \varrho(2) & \tau(2) \end{pmatrix}\,. \tag{2.440}$$

$\tau$ and $\varrho$ are the complex transmission and reflection coefficients, respectively. Due to

$$\hat{U}\hat{U}^{\dagger} = 1\,, \tag{2.441}$$

we have to require

$$|\tau(i)|^2 + |\varrho(i)|^2 = 1 \quad (i=1,2)\,, \tag{2.442}$$

$$\tau(1)\varrho^*(2) + \varrho(1)\tau^*(2) = 0\,. \tag{2.443}$$

A typical single-layer beam splitter can be described by

$$\tau(j) = \cos(\alpha/2)\,, \quad \varrho(j) = \mathrm{i}\sin(\alpha/2) \quad (j=1,2)\,. \tag{2.444}$$

By comparison with Sect. 2.2.3.2 we see that this device implements the unitary operator

$$\hat{U}^x(\alpha) = \mathrm{e}^{\mathrm{i}\alpha\hat{\lambda}_x/2}\,. \tag{2.445}$$

There are multilayer beam splitters which implement $\hat{U}^y(\beta)$ (cf. [116]). $\hat{U}^z(\gamma)$ is easily realized by phase shifters affecting the two states differently. $\alpha = \pi/2$ defines the 50–50 beam splitter with $R = T = 1/2$. Obviously, the output $\psi_i(\alpha)$ is a continuous (oscillatory) function of the parameter $\alpha$ with $\hat{U}(\alpha = 0) = \hat{1}$. Those devices operate as special quantum gates.

The projection postulate then reads with

$$\hat{P}_{ii}(\alpha) = |i(\alpha)\rangle \langle i(\alpha)| = \hat{U}^\dagger(\alpha)\hat{P}_{ii}\hat{U}(\alpha) \; : \tag{2.446}$$

$$\hat{\rho}'(\alpha) = \frac{1}{p_i(\alpha)}\hat{P}_{ii}(\alpha)\hat{\rho}\hat{P}_{ii}(\alpha) \,, \tag{2.447}$$

where

$$p_i(\alpha) = \mathrm{Tr}\left\{\hat{P}_{ii}(\alpha)\hat{\rho}\right\} \,. \tag{2.448}$$

Alternatively, this can be interpreted as a measurement projection of the transformed density operator $\hat{\rho}(\alpha)$,

$$\hat{\rho}(\alpha) = \hat{U}^\dagger(\alpha)\hat{\rho}\hat{U}(\alpha) \,, \tag{2.449}$$

with respect to the original measurement basis $|i(\alpha = 0)\rangle = |i\rangle$, and with subsequent back transformation:

$$\hat{\rho}'(\alpha) = \hat{U}(\alpha)\frac{1}{p_i(\alpha)}\hat{P}_{ii}\hat{\rho}(\alpha)\hat{P}_{ii}\hat{U}^\dagger(\alpha) \,. \tag{2.450}$$

Such a scheme can also be implemented dynamically by applying appropriate pulses before and after measurement.

**2.3.6.4 The Stern-Gerlach Beam Splitter.** The Stern-Gerlach experiment is illustrated in Fig. 2.20. As shown, a beam of neutral atoms crosses a magnetic field $\boldsymbol{B}_0$ so that the Hamiltonian for a bound electron with spin $\hat{\boldsymbol{L}}$ ($l = 1/2$) at position $\boldsymbol{R}$ (centre of mass, taken here as a classical variable) reads:

$$\hat{H}(\boldsymbol{R}) = -\frac{g_s \boldsymbol{B}_0(\boldsymbol{R})}{m_e}\hat{\boldsymbol{L}} \,. \tag{2.451}$$

As the atom is supposed to be neutral, there is no conflicting *Lorentz force*. Using the *Bohr magneton*

$$\mu_\mathrm{B} = \frac{g_s \hbar}{2m_e} \,, \tag{2.452}$$

wherein $g_s$ denotes the Landé $g$ factor, $m_e$ the mass of an electron, and $\hbar$ Planck's constant, and assuming a magnetic field in $z$ direction, (2.451) reduces to

$$\hat{H} = -\mu_\mathrm{B} B_0 \hat{\sigma} = -\mu_\mathrm{B} B_z \hat{\sigma}_z \,. \tag{2.453}$$

If $\partial B_z/\partial z \neq 0$, a force along the $z$ coordinate results, depending on $\hat{\sigma}_z$. This force generates a deflection of the beam which can be measured in real space.

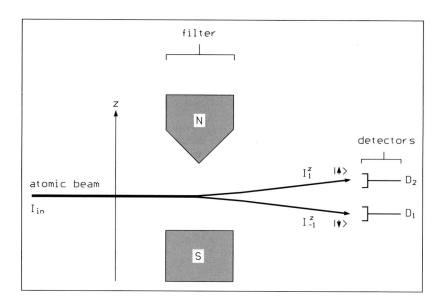

**Fig. 2.20.** The Stern-Gerlach apparatus with a magnetic field gradient in $z$ direction. It splits the input beam into two beams with polarization $P_z = \mp 1$, respectively

To measure $\hat{\sigma}_x$ $(\hat{\sigma}_y)$ one has to change the orientation of the Stern-Gerlach apparatus to have the magnetic field gradient pointing in the $x$ $(y)$ direction. The resulting information retrieval is discussed now in terms of the projection postulate. The initial state $\hat{\rho}$ is given by the matrix (in $\hat{\sigma}_z$ representation)

$$(\rho_{ij}) = \frac{1}{2} \begin{pmatrix} 1 - P_z & P_x + iP_y \\ P_x - iP_y & 1 + P_z \end{pmatrix} . \tag{2.454}$$

*Measurement of $P_z$.* The measurement basis is $|\mp 1\rangle$, the eigenstates of $\hat{\sigma}_z$, where

$$\hat{P}_{-1-1} = |-1\rangle\langle -1| , \quad \hat{P}_{11} = |1\rangle\langle 1| . \tag{2.455}$$

The atomic beam intensity into state $|-1\rangle$ (i.e. into the detector $D_1$) is

$$I_1^z = I_{\text{in}} \text{Tr}\left\{ \hat{P}_{-1-1}\hat{\rho} \right\} = \frac{I_{\text{in}}}{2} (1 - P_z) , \tag{2.456}$$

into state $|1\rangle$ (detector $D_2$)

$$I_1^z = I_{\text{in}} \text{Tr}\left\{ \hat{P}_{11}\hat{\rho} \right\} = \frac{I_{\text{in}}}{2} (1 + P_z) , \tag{2.457}$$

so that

$$\boxed{P_z = \frac{I_1^z - I_{-1}^z}{I_{\text{in}}} .} \tag{2.458}$$

*Measurement of $P_x$.* The measurement basis is now

$$\left|e_{\mp 1}^x\right\rangle = \frac{1}{\sqrt{2}}\left(|1\rangle \mp |-1\rangle\right) , \tag{2.459}$$

the eigenstates of $\hat{\sigma}_x$. The corresponding projection operators have the matrix representation

$$P_{-1-1}^x = \frac{1}{2}\begin{pmatrix} 1 & -1 \\ -1 & 1 \end{pmatrix} , \tag{2.460}$$

$$P_{11}^x = \frac{1}{2}\begin{pmatrix} 1 & 1 \\ 1 & 1 \end{pmatrix} . \tag{2.461}$$

The atomic beam intensity into state $\left|e_{-1}^x\right\rangle$ then is

$$\begin{aligned}
I_{-1}^x &= I_{\text{in}} \operatorname{Tr}\left\{\hat{P}_{-1-1}^x \hat{\rho}\right\} \\
&= \frac{I_{\text{in}}}{4} \operatorname{Tr}\left\{\begin{pmatrix} 1 & -1 \\ -1 & 1 \end{pmatrix}\begin{pmatrix} 1 - P_z & P_x + iP_y \\ P_x - iP_y & 1 + P_z \end{pmatrix}\right\} \\
&= \frac{I_{\text{in}}}{2}\left(1 - P_x\right) ,
\end{aligned} \tag{2.462}$$

into state $\left|e_1^x\right\rangle$

$$I_1^x = I_{\text{in}} \operatorname{Tr}\left\{\hat{P}_{11}^x \hat{\rho}\right\} = \frac{I_{\text{in}}}{2}\left(1 + P_x\right) , \tag{2.463}$$

so that

$$\boxed{P_x = \frac{I_1^x - I_{-1}^x}{I_{\text{in}}} .} \tag{2.464}$$

*Measurement of $P_y$.* In this case the measurement basis is

$$\left|e_{\mp 1}^y\right\rangle = \frac{1}{\sqrt{2}}\left(|1\rangle \pm i|-1\rangle\right) , \tag{2.465}$$

i.e. the eigenstates of $\hat{\sigma}_y$. The projection operators are

$$P_{-1-1}^y = \frac{1}{2}\begin{pmatrix} 1 & i \\ -i & 1 \end{pmatrix} , \tag{2.466}$$

$$P_{11}^y = \frac{1}{2}\begin{pmatrix} 1 & -i \\ i & 1 \end{pmatrix} , \tag{2.467}$$

so that the atomic beam intensity into state $\left|e_{-1}^y\right\rangle$ reads

$$\begin{aligned}
I_{-1}^y &= I_{\text{in}} \operatorname{Tr}\left\{\hat{P}_{-1-1}^y \hat{\rho}\right\} \\
&= \frac{I_{\text{in}}}{4} \operatorname{Tr}\left\{\begin{pmatrix} 1 & i \\ -i & 1 \end{pmatrix}\begin{pmatrix} 1 - P_z & P_x + iP_y \\ P_x - iP_y & 1 + P_z \end{pmatrix}\right\} \\
&= \frac{I_{\text{in}}}{2}\left(1 + P_y\right) ,
\end{aligned} \tag{2.468}$$

and into state $|e_1^y\rangle$

$$I_1^y = I_{\rm in} \operatorname{Tr}\left\{\hat{P}_{11}^y \hat{\rho}\right\} = \frac{I_{\rm in}}{2}\left(1 - P_y\right) , \qquad (2.469)$$

so that, finally,

$$\boxed{P_y = \frac{I_{-1}^y - I_1^y}{I_{\rm in}} .} \qquad (2.470)$$

The coherence vector $\boldsymbol{P}$ can thus be expressed in terms of relative intensities of three subsequent measurements with respect to 3 orthogonal field gradients. It is obvious that the three measurements cannot be performed on the same individual quantum objects: the complete measurement is based on the existence of a homogeneous ensemble, constant over the time of the measurement. If

$$|\boldsymbol{P}|^2 = P_x^2 + P_y^2 + P_z^2 < 1 , \qquad (2.471)$$

we have partial polarization.

The three sets of projection operators defined by

$$\hat{P}_{11}(x) = \frac{1}{2}\left(\hat{P}_{11} + \hat{P}_{22}\right) + \frac{1}{2}\left(\hat{P}_{12} + \hat{P}_{21}\right) ,$$
$$\hat{P}_{22}(x) = \frac{1}{2}\left(\hat{P}_{11} + \hat{P}_{22}\right) - \frac{1}{2}\left(\hat{P}_{12} + \hat{P}_{21}\right) , \qquad (2.472)$$

$$\hat{P}_{11}(y) = \frac{1}{2}\left(\hat{P}_{11} + \hat{P}_{22}\right) + \frac{i}{2}\left(\hat{P}_{21} - \hat{P}_{12}\right) ,$$
$$\hat{P}_{22}(y) = \frac{1}{2}\left(\hat{P}_{11} + \hat{P}_{22}\right) - \frac{i}{2}\left(\hat{P}_{21} - \hat{P}_{12}\right) , \qquad (2.473)$$

$$\hat{P}_{11}(z) = \hat{P}_{11} ,$$
$$\hat{P}_{22}(z) = \hat{P}_{22} \qquad (2.474)$$

can be interpreted to result from different unitary transformations of the original measurement states as discussed in Sect. 2.3.6.3: let

$$\hat{P}_{ii}(\alpha) = \hat{U}(\alpha)\hat{P}_{ii}\hat{U}^\dagger(\alpha) \quad (i = 1, 2) , \qquad (2.475)$$

$$\hat{U}(z) = \hat{1} . \qquad (2.476)$$

The associated beam intensities then are

$$I_i(\alpha) = I_{\rm in}\operatorname{Tr}\left\{\hat{P}_{ii}(\alpha)\hat{\rho}\right\} = I_{\rm in}\operatorname{Tr}\left\{\hat{P}_{ii}\hat{\rho}(\alpha)\right\} , \qquad (2.477)$$

$$\hat{\rho}(\alpha) = \hat{U}^\dagger(\alpha)\hat{\rho}\hat{U}(\alpha) , \qquad (2.478)$$

which is equivalent to a rotation of $\boldsymbol{\lambda}(\alpha)$.

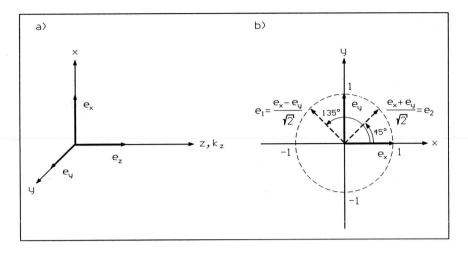

**Fig. 2.21.** Light field: real space reference frame (part a)) and alternative pairs of polarization unit vectors (part b))

**2.3.6.5 Beam Splitter in Polarization Space.** Electromagnetic fields are characterized, inter alia, by their respective polarization. Quantum mechanically, the polarization states are related to the photon spin. Though this spin is $l = 1$, the transverse condition, $\text{div}\,\boldsymbol{E} = 0$, makes this spin isomorphic to spin $l = 1/2$. Therefore, the measurement strategy discussed in the preceding section can be transcribed to the present case.

*Classical Light Polarization.* Consider two linear polarized electric fields described by (cf. Fig. 2.21)

$$\boldsymbol{E}_x \sim \boldsymbol{e}_x \exp\left[\text{i}\left(\boldsymbol{k}\boldsymbol{r} - \omega t\right)\right] , \tag{2.479}$$

$$\boldsymbol{E}_y \sim \boldsymbol{e}_y \exp\left[\text{i}\left(\boldsymbol{k}\boldsymbol{r} - \omega t + \varphi_y\right)\right] , \tag{2.480}$$

wherein $\boldsymbol{e}_x$, $\boldsymbol{e}_y$ denote the unit vectors in $x$ and $y$ direction, respectively, $\boldsymbol{k} = (0, 0, k_z)$ denotes the wave vector, $\boldsymbol{r}$ the position vector, $\omega$ the frequency, $t$ the time, and $\varphi_y$ denotes a phase. Then, the plane of polarization is the $x, y$ plane perpendicular to the direction of propagation of the light wave.

The superposition then reads

$$\boldsymbol{E} = \boldsymbol{E}_x + \boldsymbol{E}_y \sim \frac{\boldsymbol{e}_x}{\sqrt{2}} \exp\left[\text{i}\left(\boldsymbol{k}\boldsymbol{r} - \omega t\right)\right] +$$
$$\frac{\boldsymbol{e}_y}{\sqrt{2}} \exp\left[\text{i}\left(\boldsymbol{k}\boldsymbol{r} - \omega t + \varphi_y\right)\right] . \tag{2.481}$$

For phase angles $\varphi_y = 0$ ($e^{\text{i}\varphi_y} = 1$), $\varphi = \pi$ ($e^{\text{i}\varphi_y} = -1$), linear polarized waves result (compare Fig. 2.22, part a))

$$\boldsymbol{E} \sim \frac{1}{\sqrt{2}} \left(\boldsymbol{e}_x \pm \boldsymbol{e}_y\right) \exp\left[\text{i}\left(\boldsymbol{k}\boldsymbol{r} - \omega t\right)\right] . \tag{2.482}$$

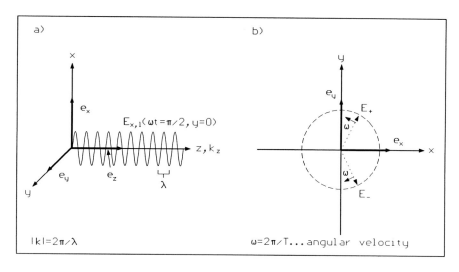

**Fig. 2.22.** Linear (part a)) and circular polarized light (part b))

The observable electric field then is described by

$$\text{Re}\boldsymbol{E} \sim \frac{1}{\sqrt{2}}\left(\boldsymbol{e}_x \pm \boldsymbol{e}_y\right) \cos\left(\boldsymbol{kr} - \omega t\right) = \boldsymbol{e}_{1,2} \cos\left(\boldsymbol{kr} - \omega t\right) \qquad (2.483)$$

(the vectors $\boldsymbol{e}_x \pm \boldsymbol{e}_y$ are illustrated in Fig. 2.21, part b)). For the phase angles $\varphi_y = \pm\pi/2$ ($e^{\pm i\varphi} = \pm i$) one finds circular polarized waves (compare (2.22), part b))

$$\boldsymbol{E} \sim \frac{1}{\sqrt{2}}\left(\boldsymbol{e}_x \pm i\boldsymbol{e}_y\right) \exp\left[i\left(\boldsymbol{kr} - \omega t\right)\right] , \qquad (2.484)$$

$$\text{Re}\boldsymbol{E} \sim \frac{1}{\sqrt{2}}\boldsymbol{e}_x \cos\left(\boldsymbol{kr} - \omega t\right) \mp \frac{1}{\sqrt{2}}\boldsymbol{e}_y \sin\left(\boldsymbol{kr} - \omega t\right) . \qquad (2.485)$$

At a fixed point $\boldsymbol{kr} = 0$ the circular polarization amounts to a right/left rotation in the $x, y$ plane:

$$\text{Re}\boldsymbol{E} \sim \frac{1}{\sqrt{2}}\boldsymbol{e}_x \cos\left(\omega t\right) \pm \frac{1}{\sqrt{2}}\boldsymbol{e}_y \sin\left(\omega t\right) := \boldsymbol{E}_\pm . \qquad (2.486)$$

The vectors $\boldsymbol{E}_\pm$ are illustrated in Fig. 2.22, part b).

Introducing the abbreviations

$$e(\pm) = \frac{1}{\sqrt{2}}\left(\boldsymbol{e}_x \pm i\boldsymbol{e}_y\right) , \qquad (2.487)$$

the circular polarized electric field (2.482) reduces to

$$\boldsymbol{E} \sim e(\pm) \exp\left[i\left(\boldsymbol{kr} - \omega t\right)\right] , \qquad (2.488)$$

where the complex amplitudes $e(\pm)$ define the two circular polarized *field states*.

The inverse relations to (2.487) are

$$
\begin{aligned}
e_x &= \frac{1}{\sqrt{2}} \left[ e(+) + e(-) \right] , \\
e_y &= -\frac{i}{\sqrt{2}} \left[ e(+) - e(-) \right] .
\end{aligned}
\tag{2.489}
$$

Inserting (2.489) into the linear polarization vectors

$$
\begin{aligned}
e_2 &= \frac{1}{\sqrt{2}} \left( e_x + e_y \right) , \\
e_1 &= \frac{1}{\sqrt{2}} \left( e_x - e_y \right) ,
\end{aligned}
\tag{2.490}
$$

one obtains

$$
\begin{aligned}
e_2 &= \frac{1}{\sqrt{2}} \left[ e(+) + i e(-) \right] , \\
e_1 &= \frac{1}{\sqrt{2}} \left[ e(+) - i e(-) \right] .
\end{aligned}
\tag{2.491}
$$

*Photon Spin.* The particles of the field, the photons, have two possible spin orientations, where the corresponding *spin states* will be denoted by $|1\rangle$ and $|-1\rangle$. These 2-dimensional eigenvectors represent a 2-dimensional Hilbert space $(\dim(\mathbb{H}) = n = 2)$. Identifying the circular polarization states $e(+)$, $e(-)$ with the eigenstates $|1\rangle$, $|-1\rangle$ of the photon spin $\hat{\sigma}_z$, one obtains the superpositions according to (2.489):

$$
\begin{aligned}
|e_x\rangle &= \frac{1}{\sqrt{2}} \left( |1\rangle + |-1\rangle \right) , \\
|e_y\rangle &= -\frac{i}{\sqrt{2}} \left( |1\rangle - |-1\rangle \right) ,
\end{aligned}
\tag{2.492}
$$

which are eigenfunctions of $\hat{\sigma}_x$, and, according to (2.491),

$$
\begin{aligned}
|e_2\rangle &= \frac{1}{\sqrt{2}} \left( |1\rangle + i |-1\rangle \right) , \\
|e_1\rangle &= \frac{1}{\sqrt{2}} \left( |1\rangle - i |-1\rangle \right) ,
\end{aligned}
\tag{2.493}
$$

which are eigenfunctions of $\hat{\sigma}_y$ (see (2.31)–(2.32)). The photon spin is thus isomorphic to the spin 1/2 algebra, even though the photon spin is 1 (boson type). $\hat{\sigma}_x$, $\hat{\sigma}_y$, $\hat{\sigma}_z$ have a different meaning: $\hat{\sigma}_z$ describes circular polarization, $\hat{\sigma}_x$ and $\hat{\sigma}_y$ two alternative linear polarization types in the plane perpendicular to the propagation of the light.

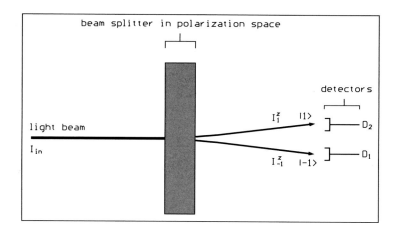

**Fig. 2.23.** Splitting into circular polarized field states

*Measurement of $P_z$.* In order to measure circular polarized field states, a circular polarizing filter has to be used. Figure 2.23 illustrates the experiment, which is analogous to the Stern-Gerlach experiment. (It is not a polarizing filter in the technical sense, where one of the two polarization modes is absorbed.) Due to this polarizing filter, the light beam splits off into two parts with right $(+)$ and left $(-)$ circular polarization so that the intensities $I_1^z$, $I_{-1}^z$ of the partial beams in $z$ direction (i.e. the intensities of two different ensembles of spin states) can be measured:

$$P_z = \frac{I_1^z - I_{-1}^z}{I_{in}} = \frac{I(\text{right}) - I(\text{left})}{I_{in}} \ . \tag{2.494}$$

Correspondingly we find, using appropriate positions of linear polarizing filters (cf. Fig. 2.21):

$$P_x = \frac{I_1^x - I_{-1}^x}{I_{in}} = \frac{I(0°) - I(90°)}{I_{in}} \ ,$$
$$P_y = \frac{I_{-1}^y - I_1^y}{I_{in}} = \frac{I(45°) - I(135°)}{I_{in}} \ . \tag{2.495}$$

The 4 observables $\eta_0 = I_{in}$, $\eta_1 = P_z$, $\eta_2 = P_x$, $\eta_3 = P_y$ are known as *Stokes parameters* (cf. [106]).

**2.3.6.6 Measurements in $SU(n)$, $n > 2$.** In the case of electron or photon spin experiments (or any other pseudospin 1/2 experiments) we need to consider only a 2-dimensional Hilbert space. However, in *scattering experiments*, for example, groups of higher order readily become relevant. In Fig. 2.24 two such experiments are sketched. The lower part illustrates the scattering of a particle beam at a known *scattering potential*, where the scattering potential represents the filter of the measuring system. The number of *detection channels* defines the effective dimension of this scenario. The upper part

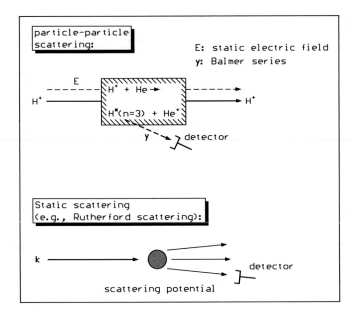

**Fig. 2.24.** Scattering experiments

shows a particle–particle scattering trace defined by (cf. [7])

$$H^+ + He \rightarrow H^*(n = 3) + He^+ \ ,$$
$$H^* \rightarrow H + \gamma \ ,$$

(2.496)

wherein H denotes Hydrogen, He Helium, and $\gamma$ a photon. The symbol $*$ indicates excited states, and the symbol $+$ a positive ion. $n$ denotes the *principal quantum number* of the electron system.

During the scattering process defined by (2.496) an *electron transfer* takes place, i.e. hydrogen ions $H^+$ receive electrons from helium atoms He. The resulting hydrogen atoms are in excited states $H^*(n = 3)$, where possible electron states are defined by the quantum numbers $n = 3, l \leq n-1 = 0, 1, 2,$ $m_l = -l \ldots +l$. The 9 angular momentum state vectors $|0,0\rangle, |1,-1\rangle, |1,0\rangle,$ $|1,1\rangle, |2,-2\rangle, |2,-1\rangle, |2,0\rangle, |2,1\rangle, |2,2\rangle$ represented by

$$\hat{1} = \sum_{l=0}^{2} \sum_{m_l=-l}^{+l} |lm_l\rangle \langle lm_l|$$

(2.497)

serve as possible initial states for spontaneous emission (Balmer series). This Hilbert space can be represented by $SU(9)$ with

$$s = n^2 - 1 = 80$$

(2.498)

generating operators; the density matrix $\rho_{lm_l;l'm_l'}$ has 80 parameters. Non-diagonal elements describe coherences including Zeeman coherence (for fixed

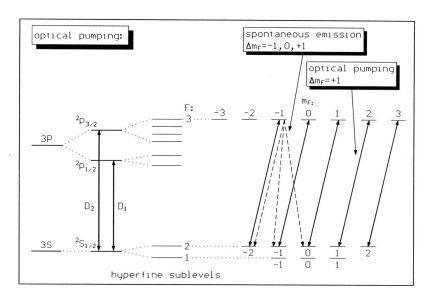

**Fig. 2.25.** Optical pumping

$l$) and angular momentum coherence (different $l$). Taking symmetry conditions into account, the set of non-vanishing parameters can usually be reduced (cf. [7]).

*Optical Pumping.* Another type of experiment is shown in Fig. 2.25. Due to interactions of electrons and protons, *hyperfine sublevels* exist in sodium (Na). Exposed to laser light, transitions from the hyperfine levels $F = 2$, $m_2 = -2, -1, 0, 1, 2$ into the hyperfine levels $F = 3$, $m_3 = -3, -2, -1, 0, 1, 2, 3$ with the *selection rule* $\Delta m_F = +1$ take place. (In Fig. 2.25 these allowed transitions are indicated.) However, this pumping process is superimposed by a *relaxation process* with the selection rule $\Delta m_F = -1, 0, +1$ (spontaneous emission). (In Fig. 2.25 decay from one state is shown.) These two combined processes then cause a change of the population: after some time the state $F = 2$, $m_2 = 2$ is populated almost exclusively.

The values $m_2 = -2, -1, 0, 1, 2$ of $F = 2$ define a 5-dimensional Hilbert space so that with

$$s = n^2 - 1 = 24 \tag{2.499}$$

the respective density operator reads in $SU(5)$:

$$\hat{\rho} = \frac{1}{5}\hat{1} + \frac{1}{2}\sum_{j=1}^{24}\lambda_j\hat{\lambda}_j \ . \tag{2.500}$$

Initially $\hat{\rho}$ will be a mixed state; the entropy of this state can be reduced to zero by means of pumping. This is a source of almost completely polarized atoms.

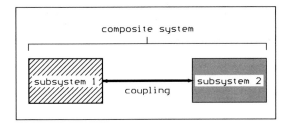

**Fig. 2.26.** The structure of a composite system consisting of two subsystems

## 2.4 Composite Systems: Two Nodes

Contents: product representations, reduced density operator, correlations, correlations due to the Pauli principle, EPR correlations (entangled states), Bell inequality, quantum cryptography.

### 2.4.1 Product Space

In this section a mathematical scheme appropriate to describe *composite systems* consisting of specific subsystems ("nodes") is introduced. Basic definitions and properties are given. *Product operators* forming an operator basis are considered. 2-node coherence (*entanglement*) is shown to supplement the notion of one-node coherence.

The term *node* refers to a specific part of the system with index $i$. In typical situations $i$ denotes a specific localization. Only in special situations will a node coincide with an elementary particle like a photon, an electron spin, or a mode index. Interactions between the nodes define the "edges" within the network.

**2.4.1.1 Product Hilbert Space.** Consider systems $\nu$ defined by sets of state vectors $|l(\nu)\rangle$ with

$$\hat{1}(\nu) = \sum_l |l(\nu)\rangle \langle l(\nu)| , \qquad (2.501)$$

where the state vectors $|l(\nu)\rangle$ span the respective Hilbert space $\mathbb{H}(\nu)$ ($\nu = 1, 2$) with dimension

$$\dim(\mathbb{H}(\nu)) = n_\nu . \qquad (2.502)$$

A system composed of two subsystems defined in $\mathbb{H}(1)$ and $\mathbb{H}(2)$ (compare with Fig. 2.26) then lives in the product Hilbert space

$$\mathbb{H} = \mathbb{H}(1) \otimes \mathbb{H}(2) \qquad (2.503)$$

with dimension

$$n = \dim(\mathbb{H}) = n_1 n_2 . \qquad (2.504)$$

Here, $\otimes$ denotes the *direct product* of two Hilbert spaces, and

$$\hat{1} = \hat{1}(1) \otimes \hat{1}(2) = \sum_{l,m} |l(1) m(2)\rangle \langle l(1) m(2)| \tag{2.505}$$

defines completeness in $\mathbb{H}$, where

$$|l(1) m(2)\rangle = |l(1)\rangle \otimes |m(2)\rangle \ . \tag{2.506}$$

These state vectors obey the orthonormality relation

$$\langle l(1) m(2)| l'(1) m'(2) \rangle = \delta_{ll'} \delta_{mm'} \ . \tag{2.507}$$

The direct product is *commutative*, i.e. $|l(1)\rangle \otimes |m(2)\rangle = |m(2)\rangle \otimes |l(1)\rangle$ holds.

**2.4.1.2 Subsystem Operators in $\mathbb{H}$.** Consider operators $\hat{A}(1)$ which are defined in Hilbert space $\mathbb{H}(1)$, and operators $\hat{B}(2)$ defined in $\mathbb{H}(2)$, with eigenvalue equations

$$\begin{aligned} \hat{A}(1) |l(1)\rangle &= |l'(1)\rangle \ , \\ \hat{B}(2) |m(2)\rangle &= |m'(2)\rangle \ . \end{aligned} \tag{2.508}$$

In the product Hilbert space $\mathbb{H}$ these operators are

$$\hat{A} = \hat{A}(1) \otimes \hat{1}(2) = \hat{1}(2) \otimes \hat{A}(1) \ , \tag{2.509}$$

$$\hat{B} = \hat{1}(1) \otimes \hat{B}(2) = \hat{B}(2) \otimes \hat{1}(1) \tag{2.510}$$

so that

$$\left[\hat{A}, \hat{B}\right]_{-} = 0 \ , \tag{2.511}$$

and (2.508) becomes

$$\begin{aligned} \hat{A} |l(1) m(2)\rangle &= |l'(1) m(2)\rangle \ , \\ \hat{B} |l(1) m(2)\rangle &= |l(1) m'(2)\rangle \ . \end{aligned} \tag{2.512}$$

If $A(1)_{ll'}$, $B(2)_{mm'}$ are matrix elements in the subspaces $\mathbb{H}_1$, $\mathbb{H}_2$, then

$$\begin{aligned} \langle l(1) m(2)| \hat{A} | l'(1) m'(2) \rangle &= A(1)_{ll'} \delta_{mm'} \ , \\ \langle l(1) m(2)| \hat{B} | l'(1) m'(2) \rangle &= \delta_{ll'} B(2)_{mm'} \ . \end{aligned} \tag{2.513}$$

Due to (2.513), the trace-relations

$$\begin{aligned} \text{Tr}_2 \left\{\hat{A}\right\} &:= \sum_m \langle l(1) m(2)| \hat{A} | l'(1) m(2)\rangle = A(1)_{ll'} n_2 \ , \\ \text{Tr}_1 \left\{\hat{B}\right\} &:= \sum_l \langle l(1) m(2)| \hat{B} | l'(1) m(2)\rangle = n_1 B(2)_{mm'} \end{aligned} \tag{2.514}$$

and thus

$$\begin{aligned} \text{Tr} \left\{\hat{A}\right\} &= \sum_l A(1)_{ll} n_2 = \text{Tr}_1 \left\{\hat{A}(1)\right\} n_2 \ , \\ \text{Tr} \left\{\hat{B}\right\} &= \sum_m n_1 B(2)_{mm} = n_1 \text{Tr}_2 \left\{\hat{B}(2)\right\} \end{aligned} \tag{2.515}$$

can be derived. Here, $\text{Tr}_\nu$ means trace operation in subspace $\nu$ only.

**2.4.1.3 Product Operators.** Based on the subsystem operators we can form new product operators:

$$\hat{A}(1,2) = \hat{A}(1) \otimes \hat{B}(2) = \left[\hat{A}(1) \otimes \hat{1}(2)\right]\left[\hat{1}(1) \otimes \hat{B}(2)\right] \tag{2.516}$$

implying

$$\left[\hat{A}(1) \otimes \hat{B}(2)\right]\left[\hat{C}(1) \otimes \hat{D}(2)\right] = \hat{A}(1)\hat{C}(1) \otimes \hat{B}(2)\hat{D}(2) \tag{2.517}$$

and

$$\left[\hat{A}(1) \otimes \hat{B}(2), \hat{C}(1) \otimes \hat{D}(2)\right]_{-} = \hat{A}\hat{C} \otimes \hat{B}\hat{D} - \hat{C}\hat{A} \otimes \hat{D}\hat{B} . \tag{2.518}$$

Matrix elements for the product operator (2.516) are then

$$\langle l(1) m(2) | \hat{A}(1) \otimes \hat{B}(2) | l'(1) m'(2) \rangle$$
$$= \left\langle l(1) \left| \hat{A} \right| l'(1) \right\rangle \left\langle m(2) \left| \hat{B} \right| m'(2) \right\rangle$$
$$= A(1)_{ll'} B(2)_{mm'} , \tag{2.519}$$

i.e. they consist of products of matrix elements $A(1)_{ll'}$, $B(2)_{mm'}$ defined in either subspace. Based on these matrix elements, one obtains

$$\mathrm{Tr}\left\{\hat{A}(1) \otimes \hat{B}(2)\right\} = \sum_{l,m} A(1)_{ll} B(2)_{mm}$$
$$= \mathrm{Tr}_1\left\{\hat{A}(1)\right\} \mathrm{Tr}_2\left\{\hat{B}(2)\right\} , \tag{2.520}$$

while

$$\mathrm{Tr}_1\left\{\hat{A}(1) \otimes \hat{B}(2)\right\} = \mathrm{Tr}_1\left\{\hat{A}(1)\right\} \hat{B}(2) ,$$
$$\mathrm{Tr}_2\left\{\hat{A}(1) \otimes \hat{B}(2)\right\} = \hat{A}(1) \mathrm{Tr}_2\left\{\hat{B}(2)\right\} . \tag{2.521}$$

From (2.520) and (2.71) it follows that

$$\mathrm{Tr}\left\{\hat{\lambda}_j(1) \otimes \hat{\lambda}_k(2)\right\} = \mathrm{Tr}_1\left\{\hat{\lambda}_j(1)\right\} \mathrm{Tr}_2\left\{\hat{\lambda}_k(2)\right\} = 0 , \tag{2.522}$$

from (2.517) and (2.71) one finds

$$\mathrm{Tr}\left\{\left[\hat{\lambda}_j(1) \otimes \hat{\lambda}_k(2)\right]\left[\hat{\lambda}_l(1) \otimes \hat{\lambda}_m(2)\right]\right\}$$
$$= \mathrm{Tr}_1\left\{\hat{\lambda}_j(1)\hat{\lambda}_l(1)\right\} \mathrm{Tr}_2\left\{\hat{\lambda}_k(2)\hat{\lambda}_m(2)\right\}$$
$$= 4\delta_{jl}\delta_{km} . \tag{2.523}$$

**2.4.1.4 Representation in $SU(n_1) \otimes SU(n_2)$.** Consider the set of operators

$$\left\{\frac{1}{\sqrt{n_\nu}}\hat{1}(\nu), \hat{\lambda}_j(\nu)\right\} , \quad \nu = 1, 2 , \tag{2.524}$$

where $\hat{\lambda}_j(\nu)$ denote the generating operators of the special unitary group $SU(n_\nu)$. The product operators then represent a complete *operator basis* (or a complete *matrix basis* if a matrix representation is considered):

$$\left\{ \frac{1}{\sqrt{n_1 n_2}}\hat{1}, \hat{\lambda}_j(1) \otimes \frac{1}{\sqrt{n_2}}\hat{1}(2), \frac{1}{\sqrt{n_1}}\hat{1}(1) \otimes \hat{\lambda}_j(2), \hat{\lambda}_j(1) \otimes \hat{\lambda}_k(2) \right\} , \quad (2.525)$$

where the number of elements is

$$1 + s_1 + s_2 + s_1 s_2 = 1 + \left(n_1^2 - 1\right) + \left(n_2^2 - 1\right) + \left(n_1^2 - 1\right)\left(n_2^2 - 1\right)$$
$$= n_1^2 n_2^2$$
$$= n^2 , \quad (2.526)$$

as required. Based on the trace properties given above we find in complete analogy with (2.294):

$$\boxed{\begin{aligned} \hat{A} &= \frac{1}{n_1 n_2}\mathrm{Tr}\left\{\hat{A}\right\}\hat{1} + \frac{1}{2n_2}\sum_{j=1}^{s_1} \mathcal{A}_j(1)\left[\hat{\lambda}_j(1) \otimes \hat{1}(2)\right] \\ &\quad + \frac{1}{2n_1}\sum_{k=1}^{s_2} \mathcal{A}_k(2)\left[\hat{1}(1) \otimes \hat{\lambda}_k(2)\right] \\ &\quad + \frac{1}{4}\sum_{j=1}^{s_1}\sum_{k=1}^{s_2} \mathcal{A}_{jk}(1,2)\left[\hat{\lambda}_j(1) \otimes \hat{\lambda}_k(2)\right] \end{aligned}} \quad (2.527)$$

with the coefficients

$$\mathcal{A}_j(1) = \mathrm{Tr}\left\{\hat{A}\left[\hat{\lambda}_j(1) \otimes \hat{1}(2)\right]\right\} := \mathcal{A}\begin{pmatrix} 1 \\ j \end{pmatrix} ,$$
$$\mathcal{A}_k(2) = \mathrm{Tr}\left\{\hat{A}\left[\hat{1}(1) \otimes \hat{\lambda}_k(2)\right]\right\} := \mathcal{A}\begin{pmatrix} 2 \\ k \end{pmatrix} , \quad (2.528)$$

$$\mathcal{A}_{jk}(1,2) = \mathrm{Tr}\left\{\hat{A}\left[\hat{\lambda}_j(1) \otimes \hat{\lambda}_k(2)\right]\right\} := \mathcal{A}\begin{pmatrix} 1 & 2 \\ j & k \end{pmatrix} = \mathcal{A}_{kj}(2,1) . \quad (2.529)$$

*Example 2.4.1.* In the special case of a subsystem operator

$$\hat{A} = \hat{A}(1) \otimes \hat{1}(2) , \quad \mathrm{Tr}\left\{\hat{A}\right\} = \mathrm{Tr}_1\left\{\hat{A}(1)\right\} n_2 \quad (2.530)$$

(cf. (2.515)), the traces (2.528), (2.529) are replaced by

$$\mathcal{A}_j(1) = \mathrm{Tr}_1\left\{\hat{A}(1)\hat{\lambda}_j(1)\right\} n_2 , \quad \mathcal{A}_j(2) = 0 , \quad \mathcal{A}_{j,k}(1,2) = 0 , \quad (2.531)$$

and the operator representation (2.527) reduces to

$$\hat{A} = \frac{1}{n_1}\mathrm{Tr}_1\left\{\hat{A}(1)\right\}\left[\hat{1}(1) \otimes \hat{1}(2)\right] + \frac{1}{2}\sum_{j=1}^{s_1} \mathcal{A}_j(1)\left[\hat{\lambda}_j(1) \otimes \hat{1}(2)\right]$$
$$= \hat{A}(1) \otimes \hat{1}(2) . \quad (2.532)$$

*Example 2.4.2.* In particular, additive operators

$$\hat{A} = \hat{A}(1) \otimes \hat{1}(2) + \hat{1}(1) \otimes \hat{A}(2) := \hat{A}_1 + \hat{A}_2 \qquad (2.533)$$

have the representation

$$\boxed{\begin{aligned}
\mathcal{A}_j(1) &= \mathrm{Tr}_1\left\{\hat{A}(1)\hat{\lambda}_j(1)\right\} n_2 \ , \\
\mathcal{A}_k(2) &= \mathrm{Tr}_2\left\{\hat{A}(2)\hat{\lambda}_k(2)\right\} n_1 \ , \\
\mathcal{A}_{jk}(1,2) &= 0 \ .
\end{aligned}} \qquad (2.534)$$

Note that – inserting these terms into (2.527) – the dimension factors $n_i$ cancel.

*Example 2.4.3.* Let us, finally, consider the product operator

$$\hat{A} = \hat{A}(1) \otimes \hat{B}(2) \ . \qquad (2.535)$$

When put into (2.528), (2.529), one obtains

$$\begin{aligned}
\mathcal{A}_j(1) &= \mathrm{Tr}\left\{\hat{A}\left[\hat{\lambda}_j(1) \otimes \hat{1}(2)\right]\right\} \\
&= \mathrm{Tr}\left\{\hat{A}(1)\hat{\lambda}_j(1) \otimes \hat{B}(2)\right\} \\
&= \mathrm{Tr}_1\left\{\hat{A}(1)\hat{\lambda}_j(1)\right\} \mathrm{Tr}_2\left\{\hat{B}(2)\right\} \ , 
\end{aligned} \qquad (2.536)$$

$$\begin{aligned}
\mathcal{A}_{jk}(1,2) &= \mathrm{Tr}\left\{\hat{A}\left[\hat{\lambda}_j(1) \otimes \hat{\lambda}_k(2)\right]\right\} \\
&= \mathrm{Tr}\left\{\hat{A}(1)\hat{\lambda}_j(1) \otimes \hat{B}(2)\hat{\lambda}_k(2)\right\} \\
&= \mathrm{Tr}_1\left\{\hat{A}(1)\hat{\lambda}_j(1)\right\} \mathrm{Tr}_2\left\{\hat{B}(2)\hat{\lambda}_k(2)\right\} \ .
\end{aligned} \qquad (2.537)$$

In the same way, $\mathcal{A}_j(2)$ can be calculated so that (2.527) reads

$$\begin{aligned}
\hat{A} = {}& \frac{1}{n}\mathrm{Tr}_1\left\{\hat{A}(1)\right\} \mathrm{Tr}_2\left\{\hat{B}(2)\right\} \hat{1} + \\
& \frac{1}{2n_2}\mathrm{Tr}_2\left\{\hat{B}(2)\right\} \sum_{j=1}^{s_1} \mathrm{Tr}_1\left\{\hat{A}(1)\lambda_j(1)\right\} \left[\hat{\lambda}_j(1) \otimes \hat{1}(2)\right] + \\
& \frac{1}{2n_1}\mathrm{Tr}_1\left\{\hat{A}(1)\right\} \sum_{k=1}^{s_2} \mathrm{Tr}_2\left\{\hat{B}(2)\lambda_k(2)\right\} \left[\hat{1}(1) \otimes \hat{\lambda}_k(2)\right] + \\
& \frac{1}{4} \sum_{j=1}^{s_1}\sum_{k=1}^{s_2} \mathrm{Tr}_1\left\{\hat{A}(1)\hat{\lambda}_j(1)\right\} \mathrm{Tr}_2\left\{\hat{B}(2)\hat{\lambda}_k(2)\right\} \\
& \left[\hat{\lambda}_j(1) \otimes \hat{\lambda}_k(2)\right] \ .
\end{aligned} \qquad (2.538)$$

### 2.4.1.5 Product Generators Expressed by Projection Operators.

Just like the generators $\hat{\lambda}_i$ (cf. (2.80)), the products $\hat{\lambda}_i(1) \otimes \hat{\lambda}_j(2)$ can also be expressed by projectors. We consider here projectors in product space. Then we find from (2.80) ($j < k$, $l < m$):

$$
\begin{aligned}
\hat{u}_{jk}(1) \otimes \hat{u}_{lm}(2) &= \hat{P}_{jk}(1) \otimes \hat{P}_{lm}(2) + \hat{P}_{jk}(1) \otimes \hat{P}_{ml}(2) + \\
&\quad \hat{P}_{kj}(1) \otimes \hat{P}_{lm}(2) + \hat{P}_{kj}(1) \otimes \hat{P}_{ml}(2) \\
&:= \hat{P}_{jl,km} + \hat{P}_{jm,kl} + \hat{P}_{kl,jm} + \hat{P}_{km,jl} .
\end{aligned} \tag{2.539}
$$

Similarly, we get

$$
\begin{aligned}
\hat{v}_{jk}(1) \otimes \hat{v}_{lm}(2) &= -\hat{P}_{jk}(1) \otimes \hat{P}_{lm}(2) + \hat{P}_{jk}(1) \otimes \hat{P}_{ml}(2) + \\
&\quad \hat{P}_{kj}(1) \otimes \hat{P}_{lm}(2) - \hat{P}_{kj}(1) \otimes \hat{P}_{ml}(2) \\
&:= -\hat{P}_{jl,km} + \hat{P}_{jm,kl} + \hat{P}_{kl,jm} - \hat{P}_{km,jl} ,
\end{aligned} \tag{2.540}
$$

$$
\begin{aligned}
\hat{u}_{jk}(1) \otimes \hat{v}_{jk}(2) &= i \left[ \hat{P}_{jk}(1) \otimes \hat{P}_{lm}(2) - \hat{P}_{jk}(1) \otimes \hat{P}_{ml}(2) \right] + \\
&\quad \hat{P}_{kj}(1) \otimes \hat{P}_{lm}(2) - \hat{P}_{kj}(1) \otimes \hat{P}_{ml}(2) \\
&= i \left( \hat{P}_{jl,km} - \hat{P}_{jm,kl} + \hat{P}_{kl,jm} - \hat{P}_{km,jl} \right) .
\end{aligned} \tag{2.541}
$$

For $SU(2)$ we require

$$
l = -1 , \quad m = 1 , \quad j = -1 , \quad k = 1 \tag{2.542}
$$

so that, with the product state identification

$$
\begin{aligned}
|-1,-1\rangle &= |1\rangle , \\
|-1,1\rangle &= |2\rangle , \\
|1,-1\rangle &= |3\rangle , \\
|1,1\rangle &= |4\rangle ,
\end{aligned} \tag{2.543}
$$

we get

$$
\boxed{\hat{u}_{12}(1) \otimes \hat{u}_{12}(2) = \hat{P}_{14} + \hat{P}_{23} + \hat{P}_{32} + \hat{P}_{41}} \tag{2.544}
$$

and

$$
\boxed{\hat{v}_{12}(1) \otimes \hat{v}_{12}(2) = -\hat{P}_{14} + \hat{P}_{23} + \hat{P}_{32} - \hat{P}_{41} .} \tag{2.545}
$$

Exchange of node indices 1 and 2 implies exchange of the two states $|2\rangle$ and $|3\rangle$. These results are easily generalized to higher dimensions. Mixed products are

$$
\boxed{
\begin{aligned}
\hat{u}_{12}(1) \otimes \hat{v}_{12}(2) &= i \left( \hat{P}_{14} - \hat{P}_{23} + \hat{P}_{32} - \hat{P}_{41} \right) , \\
\hat{u}_{12}(1) \otimes \hat{w}_1(2) &= \hat{P}_{24} + \hat{P}_{42} - \hat{P}_{13} - \hat{P}_{31} , \\
\hat{v}_{12}(1) \otimes \hat{w}_1(2) &= i \left( \hat{P}_{24} + \hat{P}_{31} - \hat{P}_{42} - \hat{P}_{13} \right) .
\end{aligned}
} \tag{2.546}
$$

Finally, with

$$\hat{w}_1 = \hat{P}_{11} - \hat{P}_{-1-1} \,, \tag{2.547}$$

one obtains

$$\hat{w}_1(1) \otimes \hat{w}_1(2) = \hat{P}_{11}(1) \otimes \hat{P}_{11}(2) - \hat{P}_{11}(1) \otimes \hat{P}_{-1-1}(2) - \\ \hat{P}_{-1-1}(1) \otimes \hat{P}_{11}(2) + \hat{P}_{-1-1}(1) \otimes \hat{P}_{-1-1}(2) \tag{2.548}$$

so that

$$\boxed{\hat{w}_1(1) \otimes \hat{w}_1(2) = \hat{P}_{11} - \hat{P}_{22} - \hat{P}_{33} + \hat{P}_{44} \,.} \tag{2.549}$$

## 2.4.2 Hamilton Models: Pair Interactions

The Hamilton models studied in Sect. 2.2.5 are now extended to models of composite systems. Fundamental interactions are of the 2-particle type (interactions in pairs). In this section, we restrict ourselves to $2 \otimes 2$-level systems specified by

$$\boxed{\begin{aligned} \hat{H} = \frac{1}{4}\mathrm{Tr}\left\{\hat{H}\right\}\hat{1} &+ \frac{1}{4}\sum_{j=1}^{3}\mathcal{H}_j(1)\left[\hat{\lambda}_j(1)\otimes\hat{1}(2)\right] \\ &+ \frac{1}{4}\sum_{k=1}^{3}\mathcal{H}_k(2)\left[\hat{1}(1)\otimes\hat{\lambda}_k(2)\right] \\ &+ \frac{1}{4}\sum_{j=1}^{3}\sum_{k=1}^{3}\mathcal{H}_{jk}(1,2)\left[\hat{\lambda}_j(1)\otimes\hat{\lambda}_k(2)\right] \,. \end{aligned}} \tag{2.550}$$

Typically, the total Hamiltonian can be decomposed as

$$\hat{H} = \hat{H}_1 + \hat{H}_2 + \hat{H}(1,2) \,, \tag{2.551}$$

where the subsystem Hamiltonians $\hat{H}_1$, $\hat{H}_2$ describe two 2-level systems according to (2.183):

$$\hat{H}_1 = \hat{H}(1) \otimes \hat{1}(2) \,, \tag{2.552}$$

$$\hat{H}_2 = \hat{1}(1) \otimes \hat{H}(2) \tag{2.553}$$

with

$$\begin{aligned} \hat{H}(1) &= \sum_{i=1}^{2} E_i \hat{P}_{ii}(1) \\ &= \frac{1}{2}\left[\left(E_1^{(1)} + E_2^{(1)}\right)\hat{1}(1) + \left(E_2^{(1)} - E_1^{(1)}\right)\hat{w}_1(1)\right] \,, \tag{2.554} \end{aligned}$$

$$\hat{H}(2) = \sum_{j=1}^{2} E_j \hat{P}_{jj}(2)$$

$$= \frac{1}{2} \left[ \left( E_1^{(2)} + E_2^{(2)} \right) \hat{1}(2) + \left( E_2^{(2)} - E_1^{(2)} \right) \hat{w}_1(2) \right] . \tag{2.555}$$

The non-interacting part, $\hat{H}_1 + \hat{H}_2$, contributes to the $SU(2)$ vectors $\mathcal{H}_j(1)$, $\mathcal{H}_k(2)$ as specified in (2.534).

For the interaction (as for the local properties $\mathcal{H}_l$) only a finite number of possibilities exist. We restrict ourselves to three basic types symmetric in both subsystems. They can be realized in physically very different ways. The parameters involved typically follow from a theoretical analysis taking into account, for example, the geometric structure of the network.

**2.4.2.1 Diagonal Interaction.** Let us start from

$$\hat{H}(1,2) = \sum_{i,j=1}^{2} \hat{P}_{ii}(1) \otimes \hat{P}_{jj}(2) h_{ij} , \tag{2.556}$$

where the operators $\hat{P}_{ii}(1)$, $\hat{P}_{jj}(2)$ are projection operators defined by $\hat{P}_{ii}(1) = |i(1)\rangle \langle i(1)|$, $\hat{P}_{jj}(2) = |j(2)\rangle \langle j(2)|$. Identifying the Hamiltonian $\hat{H}(1,2)$ with the operator $\hat{A}$ of (2.527), (2.556) can be rewritten in $SU(2) \otimes SU(2)$ as:

$$\hat{H}(1,2) = \frac{1}{4}\mathcal{H}_0' \hat{1}(1) \otimes \hat{1}(2) + \frac{1}{4}\mathcal{H}_3'(2)\hat{1}(1) \otimes \hat{w}_1(2) +$$
$$\frac{1}{4}\mathcal{H}_3'(1)\hat{w}_1(1) \otimes \hat{1}(2) + \frac{1}{4}\mathcal{H}_{33}(1,2)\hat{w}_1(1) \otimes \hat{w}_1(2) , \tag{2.557}$$

where the notation $\hat{u}_{12}$, $\hat{v}_{12}$, $\hat{w}_1$ has been used for the generating operators $\hat{\lambda}_j$. The coefficients not equal to zero are

$$\mathcal{H}_0' = \text{Tr}\left\{ \hat{H}(1,2) \right\} = h_{11} + h_{21} + h_{12} + h_{22} ,$$

$$\mathcal{H}_3'(2) = -h_{11} - h_{21} + h_{12} + h_{22} , \tag{2.558}$$

$$\mathcal{H}_3'(1) = -h_{11} + h_{21} - h_{12} + h_{22}$$

and

$$\boxed{\mathcal{H}_{33}(1,2) = h_{11} - h_{21} - h_{12} + h_{22} .} \tag{2.559}$$

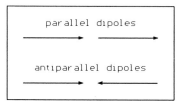

Fig. 2.27. Parallel and antiparallel dipoles

The Hamiltonian $\hat{H}(1,2)$ thus also contributes to the local vectors $\mathcal{H}_i(\nu)$; it is defined in the $\hat{w}$ subspace only.

*Example 2.4.4.* Consider two 2-level systems embedded in a rigid solid state matrix. If the centre-of-mass position of the subsystem $\nu$ is denoted by $\boldsymbol{R}_\nu$, its diagonal dipole moment in state $i$ by $\boldsymbol{d}_i(\nu)$, and $\boldsymbol{R}_{12} = \boldsymbol{R}_1 - \boldsymbol{R}_2$, one finds (cf. [104])

$$h_{ij} = \frac{1}{4\pi\epsilon_r\epsilon_0 |\boldsymbol{R}_{12}|^3} \left\{ \boldsymbol{d}_i(1)\boldsymbol{d}_j(2) - \right.$$
$$\left. \frac{3}{|\boldsymbol{R}_{12}|^2} [\boldsymbol{R}_{12}\boldsymbol{d}_i(1)] [\boldsymbol{R}_{12}\boldsymbol{d}_j(2)] \right\}. \tag{2.560}$$

$\epsilon_r$ is the relative dielectric constant of the material, $\epsilon_0$ the dielectric constant.

Restricting ourselves to dipole–dipole interactions caused by parallel and antiparallel dipoles (see Fig. 2.27), i.e. using the scheme

$$\boxed{\begin{aligned} h_{11} = h_{22} &:= -\frac{1}{2}\hbar C_{\mathrm{R}} \text{ (parallel dipoles)} , \\ h_{12} = h_{21} &:= +\frac{1}{2}\hbar C_{\mathrm{R}} \text{ (antiparallel dipoles)} , \end{aligned}} \tag{2.561}$$

the coefficients defined by (2.558) reduce to

$$\begin{aligned} \mathcal{H}_0' &= 0 , \\ \mathcal{H}_3'(2) &= 0 , \\ \mathcal{H}_3'(1) &= 0 , \\ \mathcal{H}_{33}(1,2) &= -2\hbar C_{\mathrm{R}} . \end{aligned} \tag{2.562}$$

Equation (2.562) in (2.557) shows that this interaction Hamiltonian in $SU(2) \otimes SU(2)$ is specified by the matrix

$$\boxed{(\mathcal{H}_{jk}(1,2)) = 2 \begin{pmatrix} 0 & 0 & 0 \\ 0 & 0 & 0 \\ 0 & 0 & -\hbar C_{\mathrm{R}} \end{pmatrix}} . \tag{2.563}$$

Using (2.562) and

$$\begin{aligned} \mathcal{H}_j(1) &= \mathrm{Tr}\left\{ \hat{H}\left[\hat{\lambda}_j(1) \otimes \hat{1}(2)\right] \right\} = 2\mathrm{Tr}_1\left\{ \hat{H}(1)\hat{\lambda}_j(1) \right\} \\ &= 2\left(E_2^{(1)} - E_1^{(1)}\right) , \end{aligned} \tag{2.564}$$

$$\begin{aligned} \mathcal{H}_k(2) &= \mathrm{Tr}\left\{ \hat{H}\left[\hat{1}(1) \otimes \hat{\lambda}_k(2)\right] \right\} = 2\mathrm{Tr}_2\left\{ \hat{H}(2)\hat{\lambda}_k(2) \right\} \\ &= 2\left(E_2^{(2)} - E_1^{(2)}\right) , \end{aligned} \tag{2.565}$$

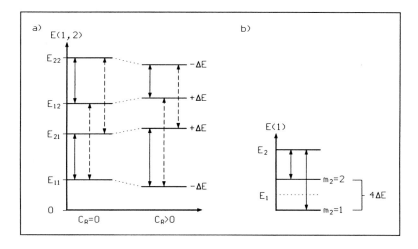

**Fig. 2.28.** Energy spectrum and energy shift in the case of diagonal interactions. a) Total spectrum. b) Effective spectrum of subsystem 1 (the transition frequency depends on the state of subsystem 2)

one obtains for the total Hamiltonian in $SU(2) \otimes SU(2)$:

$$
\begin{aligned}
\hat{H} = \text{constant} &+ \frac{1}{2}\left(E_2^{(1)} - E_1^{(1)}\right)\hat{w}_1(1) \otimes \hat{1}(2) \\
&+ \frac{1}{2}\left(E_2^{(2)} - E_1^{(2)}\right)\hat{1}(1) \otimes \hat{w}_1(2) \\
&- \frac{\hbar C_R}{2}\hat{w}_1(1) \otimes \hat{w}_1(2) \; .
\end{aligned}
\tag{2.566}
$$

The generating operators $\hat{w}_1(1)$, $\hat{w}_1(2)$ can be identified with $\hat{\sigma}_3(1)$, $\hat{\sigma}_3(2)$.

The eigenfunctions of this Hamiltonian are still the original product states $|m_1, m_2\rangle$, $m_i = 1, 2$. This is why we call this interaction *diagonal*. In Fig. 2.28, part a), we sketch the resulting spectrum $E_{m_1 m_2}$ (left for $C_R = 0$, right for $C_R > 0$). The energy shifts are $\Delta E = \hbar C_R/2$. The arrows indicate transition energies within subsystem 1 and 2 (broken line), respectively. We see that, due to the interaction, transitions within one subsystem depend on the state ($m$) of its neighbour (Fig. 2.28, part b)). As will be discussed later, this simple property has interesting dynamical consequences.

*Example 2.4.5.* The interaction (2.566) also applies to cavity electrodynamics. The coupling between a 2-level atom and a single cavity mode (vacuum and 1-photon state, say) is modelled as (cf. [23]):

$$
\hat{H}_{\text{off}} = \Delta E \hat{P}_{22}(1) \otimes \hat{P}_{22}(2) = \Delta E \hat{a}^+ \hat{a} \, |2(2)\rangle \langle 2(2)| \; ,
\tag{2.567}
$$

where $\Delta E$ is the change in atomic level spacing per photon in the cavity, which has the form (2.556) with

$$h_{ij} = \begin{cases} \Delta E & \text{for} \quad i = j = 2 \\ 0 & \text{for} \quad \text{otherwise} \end{cases} . \tag{2.568}$$

This interaction dominates under off-resonance conditions. This *off-resonant interaction* may be contrasted with the *Förster energy exchange mechanism* to be discussed next.

### 2.4.2.2 Förster Model. Consider the Hamiltonian

$$\begin{aligned} \hat{H}(1,2) &= \hat{V}_1 + \hat{V}_2 \\ &= \frac{1}{2}\gamma\hat{\sigma}^+(1) \otimes \hat{\sigma}^-(2) + \frac{1}{2}\gamma^*\hat{\sigma}^-(1) \otimes \hat{\sigma}^+(2) , \end{aligned} \tag{2.569}$$

where

$$\hat{\sigma}^\pm = \frac{1}{2}\left(\hat{\sigma}_x \pm i\hat{\sigma}_y\right) \tag{2.570}$$

and

$$\gamma = |\gamma|\, e^{i\varphi} . \tag{2.571}$$

With

$$\begin{aligned} 8\hat{V}_1 &= 4\gamma\hat{\sigma}^+(1) \otimes \hat{\sigma}^-(2) \\ &= \gamma\left[\hat{\sigma}_x(1) + i\hat{\sigma}_y(1)\right] \otimes \left[\hat{\sigma}_x(2) - i\hat{\sigma}_y(2)\right] \\ &= \gamma\left[\hat{\sigma}_x(1) \otimes \hat{\sigma}_x(2) + i\hat{\sigma}_y(1) \otimes \hat{\sigma}_x(2) - i\hat{\sigma}_x(1) \otimes \hat{\sigma}_y(2) + \right. \\ &\qquad \left. \hat{\sigma}_y(1) \otimes \hat{\sigma}_y(2)\right] , \end{aligned} \tag{2.572}$$

$$\begin{aligned} 8\hat{V}_2 &= 4\gamma^*\hat{\sigma}^-(1) \otimes \hat{\sigma}^+(2) \\ &= \gamma^*\left[\hat{\sigma}_x(1) - i\hat{\sigma}_y(1)\right] \otimes \left[\hat{\sigma}_x(2) + i\hat{\sigma}_y(2)\right] \\ &= \gamma^*\left[\hat{\sigma}_x(1) \otimes \hat{\sigma}_x(2) - i\hat{\sigma}_y(1) \otimes \hat{\sigma}_x(2) + i\hat{\sigma}_x(1) \otimes \hat{\sigma}_y(2) + \right. \\ &\qquad \left. \hat{\sigma}_y(1) \otimes \hat{\sigma}_y(2)\right] , \end{aligned} \tag{2.573}$$

we find by comparison with

$$\hat{H}(1,2) = \frac{1}{4}\sum_{j,k}\mathcal{H}_{jk}(1,2)\left[\hat{\sigma}_j(1) \otimes \hat{\sigma}_k(2)\right] \tag{2.574}$$

the coefficients of the product operators $\hat{\sigma}_j(1) \otimes \hat{\sigma}_k(2)$, $\mathcal{H}_{jk}(1,2)$:

$$\begin{aligned} (\mathcal{H}_{jk}(1,2)) &= \frac{1}{2}\begin{pmatrix} \gamma + \gamma^* & -i(\gamma - \gamma^*) & 0 \\ i(\gamma - \gamma^*) & \gamma + \gamma^* & 0 \\ 0 & 0 & 0 \end{pmatrix} \\ &= |\gamma|\begin{pmatrix} \cos\varphi & \sin\varphi & 0 \\ -\sin\varphi & \cos\varphi & 0 \\ 0 & 0 & 0 \end{pmatrix} . \end{aligned} \tag{2.575}$$

For $\gamma = -\gamma^* = i|\gamma|$ (i.e. $\varphi = \pi/2$) the coefficient matrix $(\mathcal{H}_{jk}(1,2))$ is antisymmetric. For $\gamma = \gamma^* := 2\hbar C_F$ (i.e. $\varphi = 0$) the matrix is diagonal. In

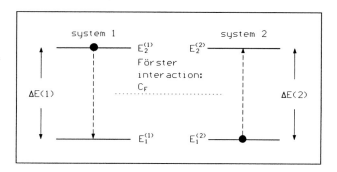

**Fig. 2.29.** Förster mechanism, in the case of resonance: $\Delta E = \Delta E(1) = \Delta E(2)$

the latter case, one obtains the $SU(2) \otimes SU(2)$ representation of the total Hamiltonian $\hat{H}$:

$$
\begin{aligned}
\hat{H} = \text{constant} &+ \frac{1}{2}\left(E_2^{(1)} - E_1^{(1)}\right)\left[\hat{\sigma}_z(1) \otimes \hat{1}(2)\right] \\
&+ \frac{1}{2}\left(E_2^{(2)} - E_1^{(2)}\right)\left[1(1) \otimes \hat{\sigma}_z(2)\right] \\
&+ \frac{\hbar C_{\mathrm{F}}}{2}\hat{\sigma}_x(1) \otimes \hat{\sigma}_x(2) + \frac{\hbar C_{\mathrm{F}}}{2}\hat{\sigma}_y(1) \otimes \hat{\sigma}_y(2) \,.
\end{aligned}
\tag{2.576}
$$

This Hamiltonian specifies the Förster coupling between two 2-level systems, which is efficient only under resonance conditions,

$$
E_2^{(1)} - E_1^{(1)} = E_2^{(2)} - E_1^{(2)} = \Delta E
\tag{2.577}
$$

(see Fig. 2.29). The resulting eigenvalue problem will be discussed in Sect. 2.4.4.3.

**2.4.2.3 Charge Transfer Coupling.** Let us consider the states of the local 4-state node of Sect. 2.2.5.6, which have been interpreted as various charge

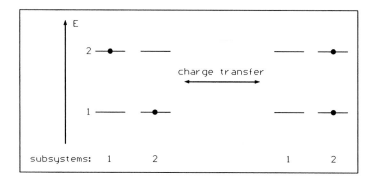

**Fig. 2.30.** Charge transfer between two subsystems

states. Interaction between such cells (in the following denoted by 1 and 2) can then be modelled as charge transfer:

$$\hat{H}(1,2) = t_{12} \sum_{\sigma=\downarrow,\uparrow} \left[ \hat{P}_{10}^{\sigma}(1) \otimes \hat{P}_{01}^{\sigma}(2) + \hat{P}_{01}^{\sigma}(1) \otimes \hat{P}_{10}^{\sigma}(2) \right]$$

$$= t_{12} \sum_{\sigma=\downarrow,\uparrow} \left[ \hat{a}_{\sigma}^{+}(1) \otimes \hat{a}_{\sigma}(2) + \hat{a}_{\sigma}(1) \otimes \hat{a}_{\sigma}^{+}(2) \right] . \tag{2.578}$$

For each spin this is a Förster-type interaction. In a closed system the total charge must be conserved, i.e.

$$N = \sum_{\sigma=\downarrow,\uparrow} \left( \hat{N}_{\downarrow}(1) + \hat{N}_{\uparrow}(2) \right) = \text{const.} . \tag{2.579}$$

Homogeneous networks of such cells constitute (one-band) *Hückel-type models*. Such models are readily generalized to more complicated level-structures. A pair of such nodes interacting via charge transfer is shown in Fig. 2.30.

Charge states play an important role for *single electron tunneling* via *isolated islands*. The disadvantage with respect to control derives from the fact that spurious charges from the environment changing $N_e$ spoil the whole repertoire of accessible states, even if the subsystems and respective energy states were fixed.

**2.4.2.4 Jaynes-Cummings Model in $SU(2) \otimes SU(2)$.** The Jaynes-Cummings model describes a 2-level system (subsystem 1) coupled to a *harmonic oscillator* (subsystem 2) (infinite number of levels) representing a single (cavity) photon mode. This model is defined by (cf. [154])

$$\boxed{\hat{H} = \frac{\hbar}{2}\omega_0\hat{\sigma}_z(1) + \frac{1}{2}\gamma\hat{a}\hat{\sigma}^{+}(1) + \frac{1}{2}\gamma^{*}\hat{a}^{+}\hat{\sigma}^{-}(1) + \hbar\omega\hat{a}^{+}\hat{a} .} \tag{2.580}$$

The 2-level subsystem is described by

$$\frac{1}{2}\left[\hat{\sigma}_x(1) - i\hat{\sigma}_y(1)\right] = \hat{\sigma}^{-}(1) = \hat{P}_{12}(1) ,$$

$$\frac{1}{2}\left[\hat{\sigma}_x(1) + i\hat{\sigma}_y(1)\right] = \hat{\sigma}^{+}(1) = \hat{P}_{21}(1) . \tag{2.581}$$

The operator $\hat{\sigma}^{+}(1)$ functions as a creation operator, and the operator $\hat{\sigma}^{-}(1)$ as an annihilation operator in Hilbert space $\mathbb{H}(1)$:

$$\langle 1| \hat{\sigma}^{-}(1) |2\rangle = 1 ,$$
$$\langle 2| \hat{\sigma}^{+}(1) |1\rangle = 1 . \tag{2.582}$$

According to Sect. 2.2.5.4, the operators $\hat{a}^{+}$, $\hat{a}$ are in the 2-dimensional subspace $|1\rangle$, $|2\rangle$ given by

$$\hat{a}^{+} = \hat{P}_{21}(2) = \sigma^{+}(2) , \tag{2.583}$$

$$\hat{a} = \hat{P}_{12}(2) = \sigma^{-}(2) , \tag{2.584}$$

$$\hat{a}^+\hat{a} = \hat{P}_{22}(2) = \frac{1}{2}\left[1 + (\sigma^z(2))\right] . \tag{2.585}$$

In this subspace the Hamiltonian $\hat{H}$ according to (2.580) reduces to

$$\hat{H} = \text{constant} + \frac{\hbar}{2}\omega_0\hat{\sigma}_z(1) + \frac{\hbar}{2}\omega\hat{\sigma}_z(2) + \frac{\gamma}{2}\hat{\sigma}^+(1)\hat{\sigma}^-(2)$$
$$+ \frac{\gamma^*}{2}\hat{\sigma}^-(1)\hat{\sigma}^+(2) , \tag{2.586}$$

i.e. the Förster-type interaction (cf. (2.569)). In this approximation an $SU(2) \otimes SU(2)$ network results as a *hybrid system* made of a 2-level atom and a single cavity photon mode (restricted to the vacuum and 1-photon state).

**2.4.2.5 Spin-boson Model.** An apparently similar model, which (contrary to the Jaynes-Cummings model) has not yet been solved exactly, is defined by (cf. [50])

$$\boxed{\hat{H} = \frac{1}{2}\hbar\omega_0\hat{\sigma}_z(1) + \hbar\omega_0\hat{a}^+\hat{a} + \frac{1}{2}\gamma\hat{\sigma}_x\left(\hat{a} + \hat{a}^+\right) .} \tag{2.587}$$

The coupling to the harmonic oscillator is here in terms of the elongation ("quadrature field"). In the truncated state space as used for (2.586) one obtains

$$\hat{H} = \text{constant} + \frac{1}{2}\hbar\omega_0\hat{\sigma}_z(1) + \hbar\omega_0\hat{\sigma}_z(2) + \frac{1}{2}\gamma\hat{\sigma}_x(1) \otimes \hat{\sigma}_x(2) . \tag{2.588}$$

This form should be compared with (2.576) for the Förster model.

### 2.4.3 Coupling Between Higher-dimensional Subsystems

The Förster coupling of (2.569) can be generalized to

$$\boxed{\hat{H} = 2\hbar \sum_{i<j}\sum_{k<l} C_{\text{F}}(ij,kl)\left[\hat{\sigma}_{ij}^+(1) \otimes \hat{\sigma}_{kl}^-(1) + \hat{\sigma}_{ij}^-(2) \otimes \hat{\sigma}_{kl}^+(2)\right] .} \tag{2.589}$$

We consider two examples for $SU(3) \otimes SU(3)$.

The first example (compare Fig. 2.31) is selective in all three possible transitions:

$$\mathcal{H}_{ij}(1,2) = 2\hbar
\begin{array}{c}
j: \quad u_{12}\ u_{13}\ u_{23}\ v_{12}\ v_{13}\ v_{23}\ w_1\ w_2 \\
\begin{pmatrix}
C_{11} & 0 & 0 & 0 & 0 & 0 & 0 & 0 \\
0 & C_{22} & 0 & 0 & 0 & 0 & 0 & 0 \\
0 & 0 & C_{33} & 0 & 0 & 0 & 0 & 0 \\
0 & 0 & 0 & C_{44} & 0 & 0 & 0 & 0 \\
0 & 0 & 0 & 0 & C_{55} & 0 & 0 & 0 \\
0 & 0 & 0 & 0 & 0 & C_{66} & 0 & 0 \\
0 & 0 & 0 & 0 & 0 & 0 & 0 & 0 \\
0 & 0 & 0 & 0 & 0 & 0 & 0 & 0
\end{pmatrix}
\begin{array}{c}
i: \\
\begin{pmatrix}
u_{12} \\ u_{13} \\ u_{23} \\ v_{12} \\ v_{13} \\ v_{23} \\ w_1 \\ w_2
\end{pmatrix}
\end{array}
\end{array} \tag{2.590}$$

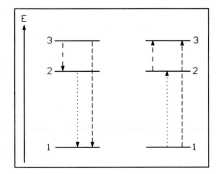

**Fig. 2.31.** Coupled 3-level nodes selective in all possible transitions

with

$$C_{11} = C_F(12,12), \quad C_{22} = C_F(13,13), \quad C_{33} = C_F(23,23),$$
$$C_{44} = C_F(12,12), \quad C_{55} = C_F(13,13), \quad C_{66} = C_F(23,23).$$

(2.591)

In the second example (cf. Fig. 2.32) two different transitions within subsystem 1 couple to one transition in subsystem 2:

$$\mathcal{H}_{ij}(1,2) = 2\hbar \begin{array}{c} j: \\ \begin{pmatrix} C_{11} & 0 & 0 & 0 & 0 & 0 & 0 & 0 \\ C_{21} & 0 & 0 & 0 & 0 & 0 & 0 & 0 \\ 0 & 0 & 0 & 0 & 0 & 0 & 0 & 0 \\ 0 & 0 & 0 & C_{44} & 0 & 0 & 0 & 0 \\ 0 & 0 & 0 & C_{54} & 0 & 0 & 0 & 0 \\ 0 & 0 & 0 & 0 & 0 & 0 & 0 & 0 \\ 0 & 0 & 0 & 0 & 0 & 0 & 0 & 0 \\ 0 & 0 & 0 & 0 & 0 & 0 & 0 & 0 \end{pmatrix} \end{array} \begin{array}{c} i: \\ \begin{pmatrix} u_{12} \\ u_{13} \\ u_{23} \\ v_{12} \\ v_{13} \\ v_{23} \\ w_1 \\ w_2 \end{pmatrix} \end{array}$$

$$\begin{array}{cccccccc} u_{12} & u_{13} & u_{23} & v_{12} & v_{13} & v_{23} & w_1 & w_2 \end{array}$$

(2.592)

with

$$C_{11} = C_F(12,12), \quad C_{21} = C_F(13,12),$$
$$C_{44} = C_F(12,12), \quad C_{54} = C_F(13,12).$$

(2.593)

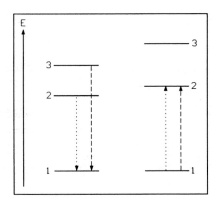

**Fig. 2.32.** Coupled 3-level nodes: two transitions within subsystem 1 couple to one transition in subsystem 2

Note that the different rows of matrices correspond to the different coherence vector entries $i$ of subsystem 1, the various columns to index $j$ of subsystem 2. This scheme is easily generalized to even higher dimensions.

We can also generalize the diagonal interaction

$$\hat{H}(1,2) = \sum_{i,j=1}^{n} \hat{P}_{ii}(1) \otimes \hat{P}_{jj}(2) h_{ij} \qquad (2.594)$$

(cf. (2.556)), which – based on

$$\hat{P}_{mm} = \frac{1}{n}\hat{1} + \frac{1}{2}\sum_{l=1}^{n-1} w_l^{(m)} \hat{w}_l \qquad (2.595)$$

(cf. (2.111)) and equation (2.103) – can be transformed into the $SU(n) \otimes SU(n)$ form. For $n = 3$ we find with

$$\hat{P}_{11} = \frac{1}{3}\hat{1} - \frac{1}{2}\hat{w}_1 - \frac{1}{2\sqrt{3}}\hat{w}_2 \ ,$$

$$\hat{P}_{22} = \frac{1}{3}\hat{1} + \frac{1}{2}\hat{w}_1 - \frac{1}{2\sqrt{3}}\hat{w}_2 \ , \qquad (2.596)$$

$$\hat{P}_{33} = \frac{1}{3}\hat{1} + \frac{1}{\sqrt{3}}\hat{w}_2$$

the terms

$$\hat{H}(1,2) = \ldots + \frac{1}{4}\hat{w}_1(1) \otimes \hat{w}_1(2)\mathcal{H}_{77}(1,2)$$

$$+ \frac{1}{4}\hat{w}_1(1) \otimes \hat{w}_2(2)\mathcal{H}_{78}(1,2)$$

$$+ \frac{1}{4}\hat{w}_2(1) \otimes \hat{w}_1(2)\mathcal{H}_{87}(1,2) +$$

$$+ \frac{1}{4}\hat{w}_2(1) \otimes \hat{w}_2(2)\mathcal{H}_{88}(1,2) \ , \qquad (2.597)$$

where

$$\begin{array}{|l|}
\hline
\mathcal{H}_{77}(1,2) = h_{11} + h_{22} - h_{12} - h_{21} \ , \\[2mm]
\mathcal{H}_{78}(1,2) = \dfrac{2}{\sqrt{3}}\left(-h_{13} + h_{23}\right) \ , \\[2mm]
\mathcal{H}_{87}(1,2) = \dfrac{2}{\sqrt{3}}\left(-h_{31} + h_{32}\right) \ , \\[2mm]
\mathcal{H}_{88}(1,2) = -\dfrac{2}{3}\left(h_{13} + h_{31} + h_{23} + h_{32}\right) + \dfrac{4}{3}h_{33} \ . \\
\hline
\end{array} \qquad (2.598)$$

$\mathcal{H}_{77}(1,2)$ is identical with $\mathcal{H}_{33}(1,2)$ for $SU(2) \otimes SU(2)$. The non-zero matrix elements of $\mathcal{H}_{ij}(1,2)$ are confined to the $w_1, w_2$ subspace.

### 2.4.4 Density Operator

In this section, state vectors, density operators or coherence vectors, and *correlation tensors* of systems composed of 2 subsytems ("nodes") are studied. Special examples of *entangled states* are introduced. The entropy of composite systems and their parts are discussed.

**2.4.4.1 Density Operator in Product Space.** Based on (2.527) the density operator can be represented as (cf. [45])

$$\hat{\rho} = \frac{1}{n_1 n_2}\hat{1} + \frac{1}{2n_2}\sum_{j=1}^{s_1}\lambda_j(1)\left[\hat{\lambda}_j(1)\otimes\hat{1}(2)\right]$$
$$+ \frac{1}{2n_1}\sum_{k=1}^{s_2}\lambda_k(2)\left[\hat{1}(1)\otimes\hat{\lambda}_k(2)\right] \tag{2.599}$$
$$+ \frac{1}{4}\sum_{j,k=1}^{s_1,s_2}K_{jk}(1,2)\left[\hat{\lambda}_j(1)\otimes\hat{\lambda}_k(2)\right]$$

with $\lambda_j(1)$, $\lambda_k(2)$ being defined by

$$\lambda_j(1) = \left\langle\hat{\lambda}_j(1)\right\rangle = \mathrm{Tr}_1\left\{\hat{\rho}\left[\hat{\lambda}_j(1)\otimes\hat{1}(2)\right]\right\} = \mathrm{Tr}_1\left\{\hat{R}(1)\hat{\lambda}_j(1)\right\} , \tag{2.600}$$

$$\lambda_k(2) = \left\langle\hat{\lambda}_k(2)\right\rangle = \mathrm{Tr}_2\left\{\hat{\rho}\left[\hat{1}(1)\otimes\hat{\lambda}_k(2)\right]\right\} = \mathrm{Tr}_1\left\{\hat{R}(2)\hat{\lambda}_k(2)\right\} . \tag{2.601}$$

The corresponding coherence vectors – which specify 1-node (or intra-node) coherence – then read

$$\boldsymbol{\lambda}(\nu) = \{\lambda_1(\nu), \lambda_2(\nu), \ldots, \lambda_{s_\nu}(\nu)\} , \quad \nu = 1, 2 . \tag{2.602}$$

2-node coherence is described by the *correlation tensor* $\mathsf{K}(1,2)$,

$$K_{jk}(1,2) = \left\langle\hat{\lambda}_j(1)\hat{\lambda}_k(2)\right\rangle = \mathrm{Tr}\left\{\hat{\rho}\left[\hat{\lambda}_j(1)\otimes\hat{\lambda}_k(2)\right]\right\} . \tag{2.603}$$

All these expectation values are real. For a product state one finds with

$$\left\langle\hat{\lambda}_j(1)\hat{\lambda}_k(2)\right\rangle = \left\langle\hat{\lambda}_j(1)\right\rangle\left\langle\hat{\lambda}_k(2)\right\rangle \tag{2.604}$$

the relations

$$K_{jk}(1,2) = \lambda_j(1)\lambda_k(2) . \tag{2.605}$$

It is, therefore, convenient to introduce the correlation tensor proper, $\mathsf{M}$, by

$$M_{jk}(1,2) = K_{jk}(1,2) - \lambda_j(1)\lambda_k(2) . \tag{2.606}$$

$\mathsf{M}$ is zero for a 2-subsystem product space. In the case of multi-node systems, a similar derivation holds. However, in such a case multi-node coherence and thus correlation tensors of higher order occur.

The tensor M allows the term *entanglement* to be defined in a precise way: in this book the term *entanglement* (between node $\nu$ and $\mu$) applies if $M(\nu, \mu) \neq 0$. Thus, a 2-subsystem product state has "no entanglement", i.e. no inter-node coherence.

**2.4.4.2 Reduced Density Operator.** Using formula (2.249) and the subsystem operators (2.509), expectation values $\overline{A(1)}$ within subsystem $\mathbb{H}_1$ can be calculated:

$$\overline{A(1)} = \mathrm{Tr}\left\{\hat{\rho}\left(\hat{A}\left(1\right) \otimes \hat{1}\left(2\right)\right)\right\} . \tag{2.607}$$

This formula can be recast into a more elementary form:

$$\overline{A(1)} =$$
$$\sum_{l,m,j,k} \langle l(1)m(2)|\,\hat{\rho}\,|j(1)k(2)\rangle\,\langle j(1)k(2)|\,\hat{A}\left(1\right)\otimes\hat{1}(2)\,|l(1)m(2)\rangle \tag{2.608}$$

so that, due to (2.513),

$$\overline{A(1)} = \sum_{l,m,j,k} \langle l(1)m(2)|\,\hat{\rho}\,|j(1)k(2)\rangle\,A(1)_{jl}\delta_{km}$$
$$= \sum_{l,m,j} \langle l(1)m(2)|\,\hat{\rho}\,|j(1)m(2)\rangle\,A(1)_{jl} . \tag{2.609}$$

If the *reduced density matrix* R(1) with the matrix elements $R(1)_{lj}$ is defined by

$$\boxed{R(1)_{lj} = \sum_{m} \langle l(1)m(2)|\,\hat{\rho}\,|j(1)m(2)\rangle = \sum_{m} \rho_{lm;jm} ,} \tag{2.610}$$

the expectation value $\overline{A(1)}$ can be expressed as

$$\boxed{\overline{A(1)} = \sum_{l,j} R(1)_{lj} A(1)_{jl} .} \tag{2.611}$$

Introducing

$$\hat{R}(1) = \mathrm{Tr}_2\left\{\hat{\rho}\right\} = \sum_{m} \langle m(2)|\,\hat{\rho}\,|m(2)\rangle \tag{2.612}$$

as the *reduced density operator*, the expectation value $\overline{A(1)}$ can, again, be written in the form (2.249),

$$\boxed{\overline{A(1)} = \mathrm{Tr}_1\left\{\hat{R}(1)\hat{A}(1)\right\} .} \tag{2.613}$$

For the product state

$$\hat{\rho} = \hat{\rho}(1) \otimes \hat{\rho}(2) \tag{2.614}$$

the reduced density operator is

$$\hat{R}(1) = \hat{\rho}(1) \tag{2.615}$$

(use (2.521) and $\mathrm{Tr}_2\{\hat{\rho}(2)\} = 1$). In the same way, the reduced density operator $\hat{R}(2)$ and representations for expectation values $\overline{A(2)}$ can be derived.

Inserting (2.599) into the reduced density operator $\hat{R}(1)$ defined by (2.612), and using (2.521), the relation

$$\boxed{\hat{R}(1) = \frac{1}{n_1}\hat{1}(1) + \frac{1}{2}\sum_{j=1}^{s_1}\lambda_j(1)\hat{\lambda}_j(1)} \tag{2.616}$$

can be derived. In the same way, the reduced density operator

$$\boxed{\hat{R}(2) = \frac{1}{n_2}\hat{1}(2) + \frac{1}{2}\sum_{k=1}^{s_2}\lambda_k(2)\hat{\lambda}_k(2)} \tag{2.617}$$

can be calculated; these formulae do not contain information about 2-node correlations.

*Global Expectation Values.* The expectation values of global quantities defined by additive operators

$$\hat{A} = \hat{A}(1) + \hat{A}(2) = \hat{A}(1) \otimes + \hat{1}(1) \otimes \hat{A}(2) \tag{2.618}$$

can be calculated from the equation

$$\begin{aligned}
\overline{A} &= \mathrm{Tr}\left\{\hat{\rho}\hat{A}\right\} = \mathrm{Tr}\left\{\hat{\rho}\left[\hat{A}(1) + \hat{A}(2)\right]\right\} \\
&= \mathrm{Tr}\left\{\hat{\rho}\hat{A}(1)\right\} + \mathrm{Tr}\left\{\hat{\rho}\hat{A}(2)\right\} \\
&= \overline{A(1)} + \overline{A(2)}
\end{aligned} \tag{2.619}$$

(cf. eqs. (2.293), (2.607)). Using the equation (2.613) and the corresponding expression for the expectation value $\overline{A(2)}$, $\overline{A}$ can be recast into

$$\overline{A} = \mathrm{Tr}_1\left\{\hat{R}(1)\hat{A}(1)\right\} + \mathrm{Tr}_2\left\{\hat{R}(2)\hat{A}(2)\right\} . \tag{2.620}$$

For identical subsystems, this expectation value reduces to

$$\overline{A} = 2\mathrm{Tr}_1\left\{\hat{R}(1)\hat{A}(1)\right\} . \tag{2.621}$$

*Non-separability.* In general it is not possible to define a state vector for individual subsystems (cf. [34]):

**Theorem 2.4.1 (d'Espagnat 1976).** *If two subsystems have been interacting in the past, then it is generally not possible to define a state vector for any of the individual subsystems.*

*Proof.* This theorem is easily verified: interactions lead to correlations. Correlations between 2 systems imply that the composite state vector can not be a single product of subsystem states:

$$|\psi\rangle = \sum_{m_1,m_2} a(m_1, m_2) |m_1\rangle \otimes |m_2\rangle \;, \quad a(m_1, m_2) \neq a(m_1)a(m_2) \;. \quad (2.622)$$

The density operator then reads

$$\hat{\rho} = |\psi\rangle\langle\psi| = \sum_{m_1,m_2,m_1',m_2'} a(m_1, m_2)a^*(m_1', m_2') |m_1 m_2\rangle\langle m_1' m_2'| \quad (2.623)$$

so that the reduced density operator $\hat{R}(1)$ is mixed, i.e. (cf. (2.612))

$$\hat{R}(1) = \mathrm{Tr}_2\{\hat{\rho}\} = \sum_{m_2}\langle m_2|\hat{\rho}|m_2\rangle = \sum_{m_2}\langle m_2|\psi\rangle\langle\psi|m_2\rangle$$

$$= \sum_{m_1,m_2,m_1'} a(m_1, m_2)a^*(m_1', m_2) |m_1\rangle\langle m_1'| \;, \quad (2.624)$$

$$\sum_{m_2} a(m_1, m_2)a^*(m_1', m_2) = \rho_{m_1,m_1'} \;. \quad (2.625)$$

Only if the two partial systems had never been interacting, i.e. if $|\psi\rangle$ was a product state implying

$$a(m_1, m_2) = a(m_1)a(m_2) \;, \quad (2.626)$$

then (2.624) reduces to

$$\hat{R}(1) = \sum_{m_1,m_1'} a(m_1)a^*(m_1') |m_1\rangle\langle m_1'| = |\phi(1)\rangle\langle\phi(1)| \;, \quad (2.627)$$

$$|\phi(1)\rangle = \sum_{m_1} a(m_1) |m_1\rangle \;, \quad (2.628)$$

i.e. a pure state.

We now discuss 2 situations in which entanglement occurs in a simple way: as eigenstates of specific global operators, and due to statistical effects (permutation symmetry required for indistinguishable particles).

**2.4.4.3 Eigenstates and Entanglement.** Just like coherence, in general, entanglement depends on the basis and decomposition into nodes. For given subsystems ($\nu = 1, 2$) the individual angular momenta are characterized by

$$\left[\hat{L}_i(\nu), \hat{L}_j(\nu')\right]_- = i\hbar\delta_{\nu\nu'} \sum_{k=1}^{3} \varepsilon_{ijk}\hat{L}_k(\nu) \quad (2.629)$$

($\varepsilon_{ijk}$ denotes the $\varepsilon$ tensor defined by (2.2)). Equation (2.629) represents a generalized form of (2.1). The eigenvalues of the operators $\hat{L}_z(\nu)$, $\hat{L}^2(\nu)$ are defined by (2.10). The product state vectors specified as

$$|l_1, l_2, m_1, m_2\rangle = |l_1, m_1\rangle \otimes |l_2, m_2\rangle \quad (2.630)$$

are eigenfunctions of $\hat{L}^2(\nu)$, $\hat{L}_z(\nu)$, $\nu = 1, 2$.

The total angular momentum of the composite system is

$$\hat{\boldsymbol{L}} = \hat{\boldsymbol{L}}_1 + \hat{\boldsymbol{L}}_2 = \hat{\boldsymbol{L}}(1) \otimes \hat{1}(2) + \hat{1}(1) \otimes \hat{\boldsymbol{L}}(2) \tag{2.631}$$

for which the commutator relations

$$\left[\hat{L}^2, \hat{L}_z\right]_- = 0 \,,$$

$$\left[\hat{L}^2, \hat{L}^2(\nu)\right]_- = 0 \,, \tag{2.632}$$

$$\left[\hat{L}_z, \hat{L}^2(\nu)\right]_- = 0$$

hold.

Let the joint eigenstates of the commuting set of operators $\hat{L}^2(1)$, $\hat{L}^2(2)$, $\hat{L}^2$, $\hat{L}_z$ be denoted by

$$|\chi\rangle = |l_1, l_2; l, m\rangle \,. \tag{2.633}$$

$|\chi\rangle$ can be represented as a superposition of the product states (see (2.630)). The respective amplitudes are known as *Clebsch-Gordon coefficients* (cf. [119]). In general, $|\chi\rangle$ will thus be an entangled state.

*Example 2.4.6.* The eigenvector for two spin $1/2$ subsystems and $l = 0, m = 0$ is

$$\boxed{\left|\tfrac{1}{2}, \tfrac{1}{2}; 0, 0\right\rangle = \tfrac{1}{\sqrt{2}} \left[\left|\tfrac{1}{2}, -\tfrac{1}{2}\right\rangle - \left|-\tfrac{1}{2}, \tfrac{1}{2}\right\rangle\right]} \tag{2.634}$$

defining the *singlet state*, and the eigenvectors ($l = 1$, $m = -1, 0, 1$)

$$\boxed{\begin{aligned}\left|\tfrac{1}{2}, \tfrac{1}{2}; 1, 0\right\rangle &= \tfrac{1}{\sqrt{2}} \left[\left|\tfrac{1}{2}, -\tfrac{1}{2}\right\rangle + \left|-\tfrac{1}{2}, \tfrac{1}{2}\right\rangle\right] \,, \\ \left|\tfrac{1}{2}, \tfrac{1}{2}; 1, -1\right\rangle &= \left|-\tfrac{1}{2}, -\tfrac{1}{2}\right\rangle \,, \\ \left|\tfrac{1}{2}, \tfrac{1}{2}; 1, +1\right\rangle &= \left|\tfrac{1}{2}, \tfrac{1}{2}\right\rangle\end{aligned}} \tag{2.635}$$

are the *triplet states*.

In this example, the l.h.s. of (2.635) and (2.634) represent state vectors in the eigenbasis of $\hat{L}^2(1)$, $\hat{L}^2(2)$, $\hat{L}^2$, $\hat{L}_z$, and the r.h.s. of (2.635) and (2.634) contain state vectors in the eigenbasis of $\hat{L}^2(1)$, $\hat{L}^2(2)$, $\hat{L}_z(1)$, $\hat{L}_z(2)$. The product states $|m_1, m_2\rangle$ form a 4-dimensional basis connected with the special unitary group $SU(2) \otimes SU(2)$. The singlet state and the first triplet state are examples of *entangled states*, i.e. total system wave functions which cannot be written as a single product of subsystem states.

Next, consider the Hamiltonian

$$\hat{H} = \hat{H}_1 + \hat{H}_2 + \hat{H}(1,2) \tag{2.636}$$

with $\hat{H}_1$, $\hat{H}_2$ given by equations (2.552), (2.553). The total Hamiltonian matrix for the Förster model is thus in the product basis $|m_1, m_2\rangle = \{|1, 1\rangle, |1, 2\rangle, |2, 1\rangle, |2, 2\rangle\}$:

$$\left\langle m_1 m_2 \left| \hat{H} \right| m_1' m_2' \right\rangle$$

$$= \begin{pmatrix} E_1^{(1)} + E_1^{(2)} & 0 & 0 & 0 \\ 0 & E_1^{(1)} + E_2^{(2)} & \hbar C_F & 0 \\ 0 & \hbar C_F & E_2^{(1)} + E_1^{(2)} & 0 \\ 0 & 0 & 0 & E_2^{(1)} + E_2^{(2)} \end{pmatrix} . \quad (2.637)$$

Eigenvalues and eigenstates are for $E_i^{(1)} = E_i^{(2)} = E_i$

$$\begin{aligned}
E_a &= 2E_1 & |\psi_a\rangle &= |1,1\rangle , \\
E_b &= E_1 + E_2 - |\hbar C_F| & |\psi_b\rangle &= \tfrac{1}{\sqrt{2}} \left( |1,2\rangle - |2,1\rangle \right) , \\
E_c &= E_1 + E_2 + |\hbar C_F| & |\psi_c\rangle &= \tfrac{1}{\sqrt{2}} \left( |1,2\rangle + |2,1\rangle \right) , \\
E_d &= 2E_2 & |\psi_d\rangle &= |2,2\rangle .
\end{aligned} \qquad (2.638)$$

The states $|\psi_b\rangle$, $|\psi_c\rangle$ are entangled. This result applies, for example, to interacting 2-level atoms or to 2-level atoms in resonance with a cavity photon mode as discussed in Sect. 2.4.2.4. The entangled atom cavity eigenstates are spectrally resolvable (*vacuum Rabi splitting*).

Operators in $SU(2) \otimes SU(2)$ which would have $|\psi_b\rangle$, $|\psi_c\rangle$ and the state $\tfrac{1}{\sqrt{2}} [|\psi_a\rangle \pm |\psi_b\rangle]$ as eigenstates are called *Bell operators*. They represent the measurement basis for the teleportation scenario (see Sect. 2.5.3.4).

For a more general interaction model (a molecule = network of two 1-electron-2-level atoms) already the ground state (of zero total spin) is found to be entangled (cf. [131]):

$$|\psi_a\rangle = \alpha \left( |1,1\rangle + |2,2\rangle \right) + \beta \left( |1,2\rangle + |2,1\rangle \right) . \qquad (2.639)$$

**2.4.4.4 Indistinguishability and Entanglement.** Consider a 2-particle Hilbert space $\mathbb{H}$ defined by

$$\mathbb{H} = \mathbb{H}(1) \otimes \mathbb{H}(2) \qquad (2.640)$$

spanned by the product states $|m(1), m(2)\rangle$. The product spaces $\mathbb{H}^{\pm}$ are defined by

$$\mathbb{H}^{\pm} : \quad |m_1, m_2\rangle^{\pm} = \frac{1}{\sqrt{2}} \left[ |m_1(1), m_2(2)\rangle \pm |m_2(1), m_1(2)\rangle \right] . \qquad (2.641)$$

$\mathbb{H}^{+}$ is the symmetric, $\mathbb{H}^{-}$ the antisymmetric subspace of $\mathbb{H}$. Due to the fact that the state vector $|m_1, m_2\rangle^{-}$ is identically zero for equal quantum numbers $m_1$, $m_2$, this state vector obeys the *Pauli principle*. Thus, this kind of state vector applies to *fermions* (particles with half-integer spin). In contrast, $|m_1, m_2\rangle^{+}$ applies to *bosons* (particles with integer spin).

For the density matrix connected with the density operator

$$\hat{\rho}_{\pm} = |m_1, m_2\rangle^{\pm} \langle m_1, m_2|^{\pm} , \qquad (2.642)$$

one obtains

$$(\rho_\pm) = \frac{1}{2}\begin{pmatrix} 0 & 0 & 0 & 0 \\ 0 & 1 & \pm 1 & 0 \\ 0 & \pm 1 & 1 & 0 \\ 0 & 0 & 0 & 0 \end{pmatrix} , \tag{2.643}$$

where the elements

$$\langle m_1(1), m_2(2)| \, \hat{\rho} \, |m_1'(1), m_2'(2)\rangle := \rho\left(m_1, m_2; m_1', m_2'\right) \tag{2.644}$$

of the density matrices (2.643) obey the symmetry relation

$$\boxed{\rho\left(m_1, m_2; m_1', m_2'\right) = \rho\left(m_2, m_1; m_2', m_1'\right)} \tag{2.645}$$

for fermions as well as for bosons. Permutation symmetry thus generates entanglement in the "particle identification number space" even in the absence of any physical interaction. Equation (2.643) is identical with the singlet state discussed before. However, contrary to the local nodes, the individual particle properties cannot be measured: allowed operators are only those symmetric in the particle identification number. This is why transitions between Fermi and Bose states do not occur (*superselection rule*). These properties resulting from the permutation symmetry are automatically taken care of in the *second quantization scheme*.

Based on density matrix (2.643), the reduced density operators are represented by

$$\hat{R}(1) = \mathrm{Tr}_2 \left\{\hat{\rho}_\pm\right\} = \frac{1}{2}\begin{pmatrix} 1 & 0 \\ 0 & 1 \end{pmatrix} = \hat{R}(2) = \mathrm{Tr}_1 \left\{\hat{\rho}_\pm\right\} . \tag{2.646}$$

These states describe single-particle properties but do not refer to a specific particle.

**2.4.4.5 Correlation Tensor of a Model State.** In general, the density operator $\hat{\rho}$ in $SU(2) \otimes SU(2)$ is given by the expression (2.599),

$$\hat{\rho} = \frac{1}{4}\hat{1} + \frac{1}{4}\sum_{i=1}^{3} \lambda_i(1)\hat{\lambda}_i(1) + \frac{1}{4}\sum_{j=1}^{3} \lambda_j(2)\hat{\lambda}_j(2) +$$

$$\frac{1}{4}\sum_{i,j=1}^{3} K_{ij}(1,2)\hat{\lambda}_i(1) \otimes \hat{\lambda}_j(2) . \tag{2.647}$$

Identifying the product states as in Sect. 2.5.1,

$$\begin{aligned} |-1, -1\rangle &= |1\rangle , \\ |-1, 1\rangle &= |2\rangle , \\ |1, -1\rangle &= |3\rangle , \\ |1, 1\rangle &= |4\rangle , \end{aligned} \tag{2.648}$$

we can write any pure state as

$$|\chi\rangle = \sum_\nu C_\nu |\nu\rangle \tag{2.649}$$

with

$$\sum_\nu |C_\nu|^2 = 1 .$$

(2.650)

The elements of the density matrix $(\rho_{ij})$ read

$$\rho_{ij} = C_i C_j^* \ (i \le j) , \quad \rho_{ji} = \rho_{ij}^*$$

(2.651)

and the density operator is given by

$$\hat\rho = \sum_{i,j} \rho_{ij} \hat P_{ij} .$$

(2.652)

The respective reduced density operator $\hat R(1)$ is

$$\hat R(1) = \mathrm{Tr}_2 \{\hat\rho\} = \sum_{m_2} \langle m_2 | \hat\rho | m_2 \rangle$$

(2.653)

with the corresponding matrix

$$R(1) = \begin{pmatrix} |C_1|^2 + |C_2|^2 & 0 \\ 0 & |C_3|^2 + |C_4|^2 \end{pmatrix} .$$

(2.654)

The matrix of $\hat R(2)$ is given by

$$R(2) = \begin{pmatrix} |C_3|^4 + |C_4|^2 & 0 \\ 0 & |C_1|^2 + |C_2|^2 \end{pmatrix} .$$

(2.655)

Observing $\mathrm{Tr}\left\{ \hat P_{ij} \hat P_{mn} \right\} = \delta_{in}\delta_{jm}$, we find (based on the results of Sect. 2.4.1.5):

$$\begin{aligned} K_{xx}(1,2) &= \mathrm{Tr}\{\hat\rho \, [\hat\sigma_x(1) \otimes \hat\sigma_x(2)]\} \\ &= \langle \chi | \hat\sigma_x(1) \otimes \hat\sigma_x(2) | \chi \rangle \\ &= \sum_{\nu,\nu'} C_\nu C_{\nu'}^* \left\langle \nu \left| \hat P_{14} + \hat P_{23} + \hat P_{32} + \hat P_{41} \right| \nu' \right\rangle \\ &= C_1 C_4^* + C_4 C_1^* + C_2 C_3^* + C_3 C_2^* . \end{aligned}$$

(2.656)

The other non-zero elements of the correlation tensor are given by

$$\begin{aligned} K_{yy}(1,2) &= \langle \chi | \hat\sigma_y(1) \otimes \hat\sigma_y(2) | \chi \rangle \\ &= -C_1 C_4^* - C_4 C_1^* + C_2 C_3^* + C_3 C_2^* , \\ K_{zz}(1,2) &= \langle \chi | \hat\sigma_z(1) \otimes \hat\sigma_z(2) | \chi \rangle \\ &= C_1 C_1^* - C_2 C_2^* - C_3 C_3^* + C_4 C_4^* . \end{aligned}$$

(2.657)

Finally, the coherence vector for a given subsystem $\nu = 1, 2$ is defined by (2.600)–(2.602):

$$\begin{aligned} \lambda_i(1) &= \left\langle \chi \left| \hat\lambda_i(1) \otimes \hat 1(2) \right| \chi \right\rangle , \\ \lambda_j(2) &= \left\langle \chi \left| \hat 1(1) \otimes \hat\lambda_j(2) \right| \chi \right\rangle . \end{aligned}$$

(2.658)

One obtains

$$
\begin{aligned}
\lambda_x(1) &= 2\mathrm{Re}\,(C_1 C_3^* + C_2 C_4^*) \ , \\
\lambda_y(1) &= -2\mathrm{Im}\,(C_1 C_3^* + C_2 C_4^*) \ , \\
\lambda_z(1) &= |C_1|^2 + |C_2|^2 - |C_3|^2 - |C_4|^2 \ , \\
\lambda_x(2) &= 2\mathrm{Re}\,(C_1 C_2^* + C_3 C_4^*) \ , \\
\lambda_y(2) &= -2\mathrm{Im}\,(C_1 C_2^* + C_3 C_4^*) \ , \\
\lambda_z(2) &= |C_1|^2 + |C_3|^2 - |C_2|^2 - |C_4|^2 \ .
\end{aligned}
\tag{2.659}
$$

There is no 1-node coherence if $C_1$ ($C_3$) and $C_2$ ($C_4$) are zero. For the singlet state $\boldsymbol{\lambda}(1) = \boldsymbol{\lambda}(2) = 0$.

The correlation tensor proper

$$
M_{ij}(1,2) = K_{ij}(1,2) - \lambda_i(1)\lambda_j(2)
\tag{2.660}
$$

is then for $C_1 = C_4 = 0$, $C_2 = \alpha$, $C_3 = \beta$:

$$
\begin{aligned}
M_{xx}(1,2) &= 1 + \alpha\beta^* + \beta\alpha^* + 4\left(|\alpha|^4 - |\alpha|^2\right) \ , \\
M_{yy}(1,2) &= 1 + \alpha\beta^* + \beta\alpha^* + 4\left(|\alpha|^4 - |\alpha|^2\right) \ , \\
M_{zz}(1,2) &= 4\left(|\alpha|^4 - |\alpha|^2\right) \ , \\
M_{ij}(1,2) &= 0 \ \ [(i,j) \neq (x,x),(y,y),(z,z)] \ .
\end{aligned}
\tag{2.661}
$$

*Example 2.4.7.* If $\alpha = \mp\beta = 1/\sqrt{2}$, we find

$$
K_{xx}(1,2) = K_{yy}(1,2) = \mp 1 \ , \quad K_{zz}(1,2) = -1
\tag{2.662}
$$

and

$$
\boldsymbol{\lambda}(1) = \boldsymbol{\lambda}(2) = 0
\tag{2.663}
$$

so that

$$
M_{ij}(1,2) = K_{ij}(1,2) \ .
\tag{2.664}
$$

Inserting the matrix elements $K_{ij}(1,2)$, and the elements of the coherence vector $\lambda_i(1)$, $\lambda_i(2)$ defined by (2.659) into (2.647), the density operator $\hat{\rho}$ for the singlet state ($\alpha = \beta$) is

$$
\hat{\rho} = \frac{1}{4}\hat{1} - \frac{1}{4}\left[\hat{\sigma}_x(1) \otimes \hat{\sigma}_x(2) + \hat{\sigma}_y(1) \otimes \hat{\sigma}_y(2) + \hat{\sigma}_z(1) \otimes \hat{\sigma}_z(2)\right] \ .
\tag{2.665}
$$

*Example 2.4.8.* For the triplet state defined by (2.635), i.e.

$$
\hat{\rho} = \left|\tfrac{1}{2},\tfrac{1}{2};1,-1\right\rangle\left\langle\tfrac{1}{2},\tfrac{1}{2};1,-1\right| = \left|-\tfrac{1}{2},-\tfrac{1}{2}\right\rangle\left\langle-\tfrac{1}{2},-\tfrac{1}{2}\right| \ ,
\tag{2.666}
$$

one obtains

$$(\rho_{ij}) = \begin{pmatrix} 0 & 0 & 0 & 0 \\ 0 & 0 & 0 & 0 \\ 0 & 0 & 0 & 0 \\ 0 & 0 & 0 & 1 \end{pmatrix}, \qquad (2.667)$$

where the elements of the coherence vectors are given by

$$\lambda_z(1) = \lambda_z(2) = -1, \quad \lambda_x(1) = \lambda_x(2) = \lambda_y(1) = \lambda_y(2) = 0, \qquad (2.668)$$

and the correlation tensor has only the one non-zero element

$$K_{zz}(1,2) = 1 = P_z(1)P_z(2), \qquad (2.669)$$

so that $M_{ij}(1,2) = 0$. Using these results, the density operator can be written in the form

$$\boxed{\hat{\rho} = \frac{1}{4}\hat{1} - \frac{1}{4}\hat{\sigma}_z(1) \otimes \hat{1}(2) - \frac{1}{4}\hat{1}(1) \otimes \hat{\sigma}_z(2) + \frac{1}{4}\hat{\sigma}_z(1) \otimes \hat{\sigma}_z(2).} \qquad (2.670)$$

**2.4.4.6 Entropy.** Let the density operator $\hat{\rho}$ of the composite system $\Sigma$ be $\hat{\rho}$, and the density operators of the subsystems be $\hat{\rho}^{(1)}$, $\hat{\rho}^{(2)}$. Then if

$$\hat{\rho} = \hat{\rho}^{(1)} \otimes \hat{\rho}^{(2)}, \qquad (2.671)$$

the entropy $S$ of the composite system reads

$$\begin{aligned} S(\hat{\rho}) &= -k\left(\mathrm{Tr}\left\{\hat{\rho}^{(1)} \otimes \hat{\rho}^{(2)} \ln \hat{\rho}^{(1)} + \hat{\rho}^{(1)} \otimes \hat{\rho}^{(2)} \ln \hat{\rho}^{(2)}\right\}\right) \\ &= -k\left(\mathrm{Tr}_1\left\{\hat{\rho}^{(1)} \ln \hat{\rho}^{(1)}\right\}\mathrm{Tr}_2\left\{\hat{\rho}^{(2)}\right\} + \right. \\ &\qquad \left. \mathrm{Tr}_1\left\{\hat{\rho}^{(1)}\right\}\mathrm{Tr}_2\left\{\hat{\rho}^{(2)} \ln \hat{\rho}^{(2)}\right\}\right) \\ &= -k\left(\mathrm{Tr}_1\left\{\hat{\rho}^{(1)} \ln \hat{\rho}^{(1)}\right\} + \mathrm{Tr}_2\left\{\hat{\rho}^{(2)} \ln \hat{\rho}^{(2)}\right\}\right), \end{aligned} \qquad (2.672)$$

where $\mathrm{Tr}\left\{\hat{\rho}^{(\nu)}\right\} = 1$ (compare with (2.252)) was used. Then applying the definition (2.340), one obtains

$$S(\hat{\rho}) = S\left(\hat{\rho}^{(1)}\right) + S\left(\hat{\rho}^{(2)}\right) = S(1) + S(2). \qquad (2.673)$$

In this case $S$ is additive. However, in the presence of inter-subsystem correlations, additivity no longer holds. In this case one finds (cf. [6])

**Theorem 2.4.2 (Araki and Lieb).**

$$\boxed{|S(1) - S(2)| \le S \le |S(1) + S(2)|.} \qquad (2.674)$$

For $S = 0$ this theorem implies

$$S(1) = S(2) \qquad (2.675)$$

no matter how the system is partitioned. Based on this theorem one can define a correlation index (cf. [11])

$$I_c = S(1) + S(2) - S \ge 0. \qquad (2.676)$$

This index is zero for product states (see (2.673)).

*Example 2.4.9.* One easily convinces oneself that the entangled state

$$|\psi\rangle = \tfrac{1}{\sqrt{2}}\left[|\tfrac{1}{2},-\tfrac{1}{2}\rangle - |-\tfrac{1}{2},\tfrac{1}{2}\rangle\right] \tag{2.677}$$

which is a pure state (entropy $S = 0$), implies reduced density matrices for each subsystem of the form

$$R(\nu) = \frac{1}{2}\begin{pmatrix} 1 & 0 \\ 0 & 1 \end{pmatrix} \tag{2.678}$$

so that $S(1) = S(2) = k\ln 2$, and

$$0 = S < S(1) + S(2). \tag{2.679}$$

**2.4.4.7 Correlation Index and Sum Rule.** Correlations proper are described by the matrix $M_{ij}(1,2)$. For qualitative purposes it would be convenient to have a simple scalar measure instead. To derive such a measure, which at the same time is part of a sum rule, we consider (cf. (2.599), $n_1 = n_2 = n$)

$$\begin{aligned}
\mathrm{Tr}\left\{\hat{\rho}^2\right\} = {} & \frac{1}{n^4}\mathrm{Tr}\left\{\hat{1}\hat{1}\right\} + \\
& \frac{1}{4n^2}\sum_{j,j'=1}^{s}\lambda_j(1)\lambda_{j'}(1)\mathrm{Tr}\left\{\hat{\lambda}_j(1)\lambda_{j'}(1)\otimes\hat{1}(2)\right\} + \\
& \frac{1}{4n^2}\sum_{k,k'=1}^{s}\lambda_k(2)\lambda_{k'}(2)\mathrm{Tr}\left\{\hat{1}(1)\otimes\lambda_k(2)\lambda_{k'}(2)\right\} + \\
& \frac{1}{16}\sum_{j,,j'k,k'=1}^{s}K_{jk}(1,2)K_{j'k'}(1,2) \\
& \mathrm{Tr}\left\{\left[\hat{\lambda}_j(1)\otimes\hat{\lambda}_k(2)\right]\left[\hat{\lambda}_{j'}(1)\otimes\hat{\lambda}_{k'}(2)\right]\right\}. \tag{2.680}
\end{aligned}$$

Here, we have applied the properties of product operators as given in Sect. 2.4.1.3. Observing

$$\begin{aligned}
\mathrm{Tr}\left\{\hat{1}\right\} &= n^2, \\
\mathrm{Tr}\left\{\hat{\lambda}_i(\nu)\hat{\lambda}_j(\nu)\right\} &= 2\delta_{ij}n \quad (\nu = 1,2), \\
\mathrm{Tr}\left\{\hat{\lambda}_j(1)\hat{\lambda}_{j'}(1)\otimes\hat{\lambda}_k(2)\hat{\lambda}_{k'}(2)\right\} &= 4\delta_{jj'}\delta_{kk'},
\end{aligned} \tag{2.681}$$

we immediately find (cf. [45])

$$\begin{aligned}
C(n,2) = \mathrm{Tr}\left\{\hat{\rho}^2\right\} = {} & \frac{1}{n^2} + \frac{1}{2n}\sum_{j=1}^{s}\lambda_j^2(1) + \frac{1}{2n}\sum_{k=1}^{s}\lambda_k^2(2) + \\
& \frac{1}{4}\sum_{j,k=1}^{s}K_{jk}(1,2)K_{jk}(1,2) \tag{2.682}
\end{aligned}$$

or (with $M_{ij}(1,2)$, $M_{ij}^T(1,2) = M_{ji}(1,2)$)

$$C(n, 2) = \frac{1}{n^2} + \frac{1}{2n} |\lambda(1)|^2 + \frac{1}{2n} |\lambda(2)|^2 +$$
$$\frac{1}{4} \left[ |\lambda(1)|^2 |\lambda(2)|^2 + \text{Tr} \{MM^T\} + 2\lambda(1)M\lambda(2) \right] . \tag{2.683}$$

Defining a correlation index $\beta$ by

$$\beta = \frac{1}{4 \left(1 - \frac{1}{n^2}\right)} \text{Tr} \{MM^T\}$$
$$= \frac{1}{4 \left(1 - \frac{1}{n^2}\right)} \sum_{j,k} M_{jk}(1, 2) M_{jk}(1, 2) , \tag{2.684}$$

we end up with the sum rule

$$C(n, 2) - \frac{1}{n^2} = \frac{1}{2n} |\lambda(1)|^2 + \frac{1}{2n} |\lambda(2)|^2 +$$
$$\frac{1}{4} |\lambda(1)|^2 |\lambda(2)|^2 + \frac{1}{2} \lambda(1)M\lambda(2) +$$
$$\left(1 - \frac{1}{n^2}\right) \beta . \tag{2.685}$$

For a pure state, $C(n, 2) = 1$, and if $|\lambda(1)|^2 = |\lambda(2)|^2 = 0$, we obtain $\beta = 1$, its maximum value. On the other hand, if $\beta = 0$, we obtain with

$$1 - \frac{1}{n^2} = \frac{2}{n} \left(1 - \frac{1}{n}\right) + \left(1 - \frac{1}{n}\right)^2 \tag{2.686}$$

the local pure states (cf. Sect. 2.3.2.1)

$$|\lambda(1)|^2 = |\lambda(2)|^2 = 2 \left(1 - \frac{1}{n}\right) . \tag{2.687}$$

This means that two quantum systems can be correlated only if they are not pure. This index $\beta$ has qualitatively similar properties as the index $I_c$ directly based on the theorem of Araki and Lieb. $\beta$ has the advantage of obeying an explicit coherence sum rule (cf. [77]).

### 2.4.5 Measurement Projections and 2-Node Coherence

In this section, measurements on composite systems involving *2-node correlations* are studied. *Indirect measurements* and *coincidence measurements* are considered. A special formulation of the *Bell inequality* is introduced.

**2.4.5.1 Subsystem Measurements.** We consider a composite system in $SU(n) \otimes SU(n)$ prepared in a state characterized by the 2 coherence vectors $\lambda(1)$, $\lambda(2)$ with elements $\lambda_i(1)$ and $\lambda_i(2)$, respectively, and the correlation matrix $M(1, 2)$ with elements $M_{ij}(1, 2)$.

Let the subsystem measurement operators be defined by the eigenvalue equations

$$\hat{A}(\nu)|a_i(\nu)\rangle = a_i(\nu)|a_i(\nu)\rangle \quad (\nu = 1, 2) \tag{2.688}$$

with

$$\hat{1}(\nu) = \sum_j |a_j(\nu)\rangle\langle a_j(\nu)| \ . \tag{2.689}$$

This complete set of orthonormalized states is taken to coincide with the basis defining the projector representation of the generators $\lambda_j(\nu)$:

$$|a_i(\nu)\rangle = |i(\nu)\rangle \ . \tag{2.690}$$

A measurement of $\hat{A}(2)$ in subsystem $\nu = 2$ can then be represented by projection operators acting in the Hilbert subspace $\mathbb{H}(2)$ only, $\hat{P}_{jj}^{(2)} = \hat{1} \otimes \hat{P}_{jj}(2)$, where

$$\hat{P}_{jj}(2) = |a_j(2)\rangle\langle a_j(2)| \ . \tag{2.691}$$

If the system is in state $\hat{\rho}$, the probability of finding the result $a_m$ in subsystem $\nu = 2$ is then given by (compare (2.414))

$$p_m(2) = \mathrm{Tr}\left\{\hat{P}_{mm}(2)\hat{\rho}\right\} = \mathrm{Tr}_2\left\{\hat{P}_{mm}(2)\hat{R}(2)\right\} \ , \tag{2.692}$$

and after measurement the system is in the state (compare (2.415))

$$\hat{\rho}' = \frac{1}{p_m(2)}\hat{P}_{mm}(2)\hat{\rho}\hat{P}_{mm}(2) \ . \tag{2.693}$$

The analogous results apply to measurements in subsystem $\nu = 1$.

In $SU(n) \otimes SU(n)$ the density operator reads

$$\begin{aligned}
\hat{\rho} = {}& \frac{1}{n^2}\hat{1}(1) \otimes \hat{1}(2) + \frac{1}{2n}\sum_{j=1}^{s}\lambda_j(1)\left[\hat{\lambda}_j(1) \otimes \hat{1}(2)\right] \\
& + \frac{1}{2n}\sum_{k=1}^{s}\lambda_k(2)\left[\hat{1}(1) \otimes \hat{\lambda}_k(2)\right] \\
& + \frac{1}{4}\sum_{j,k=1}^{s}K_{jk}(1,2)\left[\hat{\lambda}_j(1) \otimes \hat{\lambda}_k(2)\right] \ ,
\end{aligned} \tag{2.694}$$

and, according to (2.111),

$$\hat{P}_{jj}(2) = \frac{1}{n}\hat{1}(2) + \frac{1}{2}\sum_{l=1}^{n-1}w_l^{(j)}\hat{w}_l(2) \ . \tag{2.695}$$

Here, $w_l^{(j)}$ is defined by (2.103), and $\hat{w}_l(2)$ by (2.80). Applying (cf. (2.428))

$$\begin{aligned}
\hat{P}_{mm}(2)\hat{w}_l(2)\hat{P}_{mm}(2) &= w_l^{(m)}\hat{P}_{mm}(2) \ , \\
\hat{P}_{mm}(2)\hat{1}(2)\hat{P}_{mm}(2) &= \hat{P}_{mm}(2)
\end{aligned} \tag{2.696}$$

(all other generators $\hat{u}_{ij}(2)$, $\hat{v}_{ij}(2)$ transform to zero), we obtain for the density operator $\hat{\rho}'$:

$$
\begin{aligned}
p_m(2)\hat{\rho}' &= \frac{1}{n^2}\hat{1}(1) \otimes \hat{P}_{mm}(2) + \\
&\quad \frac{1}{2n}\sum_{j=1}^{s}\lambda_j(1)\left[\hat{\lambda}_j(1) \otimes \hat{P}_{mm}(2)\right] + \\
&\quad \frac{1}{2n}\sum_{l=1}^{n-1}w_l(2)w_l^{(m)}\left[\hat{1}(1) \otimes \hat{P}_{mm}(2)\right] + \\
&\quad \frac{1}{4}\sum_{j=1}^{s}\sum_{l=1}^{n-1}K_{js'+l}(1,2)w_l^{(m)}\left[\hat{1}(1) \otimes \hat{P}_{mm}(2)\right] .
\end{aligned}
\tag{2.697}
$$

$w_l(2)$ is the expectation value of $\hat{w}_l(2)$. Here $s' = n^2 - n$. Replacing $\hat{P}_{mm}(2)$ by its representation (2.695), and observing that

$$
\boxed{p_m(2) = \frac{1}{n} + \frac{1}{2}\sum_{l=1}^{n-1}w_l(2)w_l^{(m)} ,}
\tag{2.698}
$$

we finally obtain, assuming $p_m(2) \neq 0$:

$$
\begin{aligned}
\hat{\rho}' &= \frac{1}{n^2}\hat{1}(1) \otimes \hat{1}(2) + \\
&\quad \frac{1}{2n}\sum_{l'=1}^{n-1}w_{l'}^{(m)}\left[\hat{1}(1) \otimes \hat{w}_{l'}(2)\right] + \\
&\quad \frac{1}{2np_m(2)}\sum_{j=1}^{s}K_j^{(1)}(1,2)\left[\hat{\lambda}_j(1) \otimes \hat{1}(2)\right] + \\
&\quad \frac{1}{4p_m(2)}\sum_{j=1}^{s}\sum_{l=1}^{n-1}K_{jl}^{(2)}(1,2)\left[\hat{\lambda}_j(1) \otimes \hat{w}_l(2)\right]
\end{aligned}
\tag{2.699}
$$

with

$$
K_j^{(1)}(1,2) = \frac{1}{n}\lambda_j(1) + \frac{1}{2}\sum_{l=1}^{n-1}K_{js'+l}(1,2)w_l^{(m)} ,
\tag{2.700}
$$

$$
K_{jl}^{(2)}(1,2) = \frac{1}{n}\lambda_j(1)w_l^{(m)} + \frac{1}{2}\sum_{l'=1}^{n-1}K_{js'+l'}(1,2)w_{l'}^{(m)}w_l^{(m)} .
\tag{2.701}
$$

This equation may be compared with the form of (2.694). We can thus read off the coherence vectors and the correlation tensor after measurement in terms of the constants $w_l^{(m)}$, $p_m(2)$, and the corresponding terms before measurement:

$$\lambda'_{s'+l}(2) = w_l^{(m)} \ ,$$

$$\lambda'_j(1) = \frac{1}{p_m(2)} \left[ \frac{1}{n} \lambda_j(1) + \frac{1}{2} \sum_{l=1}^{n-1} K_{js'+l}(1,2) w_l^{(m)} \right] \ ,$$

$$K'_{js'+l}(1,2) = \frac{w_l^{(m)}}{p_m(2)} \left[ \frac{1}{n} \lambda_j(1) + \frac{1}{2} \sum_{l'=1}^{n-1} K_{js'+l'}(1,2) w_{l'}^{(m)} \right]$$

(2.702)

with

$$j = 1,2,\ldots,s \ , \quad l,l' = 1,2,\ldots,n-1 \ , \quad s' = n^2 - n \ . \tag{2.703}$$

Here we have observed that $\lambda_{s'+l}(2) = w_l(2)$. $m$ is the index of the measurement outcome $a_m(2)$, $m = 1,2,\ldots,n$. Alternatively, with the help of

$$K_{jl}(1,2) = M_{jl}(1,2) + \lambda_j(1)\lambda_l(2) \ , \tag{2.704}$$

we find

$$\lambda'_{s'+l}(2) = w_l^{(m)} \ ,$$

$$\lambda'_j(1) = \lambda_j(1) + \frac{1}{2p_m(2)} \sum_{l=1}^{n-1} M_{js'+l}(1,2) w_l^{(m)} \ ,$$

$$M'_{jk}(1,2) = 0$$

(2.705)

with

$$j,k = 1,2,\ldots s \ , \quad l = 1,2,\ldots n-1 \ , \quad s' = n^2 - n \ . \tag{2.706}$$

We see that the effect on the directly measured subsystem 2, $\boldsymbol{\lambda}(2) \to \boldsymbol{\lambda}'(2)$, is the same as for the single system studied in Sect. 2.3.6. The updating $\boldsymbol{\lambda}(1) \to \boldsymbol{\lambda}'(1)$ is the effect of the correlation proper, $M_{jk}(1,2)$. After measurement this correlation is set to zero.

*Example 2.4.10.* In Fig. 2.33 we contrast the direct measurement of a single system in $SU(2)$ with a subsystem measurement on a composite system in $SU(2) \otimes SU(2)$. The circles indicate the respective coherence vector spheres, projected on the $\lambda_3, \lambda_2$ plane; the pointers are the (projected) coherence vectors. Subsystem 2 could, for example, represent the number states 0 or 1 of a *luminescence photon channel* connected with a transition in a molecular subsystem 1. As illustrated, for the single $SU(2)$ system the two projections of the coherence vector are

$$\lambda \ \begin{array}{l} \nearrow \ \lambda'_- = (0,0,-1) \\ \searrow \ \lambda'_+ = (0,0,+1) \end{array} \ , \tag{2.707}$$

while for the $SU(2) \otimes SU(2)$ system the corresponding projections are

$$\boldsymbol{\lambda}(2) \ \begin{array}{l} \nearrow \ \lambda'_-(2) = \lambda_-(2) = (0,0,-1) \\ \searrow \ \lambda'_+(2) = \lambda_+(2) = (0,0,+1) \end{array} \ , \tag{2.708}$$

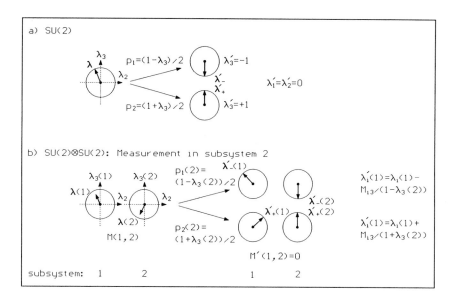

**Fig. 2.33.** Measurement in single system (part a)), and in subsystem (part b)). In both cases the initial state is supposed to be a pure state. In case b), correlations imply that the subsystem coherence vectors cannot be pure (length < 1)

$$\lambda_j(1) \left\langle \begin{array}{l} \lambda'_{j,-}(1) = \lambda_j(1) - \dfrac{1}{2p_1(2)} M_{j3} \quad \text{for} \quad \lambda'_3 = -1 \\[3mm] \lambda'_{j,+}(1) = \lambda_j(1) + \dfrac{1}{2p_2(2)} M_{j3} \quad \text{for} \quad \lambda'_3 = +1 \end{array} \right. \tag{2.709}$$

One notes that for the composite system the single-node coherence of subsystem 1 is not destroyed; part of the *non-local correlation* M has "materialized" in the updated coherence vector of subsystem 1! This *indirect measurement* (with respect to subsystem 1) is thus in terms of non-orthogonal alternatives, i.e., the corresponding projectors are non-commuting (*probability operator value measure* (POVM), see [113, 153]). Such indirect measurements are quite common: in *atom spectroscopy* the measurement on photons serves as a means to learn something about the atomic states. As M contains state properties before measurement projection, such indirect measurements do not completely destroy memories of the past.

It is now straightforward to calculate conditional probabilities for the outcome of a subsequent measurement on subsystem 1. We get for $SU(2) \otimes SU(2)$ ($|1\rangle = |-1\rangle$, $|2\rangle = |1\rangle$):

$$\begin{aligned} p_1(m_1/m_2) &= \frac{1}{2} \left[ 1 + m_1 \lambda'_3(1) \right] \\ &= \frac{1}{2} \left\{ 1 + m_1 \left[ \lambda_3(1) + m_2 \frac{M_{33}(1,2)}{1 + m_2 \lambda_3(2)} \right] \right\}, \end{aligned} \tag{2.710}$$

where $m_1 = \mp 1$, $m_2 = \mp 1$. The joint probability is

$$
\begin{aligned}
p_1(m_1, m_2) &= p_2(m_2) p_1(m_1/m_2) \\
&= \frac{1}{4} \left\{ [1 + m_1 \lambda_3(1)] [1 + m_2 \lambda_3(2)] + m_1 m_2 M_{33}(1,2) \right\} \\
&= p_1(m_1) p_2(m_2/m_1)
\end{aligned}
\tag{2.711}
$$

or, alternatively,

$$
p_1(m_1, m_2) = \frac{1}{4} [1 + m_1 \lambda_3(1) + m_2 \lambda_3(2) + m_1 m_2 K_{33}(1,2)] .
\tag{2.712}
$$

This latter relation implies

$$
K_{33}(1,2) = \sum_{m_1, m_2 = \pm 1} m_1 m_2 p(m_1, m_2) .
\tag{2.713}
$$

Finally, we consider

$$
\sum_{m_1} p(m_1, m_2) = p_2(m_2) \sum_{m_2} p_1(m_1/m_2) .
\tag{2.714}
$$

Due to the fact that in $SU(2) \otimes SU(2)$

$$
p_1(+1/m_2) + p_1(-1/m_2) = \frac{1}{2} [1 + \lambda_3'(1)] + \frac{1}{2} [1 - \lambda_3'(1)] = 1
\tag{2.715}
$$

holds, we end up with the consistency condition (valid in any $SU(n)$):

$$
\sum_{m_1} p(m_1, m_2) = p_2(m_2) .
\tag{2.716}
$$

**2.4.5.2 EPR Paradox.** 2-node coherent states can be used to generate apparently paradoxical situations. The best-known examples refer to nodes, each consisting of a single elementary particle, arranged in such a way that these particles are separated in real space (cf. [25]). Actually, the node index

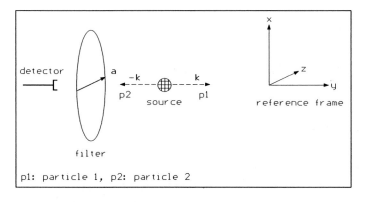

**Fig. 2.34.** Measurement of spin subsystem (2)

refers to these alternative localizations (rather than to the individual particle index, which is not measurable). Originally proposed by Einstein, Podolski and Rosen for continuous variables (*Einstein-Podolski-Rosen paradox (EPR paradox)*, cf. [40]), it has then been transcribed to discrete finite spaces, in particular, to $SU(2) \otimes SU(2)$ (by D. Bohm, [21] ).

As already noted, the 2-spin state with the zero total angular momentum (singlet state), is given by

$$|\psi\rangle = \frac{1}{\sqrt{2}}\left[|1,-1\rangle - |-1,1\rangle\right] \tag{2.717}$$

or, alternatively, by

$$\lambda(1) = \lambda(2) = 0 , \quad M_{ij}(1,2) = -\delta_{ij} . \tag{2.718}$$

This state is obviously invariant under any orthogonal translation of the basis states used for representation (presently the eigenvectors of $\hat{\sigma}_z(\nu)$). This would not be true, for example, for the triplet state with $M_{11}(1,2) = M_{22}(1,2) = 1$, $M_{33}(1,2) = -1$. Based on those properties and the subsystem measurements discussed in Sect. 2.4.5.1 we can argue as follows. Suppose that the polarization filter for subsystem 2 is set to select direction $a \parallel e_z$. If we find $\lambda_3'(2)$, we know that

$$\lambda_3'(1) = \begin{cases} -M_{33}(1,2) = +1 \\ +M_{33}(1,2) = -1 \end{cases} \quad \text{for} \quad \begin{array}{l} \lambda_3'(2) = -1 \\ \lambda_3'(2) = +1 \end{array} \tag{2.719}$$

(i.e. $\lambda_3'(1) = -\lambda_3'(2)$). If we make sure that the measurement of subsystem 2 has no influence on subsystem 1 (by separating the 2 spins in real space, see Fig. 2.34), we conclude that the subsystem 1 must have had its value $\lambda_3'(1) = \pm 1$ even before measurement. We now change the polarizer to $a \parallel e_x$, say. The same arguments apply as before, only with respect to the basis of $\sigma_x(\nu)$. We infer that $\lambda_3(1) = -\lambda_3(2)$ should have existed already before measurement. As $\lambda_3(1)$ now denotes the $x$ component of the spin 1, this would mean that we have measured its $x$ and $z$ component simultaneously, in contradiction to quantum mechanics.

The practical realization of these experiments is usually based on photon pairs, the polarization state of which is isomorphic to the spin 1/2 algebra. The advantage is that one has good 2-photon sources based on 2-photon transitions or "down-conversion" and reliable polarization filters and detectors (cf. [8]).

Coincidence measurements (cf. Fig. 2.35 for $a \parallel b$) are described by the joint probabilities (cf. (2.712))

$$p(m_1, m_2) = \frac{1}{4}\left[1 - m_1 m_2\right]. \tag{2.720}$$

The case of $a$ not parallel to $b$ will be discussed later.

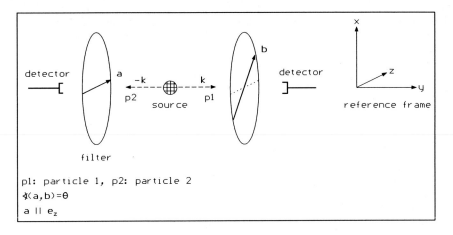

**Fig. 2.35.** Coincidence measurement

**2.4.5.3 Correlation Functions.** Consider two subsystems $\nu = 1, 2$ in $SU(n)$ with a measurement operator $\hat{A}(\nu)$ defined by $(s = n^2 - 1)$:

$$\hat{A}(\nu) = \frac{1}{n} \text{Tr}\left\{\hat{A}(\nu)\right\} \hat{1}(\nu) + \frac{1}{2} \sum_{j=1}^{s} A_j(\nu) \hat{\lambda}_j(\nu) \tag{2.721}$$

(cf. (2.294)). The respective expectation values, $\left\langle \hat{A}(\nu) \right\rangle$, are then

$$\left\langle \hat{A}(\nu) \right\rangle = \frac{1}{n} \text{Tr}\left\{\hat{A}(\nu)\right\} + \frac{1}{2} \sum_{j=1}^{s} A_j(\nu) \lambda_j(\nu) \tag{2.722}$$

(cf. (2.296)).

In the composite system, the product operator is given by

$$\hat{A}(1) \otimes \hat{A}(2) = \frac{1}{n^2} \text{Tr}\left\{\hat{A}(1)\right\} \text{Tr}\left\{\hat{A}(2)\right\} \left[\hat{1}(1) \otimes \hat{1}(2)\right] +$$

$$\frac{1}{2n} \text{Tr}\left\{\hat{A}(1)\right\} \sum_{j=1}^{s} A_j(2) \left[\hat{1}(1) \otimes \hat{\lambda}_j(2)\right] +$$

$$\frac{1}{2n} \text{Tr}\left\{\hat{A}(2)\right\} \sum_{j=1}^{s} A_j(1) \left[\hat{\lambda}_j(1) \otimes \hat{1}(2)\right] +$$

$$\frac{1}{4} \sum_{j,k=1}^{s} A_j(1) A_k(2) \left[\hat{\lambda}_j(1) \otimes \hat{\lambda}_k(2)\right] . \tag{2.723}$$

The correlation function (= expectation value of $A(1) \otimes A(2)$) can be expressed as

$$\left\langle \hat{A}(1) \otimes \hat{A}(2) \right\rangle$$

$$= \frac{1}{n^2} \text{Tr}\left\{\hat{A}(1)\right\} \text{Tr}\left\{\hat{A}(2)\right\} + \frac{1}{2n} \text{Tr}\left\{\hat{A}(1)\right\} \sum_{j=1}^{s} A_j(2)\lambda_j(2) +$$

$$\frac{1}{2n} \text{Tr}\left\{\hat{A}(2)\right\} \sum_{j=1}^{s} A_j(1)\lambda_j(1) +$$

$$\frac{1}{4} \sum_{j,k=1}^{s} K_{jk}(1,2)A_j(1)A_k(2) . \tag{2.724}$$

Comparing with (2.722), one obtains the relation

$$\boxed{\left\langle \hat{A}(1) \otimes \hat{A}(2) \right\rangle = \left\langle \hat{A}(1) \right\rangle \left\langle \hat{A}(2) \right\rangle + C_{AA}} \tag{2.725}$$

with

$$\boxed{C_{AA} = \frac{1}{4} \sum_{j,k=1}^{s} A_j(1)M_{jk}(1,2)A_k(2) ,} \tag{2.726}$$

i.e. a correlation function of second order can be decomposed into a product of correlation functions of first order (= expectation values of individual subsystems) and an additional correlation term $C_{AA}$. The scalar $C_{AA}$ can thus be calculated from the vectors $A_j(\nu)$ defining the operator $\hat{A}(\nu)$ in $SU(n)$ and the correlation tensor $M_{jk}(1,2)$ resulting from the considered state. For an uncorrelated system, this correlation term is identically zero.

*Example 2.4.11.* The polarization of an $SU(n)$ subsystem $\nu$ in the $x, z$ plane is described by

$$\hat{A}(\nu) = \hat{\sigma}_x(\nu)\sin\Theta(\nu) + \hat{\sigma}_z(\nu)\cos\Theta(\nu) \tag{2.727}$$

which implies

$$\boxed{A(\nu) = 2\left[\sin\Theta(\nu), 0, \cos\Theta(\nu)\right] .} \tag{2.728}$$

For a singlet state (EPR state) we have

$$M_{ik} = -\delta_{ik} \tag{2.729}$$

so that the correlation (2.726) reads

$$C_{AA} = C\left(\Theta(1), \Theta(2)\right) = -\sin\Theta(1)\sin\Theta(2) - \cos\Theta(1)\cos\Theta(2) . \tag{2.730}$$

Applying the trigonometric relation $\cos\alpha\cos\beta + \sin\alpha\sin\beta = \cos(\beta - \alpha)$, one finally obtains the well-known result

$$\boxed{C\left(\Theta(1), \Theta(2)\right) = -\cos\left[\Theta(2) - \Theta(1)\right]} \tag{2.731}$$

with the limiting cases

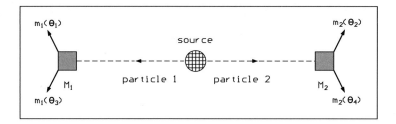

**Fig. 2.36.** Pair measurements of 2 alternative polarization directions in $x, y$ plane

$$\Theta(2) - \Theta(1) = 0 : \quad C\left(\Theta(1), \Theta(2)\right) = -1 \quad \text{(anticorrelated)} ,$$
$$\Theta(2) - \Theta(1) = \frac{\pi}{2} : \quad C\left(\Theta(1), \Theta(2)\right) = 0 \quad \text{(uncorrelated)} . \tag{2.732}$$

Such a correlation determines the inter-dependence of subsystem measurements.

**2.4.5.4 Bell Inequality.** Suppose that a source generates 2-photon states travelling along the $z$ coordinate in opposite directions. Then consider a co-incidence measurement of the polarization as shown in Fig. 2.36, where the measurements $M_1$ and $M_2$ are supposed to have no influence on each other ("locality"). Each measurement apparatus consists of a polarization filter in the $x, y$ plane with angle $\Theta$, followed by a photon counter (intensity measurement). Alternative pairs of angles, $\Theta_1$, $\Theta_3$ (index odd for system 1) and $\Theta_2$, $\Theta_4$ (index even for system 2), are indicated. The scheme

$$m_1\left(\Theta_1\right) = \pm 1 ,$$
$$m_1\left(\Theta_3\right) = \pm 1 ,$$
$$m_2\left(\Theta_2\right) = \pm 1 , \tag{2.733}$$
$$m_2\left(\Theta_4\right) = \pm 1$$

represents the possible measurement values of the single particles, $\nu = 1, 2$, as a function of the angle $\Theta$ in the $x, y$ plane.

A complete description of all 2-particle states is given by

$$Z_\alpha = \{m_1\left(\Theta_1\right), m_1\left(\Theta_3\right), m_2\left(\Theta_2\right), m_2\left(\Theta_4\right)\} = \{\pm 1, \pm 1, \pm 1, \pm 1\} ,$$
$$\alpha = 1, 2, \ldots 2^4 . \tag{2.734}$$

The scheme

$$\left.\begin{array}{c} m_1\left(\Theta_1\right) m_2\left(\Theta_2\right) \\ m_1\left(\Theta_1\right) m_2\left(\Theta_4\right) \\ m_1\left(\Theta_3\right) m_2\left(\Theta_2\right) \\ m_1\left(\Theta_3\right) m_2\left(\Theta_4\right) \end{array}\right\} = \pm 1 \tag{2.735}$$

describes all possible combined measurement results; of course, for any single pair only one out of those 4 possible angle settings can be realized. Local

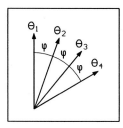

**Fig. 2.37.** Examples of measurement angles

reality, however, assumes that all these data exist, even if they cannot jointly be measured.

The variable

$$s = m_1(\Theta_1)\,m_2(\Theta_2) + m_1(\Theta_3)\,m_2(\Theta_2) + m_1(\Theta_1)\,m_2(\Theta_4) - m_1(\Theta_3)\,m_2(\Theta_4) \tag{2.736}$$

has the values $s_\alpha = \pm 2$ for all possible 2-particle states $Z_\alpha$. For an ensemble of such 2-particle states $Z_\alpha$ with weight function $f(\alpha)$, expectation values like the correlation functions

$$\boxed{C(\Theta_1,\Theta_2) = \langle m_1(\Theta_1)\,, m_2(\Theta_2)\rangle} \tag{2.737}$$

can be defined. Thus, the expectation value

$$\langle s\rangle = C(\Theta_1,\Theta_2) + C(\Theta_3,\Theta_2) + C(\Theta_1,\Theta_4) - C(\Theta_3,\Theta_4)$$
$$= \sum_\alpha f(\alpha)s_\alpha \tag{2.738}$$

has to obey the inequality

$$-2 \geq \langle s\rangle \leq 2\,. \tag{2.739}$$

Alternatively,

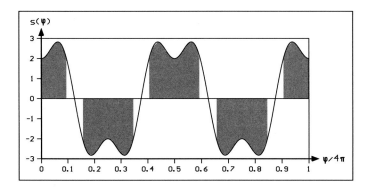

**Fig. 2.38.** Test function $s(\varphi)$. $|s(\varphi)| > 2$ violates Bell's inequality: the corresponding $\varphi$ values are in the shaded region

$$\boxed{|\langle s \rangle| = |C(\Theta_1, \Theta_2) + C(\Theta_2, \Theta_3) + C(\Theta_3, \Theta_4) - C(\Theta_4, \Theta_1)| \leq 2 \, .} \quad (2.740)$$

This inequality represents a special formulation of *Bell's inequality*. One easily shows that this inequality can be violated by quantum mechanics: apply (2.731) with

$$C(\Theta_1, \Theta_2) = - \cos[\Theta_2 - \Theta_1] \quad (2.741)$$

and select the chain of angles sketched in Fig. 2.37. Then one obtains

$$s(\varphi) = 3 \cos \varphi - \cos 3\varphi \, . \quad (2.742)$$

$s(\varphi)$ violates the limit postulated by (2.740) for, for example, $\varphi = \pi/4$ (cf. Fig. 2.38). Violations of the Bell inequalities for photon pairs first were demonstrated experimentally by Aspect et al. (cf. [8]).

**2.4.5.5 Quantum Cryptography.** (See [11, 41].) There is a wide range of different communication tasks which call for appropriate technical solutions. One is to protect information transmission from eavesdropping. If this security cannot be achieved, the next best circumstance is to guarantee that any attempt at eavesdropping will be detected by the legitimate users.

A classical channel can always be monitored by an unauthorized intruder: physically this possibility reduces to a measurement problem and, classically, measurements can be performed without any significant back action: information can be copied at will. This is not the case in the quantum regime. The quantum information channel to be discussed below is not meant to serve directly for the exchange of messages: this would be impractible. Rather, the quantum channel is reserved for the so-called *key distribution*.

The key is a set of specific parameters entering the *encrypting* and *decrypting algorithms*, which, in turn, generate the *cryptotext* to be transmitted, then, by conventional means. The security of this cryptographic technique depends on various components, one of which is the protection of the key. If the two legitimate parties, "Alice" (A) and "Bob" (B), have not anticipated their need for secret communication, they must agree on the use of a specific key in an otherwise known algorithm. This is done using the quantum channel (cf. [41]). One version of such a channel makes use of 2-particle entangled states (singlet or EPR states). It represents a first technical application of such non-classical states. The basic scheme is practically identical with the coincidence measurements sketched in Fig. 2.36; the 2 measurement devices are located with Alice ($M_1$) and Bob ($M_2$), respectively. The allowed discrete angle settings are for the two users:

$$A : \Theta_1 = 0 \quad , \quad \Theta_3 = \frac{\pi}{4} \quad , \quad \Theta_5 = \frac{\pi}{2} \quad ,$$

$$B : \Theta_2 = \frac{\pi}{4} \quad , \quad \Theta_4 = \frac{\pi}{2} \quad , \quad \Theta_6 = \frac{3\pi}{4} \, . \quad (2.743)$$

The source continuously produces photon pairs, one member reaching Alice, the other reaching Bob, who both select one of the allowed polarization angles

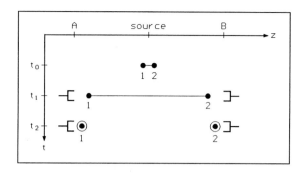

**Fig. 2.39.** Quantum cryptography. $t_0$: source generates entangled photon spins. $t_1$: photons separate in real space. $t_2$: photons reach the detectors, where they are projected in local polarization states (and destroyed)

at random. Measurement results and polarization angles are put down in local protocols A, B. After a finite series of detection events, Bob and Alice report their polarization settings in public (i.e. anybody may get this information). Then the measurement protocols are divided into two subsections:

$$1. \text{ Same orientation} = \{(\Theta_3, \Theta_2), (\Theta_5, \Theta_4)\} \tag{2.744}$$

and

$$2. \text{ Different orientation} = \left\{ \begin{matrix} (\Theta_1, \Theta_2), (\Theta_1, \Theta_4), (\Theta_1, \Theta_6), \\ (\Theta_3, \Theta_4), (\Theta_3, \Theta_6), (\Theta_5, \Theta_2), \\ (\Theta_5, \Theta_6) \end{matrix} \right\} . \tag{2.745}$$

The measurement results within the second group are announced publicly. Alice and Bob can now test whether

$$\begin{aligned} \langle s \rangle &= C(\Theta_5, \Theta_2) + C(\Theta_1, \Theta_2) + C(\Theta_5, \Theta_6) - C(\Theta_1, \Theta_6) \\ &= 3\cos(\pi/4) - \cos(3\pi/4) \\ &= -2\sqrt{2} . \end{aligned} \tag{2.746}$$

If $\langle s \rangle \neq -2\sqrt{2}$, the channel has been disturbed. If $\langle s \rangle = -2\sqrt{2}$, no one can have interfered. With high probability this holds also for the anticorrelated measurement results of group 1, which are joint pieces of information and can then be used as the key.

The remarkable feature of this communication channel is that Alice and Bob are not transmitting any information encoded in the individual photons! The *local information* comes into being only after measurement. The potential eavesdropper cannot avoid working in the same way, creating local information, where there should be none before the photons have reached their final destination (cf. Fig. 2.39). Such a scheme is definitely impossible in the classical domain. Quantum cryptographic schemes can be based also on the polarization space of single photons.

**2.4.5.6 Relativistic Constraints.** It is not quite clear to what extent discussions on relativistic effects are meaningful in the present context: the quantum theory discussed here is not a relativistic invariant formulation to begin with; it contains the speed of light at most indirectly (the travelling of photons as carriers of polarization). On the other hand, one cannot avoid discussing events in space-time, the moment one is addressing composite systems (the parts are typically separated in real space). Indirect measurements, described by projections possibly affecting subsystems miles away, therefore appear to involve "spooky" actions at a distance, not constrained by the speed of light. It has been noted with great relief that these EPR-type interactions do not allow the transmission of information, so that the relativistic maximum signal velocity is not violated (cf. [25]). Nevertheless, the projection postulate (like the dynamical measurement models of Chap. 4) is not Lorentz-invariant. This means that one can construct contradictions from the fact that simultaneity depends on the reference frame. On the other hand, the events we are talking about explicitly refer to a specific environment (cf. [66, 97]): so, there is one particular reference frame in which this environment is at rest. One must be very careful when discussing "additional" observations from other frames: on the level of actual measurements they would induce their own projections. On the level of observing (part of) the observational results, one deals with information and inference. With this in mind it appears to be possible to avoid inconsistencies in the reconstruction of reality (cf. [145]).

## 2.5 Composite Systems: Three Nodes

Contents: 3-node states, 3-node correlations, teleportation, Greenberger-Horne-Zeilinger states, measurement projections.

### 2.5.1 Product Generators Expressed by Projection Operators

The state vectors of a 3-node system are defined in the Hilbert space

$$\mathbb{H} = \mathbb{H}(1) \otimes \mathbb{H}(2) \otimes \mathbb{H}(3) , \tag{2.747}$$

where $\mathbb{H}(\nu)$ ($\nu = 1, 2, 3$) denote the Hilbert spaces of dimension $n_\nu$ of the respective subsystems. The state vectors $|m_1(1)m_2(2)m_3(3)\rangle$ form the product basis:

$$\hat{1} = \sum_{m_1, m_2, m_3} |m_1(1)m_2(2)m_3(3)\rangle \langle m_1(1)m_2(2)m_3(3)| . \tag{2.748}$$

The product generators can be evaluated as in the 2-node case (see Sect. 2.4.1.5). We restrict ourselves to $SU(2) \otimes SU(2) \otimes SU(2)$ and the product basis

$$|---\rangle = |1\rangle \ ,$$
$$|--+\rangle = |2\rangle \ ,$$
$$|-+-\rangle = |3\rangle \ ,$$
$$|-++\rangle = |4\rangle \ .$$
$$|+--\rangle = |5\rangle \ ,$$
$$|+-+\rangle = |6\rangle \ ,$$
$$|++-\rangle = |7\rangle \ ,$$
$$|+++\rangle = |8\rangle \ .$$

(2.749)

For the following one should note that products resulting from exchanged nodes are easily obtained by exchanging the state numbers:

$$\text{node } 1 \leftrightarrow 2 : \quad \begin{matrix} |3\rangle \leftrightarrow |5\rangle \\ |4\rangle \leftrightarrow |6\rangle \end{matrix} \ ,$$

$$\text{node } 1 \leftrightarrow 3 : \quad \begin{matrix} |2\rangle \leftrightarrow |5\rangle \\ |4\rangle \leftrightarrow |7\rangle \end{matrix} \ ,$$

(2.750)

$$\text{node } 2 \leftrightarrow 3 : \quad \begin{matrix} |2\rangle \leftrightarrow |3\rangle \\ |6\rangle \leftrightarrow |7\rangle \end{matrix} \ .$$

Products of $\hat{\sigma}_3$ and $\hat{1}$ are diagonal:

$$\hat{1}(1) \otimes \hat{1}(2) \otimes \hat{1}(3)$$
$$= \hat{P}_{11} + \hat{P}_{22} + \hat{P}_{33} + \hat{P}_{44} + \hat{P}_{55} + \hat{P}_{66} + \hat{P}_{77} + \hat{P}_{88} \ , \quad (2.751)$$

$$\hat{\sigma}_3(1) \otimes \hat{1}(2) \otimes \hat{1}(3)$$
$$= -\hat{P}_{11} - \hat{P}_{22} - \hat{P}_{33} - \hat{P}_{44} + \hat{P}_{55} + \hat{P}_{66} + \hat{P}_{77} + \hat{P}_{88} \ , \quad (2.752)$$

$$\hat{\sigma}_3(1) \otimes \hat{\sigma}_3(2) \otimes \hat{1}(3)$$
$$= \hat{P}_{11} + \hat{P}_{22} - \hat{P}_{33} - \hat{P}_{44} - \hat{P}_{55} - \hat{P}_{66} + \hat{P}_{77} + \hat{P}_{88} \ , \quad (2.753)$$

$$\hat{\sigma}_3(1) \otimes \hat{\sigma}_3(2) \otimes \hat{\sigma}_3(3)$$
$$= -\hat{P}_{11} + \hat{P}_{22} + \hat{P}_{33} - \hat{P}_{44} + \hat{P}_{55} - \hat{P}_{66} - \hat{P}_{77} + \hat{P}_{88} \ . \quad (2.754)$$

Products of the generators $\hat{\sigma}_1$, $\hat{\sigma}_2$ have non-zero matrix elements on the counter-diagonal:

$$\hat{\sigma}_1(1) \otimes \hat{\sigma}_1(2) \otimes \hat{\sigma}_1(3)$$
$$= \hat{P}_{18} + \hat{P}_{27} + \hat{P}_{36} + \hat{P}_{45} + \hat{P}_{54} + \hat{P}_{63} + \hat{P}_{72} + \hat{P}_{81} \ , \quad (2.755)$$

$$\hat{\sigma}_2(1) \otimes \hat{\sigma}_2(2) \otimes \hat{\sigma}_2(3)$$
$$= i \left( -\hat{P}_{18} + \hat{P}_{27} + \hat{P}_{36} - \hat{P}_{45} + \hat{P}_{54} - \hat{P}_{63} - \hat{P}_{72} + \hat{P}_{81} \right) \ , \quad (2.756)$$

$$\hat{\sigma}_1(1) \otimes \hat{\sigma}_2(2) \otimes \hat{\sigma}_2(3)$$
$$= -\hat{P}_{18} + \hat{P}_{27} + \hat{P}_{36} - \hat{P}_{45} - \hat{P}_{54} + \hat{P}_{63} + \hat{P}_{72} - \hat{P}_{81} , \qquad (2.757)$$

$$\hat{\sigma}_1(1) \otimes \hat{\sigma}_1(2) \otimes \hat{\sigma}_2(3)$$
$$= i\left(\hat{P}_{18} + \hat{P}_{27} + \hat{P}_{36} - \hat{P}_{45} + \hat{P}_{54} - \hat{P}_{63} + \hat{P}_{72} - \hat{P}_{81}\right) . \qquad (2.758)$$

Operators involving $\hat{\sigma}_1$ and $\hat{1}$ only are

$$\hat{\sigma}_1(1) \otimes \hat{1}(2) \otimes \hat{1}(3)$$
$$= \hat{P}_{15} + \hat{P}_{26} + \hat{P}_{37} + \hat{P}_{48} + \hat{P}_{51} + \hat{P}_{62} + \hat{P}_{73} + \hat{P}_{84} , \qquad (2.759)$$

$$\hat{\sigma}_1(1) \otimes \hat{\sigma}_1(2) \otimes \hat{1}(3)$$
$$= \hat{P}_{17} + \hat{P}_{28} + \hat{P}_{35} + \hat{P}_{46} + \hat{P}_{53} + \hat{P}_{64} + \hat{P}_{71} + \hat{P}_{82} , \qquad (2.760)$$

from which those replacing $\hat{1}$ by $\hat{\sigma}_3$ differ only by sign:

$$\hat{\sigma}_1(1) \otimes \hat{\sigma}_3(2) \otimes \hat{\sigma}_3(3)$$
$$= \hat{P}_{15} - \hat{P}_{26} - \hat{P}_{37} + \hat{P}_{48} + \hat{P}_{51} - \hat{P}_{62} - \hat{P}_{73} + \hat{P}_{84} , \qquad (2.761)$$

$$\hat{\sigma}_1(1) \otimes \hat{\sigma}_1(2) \otimes \hat{\sigma}_3(3)$$
$$= -\hat{P}_{17} + \hat{P}_{28} - \hat{P}_{35} + \hat{P}_{46} - \hat{P}_{53} + \hat{P}_{64} - \hat{P}_{71} + \hat{P}_{82} . \qquad (2.762)$$

Operator products involving $\hat{\sigma}_2$ and $\hat{\sigma}_3$ are

$$\hat{\sigma}_2(1) \otimes \hat{\sigma}_3(2) \otimes \hat{\sigma}_3(3)$$
$$= i\left(\hat{P}_{15} + \hat{P}_{48} + \hat{P}_{62} + \hat{P}_{73} - \hat{P}_{51} - \hat{P}_{84} - \hat{P}_{26} - \hat{P}_{37}\right) , \qquad (2.763)$$

$$\hat{\sigma}_2(1) \otimes \hat{\sigma}_2(2) \otimes \hat{\sigma}_3(3)$$
$$= \hat{P}_{17} + \hat{P}_{46} + \hat{P}_{64} + \hat{P}_{71} - \hat{P}_{28} - \hat{P}_{35} - \hat{P}_{53} - \hat{P}_{82} , \qquad (2.764)$$

and products involving $\hat{\sigma}_2$ and $\hat{1}$ are

$$\hat{\sigma}_2(1) \otimes \hat{1}(2) \otimes \hat{1}(3)$$
$$= i\left(\hat{P}_{15} + \hat{P}_{26} + \hat{P}_{37} + \hat{P}_{48} - \hat{P}_{51} - \hat{P}_{62} - \hat{P}_{73} - \hat{P}_{84}\right) , \qquad (2.765)$$

$$\hat{\sigma}_2(1) \otimes \hat{\sigma}_2(2) \otimes \hat{1}(3)$$
$$= \hat{P}_{35} + \hat{P}_{46} + \hat{P}_{53} + \hat{P}_{64} - \hat{P}_{28} - \hat{P}_{82} - \hat{P}_{17} - \hat{P}_{71} . \qquad (2.766)$$

Finally, products of 3 different operators are

$$\hat{\sigma}_1(1) \otimes \hat{\sigma}_2(2) \otimes \hat{\sigma}_3(3)$$
$$= i\left(\hat{P}_{28} + \hat{P}_{35} + \hat{P}_{64} + \hat{P}_{71} - \hat{P}_{82} - \hat{P}_{53} - \hat{P}_{46} - \hat{P}_{17}\right) , \qquad (2.767)$$

$$\hat{\sigma}_2(1) \otimes \hat{\sigma}_3(2) \otimes \hat{1}(3)$$
$$= i\left(\hat{P}_{37} + \hat{P}_{51} + \hat{P}_{48} + \hat{P}_{62} - \hat{P}_{73} - \hat{P}_{15} - \hat{P}_{84} - \hat{P}_{62}\right) , \qquad (2.768)$$

$$\hat{\sigma}_1(1) \otimes \hat{\sigma}_3(2) \otimes \hat{1}(3)$$
$$= \hat{P}_{37} + \hat{P}_{73} + \hat{P}_{48} + \hat{P}_{84} - \hat{P}_{15} - \hat{P}_{51} - \hat{P}_{26} - \hat{P}_{62} \ , \tag{2.769}$$

$$\hat{\sigma}_1(1) \otimes \hat{\sigma}_2(2) \otimes \hat{1}(3)$$
$$= \mathrm{i} \left( \hat{P}_{17} + \hat{P}_{28} + \hat{P}_{53} + \hat{P}_{64} - \hat{P}_{71} - \hat{P}_{82} - \hat{P}_{35} - \hat{P}_{46} \right) \ . \tag{2.770}$$

### 2.5.2 Hamilton Model in $SU(n_1) \otimes SU(n_2) \otimes SU(n_3)$

The representation of a Hamilton operator thus reads, generalizing the result of (2.527), (2.550):

$$
\hat{H} = \frac{1}{n_1 n_2 n_3} \mathrm{Tr} \left\{ \hat{H} \right\} \hat{1} +
$$
$$
\frac{1}{2n_2 n_3} \sum_{j=1}^{s_1} \mathcal{H}_j(1) \left[ \hat{\lambda}_j(1) \otimes \hat{1}(2) \otimes \hat{1}(3) \right] +
$$
$$
\frac{1}{2n_1 n_3} \sum_{k=1}^{s_2} \mathcal{H}_k(2) \left[ \hat{1}(1) \otimes \hat{\lambda}_k(2) \otimes \hat{1}(3) \right] +
$$
$$
\frac{1}{2n_1 n_2} \sum_{l=1}^{s_3} \mathcal{H}_l(3) \left[ \hat{1}(1) \otimes \hat{1}(2) \otimes \hat{\lambda}_l(3) \right] +
$$
$$
\frac{1}{4n_3} \sum_{j=1}^{s_1} \sum_{k=1}^{s_2} \mathcal{H}_{jk}(1,2) \left[ \hat{\lambda}_j(1) \otimes \hat{\lambda}_k(2) \otimes \hat{1}(3) \right] +
$$
$$
\frac{1}{4n_2} \sum_{j=1}^{s_1} \sum_{l=1}^{s_3} \mathcal{H}_{jl}(1,3) \left[ \hat{\lambda}_j(1) \otimes \hat{1}(2) \otimes \hat{\lambda}_l(3) \right] +
$$
$$
\frac{1}{4n_1} \sum_{k=1}^{s_2} \sum_{l=1}^{s_3} \mathcal{H}_{kl}(2,3) \left[ \hat{1}(1) \otimes \hat{\lambda}_k(2) \otimes \hat{\lambda}_l(3) \right] \ . \tag{2.771}
$$

Here we have assumed that $\hat{H}$ does not contain 3-node (3-particle) interactions, i.e. $\mathcal{H}_{jkl} = 0$.

### 2.5.3 Density Operator

In Sect. 2.4 systems composed of 2 subsystems have been considered. This section extends to *3-node systems*. The respective density operator, coherence vectors and correlation functions are considered. Applications (*teleportation, Greenberger-Horne-Zeilinger states*) will be studied.

**2.5.3.1 Representation in $SU(n_1) \otimes SU(n_2) \otimes SU(n_3)$.** The representation of the density operator (2.599) has to be generalized so that one obtains the operator

$$
\begin{aligned}
\hat{\rho} = {} & \frac{1}{n_1 n_2 n_3} \hat{1} + \frac{1}{2 n_2 n_3} \sum_{j=1}^{s_1} \lambda_j(1) \left[ \hat{\lambda}_j(1) \otimes \hat{1}(2) \otimes \hat{1}(3) \right] \\
& + \frac{1}{2 n_1 n_3} \sum_{k=1}^{s_2} \lambda_k(2) \left[ \hat{1}(1) \otimes \lambda_k(2) \otimes \hat{1}(3) \right] \\
& + \frac{1}{2 n_1 n_2} \sum_{l=1}^{s_3} \lambda_l(3) \left[ \hat{1}(1) \otimes \hat{1}(2) \otimes \lambda_l(3) \right] \\
& + \frac{1}{4 n_3} \sum_{j,k=1}^{s_1, s_2} K_{jk}(1,2) \left[ \hat{\lambda}_j(1) \otimes \hat{\lambda}_k(2) \otimes \hat{1}(3) \right] \\
& + \frac{1}{4 n_2} \sum_{j,l=1}^{s_1, s_3} K_{jl}(1,3) \left[ \hat{\lambda}_j(1) \otimes \hat{1}(2) \otimes \hat{\lambda}_l(3) \right] \\
& + \frac{1}{4 n_1} \sum_{k,l=1}^{s_2, s_3} K_{kl}(2,3) \left[ \hat{1}(1) \otimes \hat{\lambda}_k(2) \otimes \hat{\lambda}_l(3) \right] \\
& + \frac{1}{8} \sum_{j,k,l=1}^{s_1, s_2, s_3} K_{jkl}(1,2,3) \left[ \hat{\lambda}_j(1) \otimes \hat{\lambda}_k(2) \otimes \hat{\lambda}_l(3) \right] ,
\end{aligned}
\tag{2.772}
$$

which contains matrix elements $K_{jkl}(1,2,3)$. These matrix elements define a correlation tensor of third order:

$$
K_{jkl}(1,2,3) = \text{Tr} \left\{ \hat{\rho} \hat{\lambda}_j(1) \otimes \hat{\lambda}_k(2) \otimes \hat{\lambda}_l(3) \right\} .
\tag{2.773}
$$

Contrary to the representation of $\hat{H}$, these 3rd-order terms will not, in general, be equal to zero. In addition, there are now three 2-node correlation matrices $K_{ij}(\nu\nu')$, $\nu < \nu'$; 3 (local) coherence vectors $\boldsymbol{\lambda}(\nu)$ with $\lambda_j(\nu)$, $\nu = 1,2,3$. The reduced density operators are

$$
\hat{R}(1) = \text{Tr}_{2,3} \{ \hat{\rho} \} = \frac{1}{n_1} \hat{1}(1) + \frac{1}{2} \sum_{j=1}^{s_1} \lambda_j(1) \hat{\lambda}_j(1) ,
\tag{2.774}
$$

and $\hat{R}(2)$, $\hat{R}(3)$ correspondingly. The 3-node correlation tensor proper, $M_{jkl}$, is defined by (cf. (2.606))

$$
\begin{aligned}
K_{jkl}(1,2,3) = {} & \lambda_j(1) \lambda_k(2) \lambda_l(3) + \lambda_j(1) K_{kl}(2,3) + \lambda_k(2) K_{jl}(1,3) + \\
& \lambda_l(3) K_{jk}(1,2) + M_{jkl}(1,2,3) .
\end{aligned}
\tag{2.775}
$$

The representation (2.772) of $\hat{\rho}$ may be considered a kind of *cluster expansion*. We see that the "non-classical" nature of such states becomes more and more dominant with increasing number $N$ of subsystems (compare Fig. 2.40): large systems are not necessarily classical at all. On the other hand, for a closed system the basis of representation is completely ambiguous: one may always find a representation in which the respective state appears as a simple (pure) state vector. Non-classicality has no meaning without an outside (classical) world.

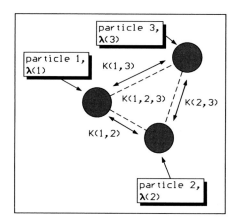

**Fig. 2.40.** State description of a 3-node system

**2.5.3.2 Sum Rule.** In complete analogy with (2.682) for $SU(n) \otimes SU(n)$, we can calculate $C(n,2)$ for $SU(n) \otimes SU(n) \otimes SU(n)$ based on $\hat{\rho}$ according to (2.772). Again, only the quadratic terms survive the trace-operation:

$$C(n,2) = \text{Tr}\left\{\hat{\rho}^2\right\}$$

$$= \frac{1}{n^6}\text{Tr}\left\{\hat{1}\right\} +$$

$$\frac{1}{4n^4}\sum_{j,j'=1}^{s}\lambda_j(1)\lambda_{j'}(1)\text{Tr}\left\{\hat{\lambda}_j(1)\hat{\lambda}_{j'}(1)\otimes\hat{1}\otimes\hat{1}\right\} +$$

$$\frac{1}{4n^4}\sum_{k,k'=1}^{s}\lambda_k(2)\lambda_{k'}(2)\text{Tr}\left\{\hat{1}\otimes\hat{\lambda}_k(2)\hat{\lambda}_{k'}(2)\otimes\hat{1}\right\} +$$

$$\frac{1}{4n^4}\sum_{l,l'=1}^{s}\lambda_l(3)\lambda_{l'}(3)\text{Tr}\left\{\hat{1}\otimes\hat{1}\otimes\hat{\lambda}_l(3)\hat{\lambda}_{l'}(3)\right\} +$$

$$\frac{1}{16n^2}\sum_{j,j',k,k'=1}^{s}\left[K_{jk}(1,2)K_{j'k'}(1,2)\right.$$

$$\left.\text{Tr}\left\{\hat{\lambda}_j(1)\hat{\lambda}_{j'}(1)\otimes\hat{\lambda}_k(2)\hat{\lambda}_{k'}(2)\otimes\hat{1}\right\}\right] +$$

$$\frac{1}{16n^2}\sum_{j,j',l,l'=1}^{s}\left[K_{jl}(1,3)K_{j'l'}(1,3)\right.$$

$$\left.\text{Tr}\left\{\hat{\lambda}_j(1)\hat{\lambda}_{j'}(1)\otimes\hat{1}\otimes\hat{\lambda}_l(3)\hat{\lambda}_{l'}(3)\right\}\right] +$$

$$\frac{1}{16n^2}\sum_{k,k',l,l'=1}^{s}\left[K_{kl}(2,3)K_{k'l'}(2,3)\right.$$

$$\mathrm{Tr}\left\{\hat{1}\otimes\hat{\lambda}_k(2)\hat{\lambda}_{k'}(2)\otimes\hat{\lambda}_l(3)\hat{\lambda}_{l'}(3)\right\}\Bigg]+$$

$$\frac{1}{64}\sum_{j,j',k,k',l,l'=1}^{s}\Bigg[K_{jkl}(1,2,3)K_{j'k'l'}(1,2,3)$$

$$\mathrm{Tr}\left\{\hat{\lambda}_j(1)\hat{\lambda}_{j'}(1)\otimes\otimes\hat{\lambda}_k(2)\hat{\lambda}_{k'}(2)\otimes\hat{\lambda}_l(3)\hat{\lambda}_{l'}(3)\right\}\Bigg].\qquad(2.776)$$

Observing that

$$\mathrm{Tr}\left\{\hat{1}\right\}=n^3\;,$$

$$\mathrm{Tr}\left\{\hat{\lambda}_j(1)\hat{\lambda}_{j'}(1)\otimes\hat{1}\otimes\hat{1}\right\}=2\delta_{jj'}n^2\;,\quad\mathrm{etc.}\;,$$

$$\mathrm{Tr}\left\{\hat{\lambda}_j(1)\hat{\lambda}_{j'}(1)\otimes\hat{\lambda}_k(2)\hat{\lambda}_{k'}(2)\otimes\hat{1}\right\}=4\delta_{jj'}\delta_{kk'}n\;,\quad\mathrm{etc.}\;,$$

$$\mathrm{Tr}\left\{\hat{\lambda}_j(1)\hat{\lambda}_{j'}(1)\otimes\hat{\lambda}_k(2)\hat{\lambda}_{k'}(2)\otimes\hat{\lambda}_l(3)\hat{\lambda}_{l'}(3)\right\}=8\delta_{jj'}\delta_{kk'}\delta_{ll'}\;,$$

$$(2.777)$$

we find based on the *sum convention* (i.e. summation with respect to equal indices is implied)

$$\boxed{\begin{aligned}C(n,2)&=\mathrm{Tr}\left\{\hat{\rho}^2\right\}\\&=\frac{1}{n^3}+\\&\quad\frac{1}{2n^2}\left[\lambda_j(1)\lambda_j(1)+\lambda_k(2)\lambda_k(2)+\lambda_l(3)\lambda_l(3)\right]+\\&\quad\frac{1}{4n}[K_{jk}(1,2)K_{jk}(1,2)+K_{jl}(1,3)K_{jl}(1,3)+\\&\qquad\quad K_{kl}(2,3)K_{kl}(2,3)]+\\&\quad\frac{1}{8}K_{jkl}(1,2,3)K_{jkl}(1,2,3)\;.\end{aligned}}\qquad(2.778)$$

$C(n,q)$ represent constants of motion (for unitary evolution).

**2.5.3.3 Model States.** The calculation of the various expectation values $K_{jk}$, $K_{jkl}$, etc. can be simplified in the product basis. We specialize in $SU(2)\otimes SU(2)\otimes SU(2)$. The local states are $|\pm\nu\rangle$, the product states are the 8 states $|m\rangle$, $m=1,2,\ldots,8$ (see Sect. 2.5.1):

$$\begin{aligned}|---\rangle&=|1\rangle\;,\\|--+\rangle&=|2\rangle\;,\\|-+-\rangle&=|3\rangle\;,\\|-++\rangle&=|4\rangle\;,\\|+--\rangle&=|5\rangle\;,\\|+-+\rangle&=|6\rangle\;,\\|++-\rangle&=|7\rangle\;,\\|+++\rangle&=|8\rangle\;.\end{aligned}\qquad(2.779)$$

They form a complete basis in $SU(8)$.

For a density operator given in terms of the projection operators

$$\hat{P}_{mn} = |m\rangle \langle n| \tag{2.780}$$

it is convenient to have the generator products in the same representation (see Sect. 2.5.1). The calculation of any expectation value can then be based on

$$\mathrm{Tr}\left\{\hat{P}_{mn}\hat{P}_{jl}\right\} = \delta_{ml}\delta_{nj} . \tag{2.781}$$

*Example 2.5.1 (Frenkel State).* Collective excited states in chains of (interacting) 2-level systems can be described by *Frenkel-exciton-type superposition states* like (cf. [103])

$$|\psi\rangle = \frac{1}{\sqrt{3}} \left(|+ - -\rangle + |- + -\rangle + |- - +\rangle\right) . \tag{2.782}$$

This is an extension of the 2-node state

$$|\psi\rangle = \frac{1}{\sqrt{2}} \left(|+-\rangle + |-+\rangle\right) . \tag{2.783}$$

In the above notation this state reads

$$|\psi\rangle = \frac{1}{\sqrt{3}} \left(|5\rangle + |3\rangle + |2\rangle\right) . \tag{2.784}$$

Based on Sect. 2.5.1 one readily shows that the only non-zero expectation values in the $SU(2) \otimes SU(2) \otimes SU(2)$ representation are

$$\lambda_3(\nu)$$
$$= \left\langle \psi \left| -\hat{P}_{11} - \hat{P}_{22} - \hat{P}_{33} - \hat{P}_{44} + \hat{P}_{55} + \hat{P}_{66} + \hat{P}_{77} + \hat{P}_{88} \right| \psi \right\rangle$$
$$= -\frac{1}{3} \ (\nu = 1, 2, 3) , \tag{2.785}$$

$$K_{22}(\mu, \nu)$$
$$= \left\langle \psi \left| \hat{P}_{35} + \hat{P}_{46} + \hat{P}_{53} + \hat{P}_{64} - \hat{P}_{28} - \hat{P}_{82} - \hat{P}_{17} - \hat{P}_{71} \right| \psi \right\rangle$$
$$= \frac{2}{3} \ (\mu < \nu = 1, 2, 3) , \tag{2.786}$$

$$K_{11}(\mu, \nu) = \frac{2}{3} , \tag{2.787}$$

$$K_{33}(\mu, \nu) = -\frac{1}{3} , \tag{2.788}$$

$$K_{113}(1, 2, 3) = K_{131}(1, 2, 3) = K_{311}(1, 2, 3) = -\frac{2}{3} , \tag{2.789}$$

$$K_{223}(1,2,3) = K_{232}(1,2,3) = K_{322}(1,2,3) = -\frac{2}{3} , \qquad (2.790)$$

$$K_{333}(1,2,3) = 1 . \qquad (2.791)$$

The correlation tensors are

$$
\begin{aligned}
M_{11}(\mu,\nu) &= K_{11}(\mu,\nu) = \frac{2}{3} , \\
M_{22}(\mu,\nu) &= K_{22}(\mu,\nu) = \frac{2}{3} , \\
M_{33}(\mu,\nu) &= -\frac{4}{9} , \\
M_{113}(1,2,3) &= K_{113}(1,2,3) - \lambda_3(3)K_{11}(1,2) = -\frac{4}{9} , \\
M_{223}(1,2,3) &= K_{223}(1,2,3) - \lambda_3(3)K_{22}(1,2) = -\frac{4}{9} , \\
M_{333}(1,2,3) &= \frac{16}{27} .
\end{aligned}
\qquad (2.792)
$$

*Example 2.5.2.* As a second example consider the pure state

$$|\psi(1,2,3)\rangle = |\psi(2,3)\rangle^- |\chi(1)\rangle \qquad (2.793)$$

with

$$|\psi(2,3)\rangle^\pm = \frac{1}{\sqrt{2}} (|+-\rangle \pm |-+\rangle) \qquad (2.794)$$

and

$$|\chi(1)\rangle = a \left( |+(1)\rangle + b \, |-(1)\rangle \right) , \quad a^2 + b^2 = 1 . \qquad (2.795)$$

Rewriting

$$
\begin{aligned}
|\psi\rangle &= \frac{a}{\sqrt{2}} |++-\rangle - \frac{a}{\sqrt{2}} |+-+\rangle + \frac{b}{\sqrt{2}} |-+-\rangle - \frac{b}{\sqrt{2}} |--+\rangle \\
&= \frac{a}{\sqrt{2}} |7\rangle - \frac{a}{\sqrt{2}} |6\rangle + \frac{b}{\sqrt{2}} |3\rangle - \frac{b}{\sqrt{2}} |2\rangle \\
&= \sum_m C_m \, |m\rangle ,
\end{aligned}
\qquad (2.796)
$$

we find

$$
\begin{aligned}
K_{333} &= \left\langle \psi \left| -\hat{P}_{11} + \hat{P}_{22} + \hat{P}_{33} - \hat{P}_{44} + \hat{P}_{55} - \hat{P}_{66} - \hat{P}_{77} + \hat{P}_{88} \right| \psi \right\rangle \\
&= b^2 - a^2 ,
\end{aligned}
\qquad (2.797)
$$

$$K_{111} = 0 ,$$

etc.

as well as

$$K_{11}(2,3) = K_{22}(2,3) = K_{33}(2,3) = -1 , \quad \mathsf{K}(1,3) = \mathsf{K}(1,2) = 0 \quad (2.798)$$

and

$$\lambda_i(2) = 0 , \quad \lambda_i(3) = 0 , \quad \lambda_3(1) = a^2 - b^2 , \quad \lambda_{1,2}(1) = 0 , \quad (2.799)$$

implying

$$M_{333} = K_{333} - \lambda_3(1)K_{33}(2,3) = 0 , \quad (2.800)$$

i.e. as can be seen already from (2.793), there are only 2-node correlations within the (2,3)-pair.

In the following, two applications of the above 3-node formalism are studied.

**2.5.3.4 Teleportation.** (See [13].) Consider first the scenario sketched in Fig. 2.41. At $t = t_0$ Alice (A) has a photon 1 in a definite polarization state (possibly not known to her), while the source creates an entangled photon pair 2,3 in a singlet state. This 3-particle state is described (in polarization subspace) by the state vector discussed in the preceding section (cf. (2.793)):

$$|\psi(1,2,3)\rangle_{t_0} = |\psi(2,3)\rangle^- |\chi(1)\rangle . \quad (2.801)$$

At $t = t_1$ photon 2 reaches A, photon 3 reaches B (details of the dynamics in real space are not considered). Then, at $t = t_2$ Alice performs a 2-photon measurement on her photon pair with respect to the 4 alternative states (measurement basis):

$$|\psi(1,2)\rangle^{\mp} = \frac{1}{\sqrt{2}} \left( |\uparrow\downarrow\rangle \mp |\downarrow\uparrow\rangle \right) ,$$

$$|\phi(1,2)\rangle^{\mp} = \frac{1}{\sqrt{2}} \left( |\uparrow\uparrow\rangle \mp |\downarrow\downarrow\rangle \right) . \quad (2.802)$$

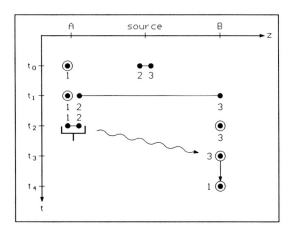

**Fig. 2.41.** Teleportation. State of particle 1 at A is teleported to B

(This basis is interpreted as the eigenstates of some fictitious *Bell operator*; with respect to the original basis these states are entangled.) In terms of these states $|\psi(1,2,3)\rangle$ can be rewritten as

$$2\,|\psi(1,2,3)\rangle = |\psi(1,2)\rangle^- |\phi(3)\rangle_1 + |\psi(1,2)\rangle^+ |\phi(3)\rangle_2 +$$
$$|\phi(1,2)\rangle^- |\phi(3)\rangle_3 + |\phi(1,2)\rangle^+ |\phi(3)\rangle_4 \qquad (2.803)$$

with

$$|\phi(3)\rangle_1 = -a\,|\!\uparrow(3)\rangle - b\,|\!\downarrow(3)\rangle \; ,$$
$$|\phi(3)\rangle_2 = -a\,|\!\uparrow(3)\rangle + b\,|\!\downarrow(3)\rangle \; ,$$
$$|\phi(3)\rangle_3 = +b\,|\!\uparrow(3)\rangle + a\,|\!\downarrow(3)\rangle \; , \qquad (2.804)$$
$$|\phi(3)\rangle_4 = -b\,|\!\uparrow(3)\rangle + a\,|\!\downarrow(3)\rangle \; .$$

In this basis $|\psi\rangle$ is a 3-node entangled state. She reports the outcome of the experiment to Bob $(t = t_3)$. As we see from (2.804), Bob's photon, due to the non-local correlation, has changed to a state $|\phi(3)\rangle_i$, with $i$ denoting one of the 4 possible outcomes of Alice's experiment. This state is related by a unitary transformation to Alice's original photon state 1:

$$|\psi(1,2)\rangle^- : \quad |\phi(3)\rangle_1 = -\begin{pmatrix} a \\ b \end{pmatrix} = -|\chi(1)\rangle$$

$$|\psi(1,2)\rangle^+ : \quad |\phi(3)\rangle_2 = -\begin{pmatrix} 1 & 0 \\ 0 & -1 \end{pmatrix} |\chi(1)\rangle \; ,$$

$$\qquad\qquad\qquad\qquad\qquad\qquad\qquad\qquad\qquad (2.805)$$

$$|\phi(1,2)\rangle^- : \quad |\phi(3)\rangle_3 = \begin{pmatrix} 0 & 1 \\ 1 & 0 \end{pmatrix} |\chi(1)\rangle \; ,$$

$$|\phi(1,2)\rangle^+ : \quad |\phi(3)\rangle_4 = \begin{pmatrix} 0 & -1 \\ 1 & 0 \end{pmatrix} |\chi(1)\rangle \; .$$

Performing the reverse transformation, Bob has "received" photon 1 without measuring it and without Alice actually transmitting it! This kind of *information transmission* has been termed *swapping*.

**2.5.3.5 Greenberger-Horne-Zeilinger States.** Consider again subsystems with two possible states. Then the special 3-particle coherent state defined by

$$|\psi\rangle = \frac{1}{\sqrt{2}}\left(|+++\rangle - |---\rangle\right) \qquad (2.806)$$

represents a so-called *Greenberger-Horne-Zeilinger state (GHZ state)*.

Using the notation of Sect. 2.5.3.3, we find

$$|\psi\rangle = \frac{1}{\sqrt{2}}|8\rangle - \frac{1}{\sqrt{2}}|1\rangle \qquad (2.807)$$

and

$$\hat{\rho} = |\psi\rangle\langle\psi| = \frac{1}{2}\left(\hat{P}_{88} + \hat{P}_{11} - \hat{P}_{81} - \hat{P}_{18}\right) . \tag{2.808}$$

One easily convinces oneself (cf. Sect. 2.3.6.2) that

$$K_{111} = -1 , \quad K_{122} = K_{212} = K_{221} = 1 , \quad K_{222} = K_{333} = 0 ,$$
$$K_{113} = K_{133} = 0 , \quad \text{etc.} , \tag{2.809}$$

$$K_{33}(1,2) = K_{33}(2,3) = K_{33}(1,3) = 1 , \tag{2.810}$$

$$\lambda_i(1) = \lambda_i(2) = \lambda_i(3) = 0 , \quad i = 1,2,3 , \tag{2.811}$$

and

$$M_{jkl} = K_{jkl} . \tag{2.812}$$

This 3-particle coherent state contradicts objective local interpretation not only in terms of ensemble measurements as in the case of 2-particle correlations (EPR correlations). Local reality, again, means that for any subsystem $\nu$ all the spin components exist: $m_i(\nu) = \pm 1$, $i = 1,2,3$, $\nu = 1,2,3$. Now, with the expectation values $K_{111}$, $K_{122}$, $K_{212}$, $K_{211}$ actually being eigenvalues of the respective density operator, the individual state (not only the ensemble) must obey the relations:

$$1 = K_{122} = m_1(1)m_2(2)m_2(3) ,$$
$$1 = K_{212} = m_2(1)m_1(1)m_2(3) , \tag{2.813}$$
$$1 = K_{221} = m_2(1)m_2(2)m_1(3) ,$$

where $m_i(\nu) = \pm 1$, $i = 1,2,3$, $\nu = 1,2,3$ according to local reality. On the other hand

$$1 = K_{122}K_{212}K_{221} = m_1(1)m_1(2)m_1(3) = K_{111} \tag{2.814}$$

as $m_2^2(\nu) = 1$. But this is a contradiction to quantum mechanics: $K_{111}$ should be $-1$. This means that the complete set $m_i(\nu)$ cannot be an element of reality.

### 2.5.4 Measurement Projections

Let us consider a measurement on subsystem 3 with respect to the observable $\hat{A}(3)$ with

$$\hat{A}(3)|a_i(3)\rangle = a_i(3)|a_i(3)\rangle \quad (i = 1,2,\dots,n) . \tag{2.815}$$

The measurement basis is, like in Sect. 2.4.5.1, assumed to coincide with the basis defining the projector representation of the generators $\hat{\lambda}(3)$, i.e.

$$|a_i(3)\rangle = |i(3)\rangle . \tag{2.816}$$

If the total system is initially in state $\hat{\rho}$, the probability to find the result $a_m(3)$ is then given by

$$p_m(3) = \text{Tr}\left\{\hat{P}_{mm}(3)\hat{\rho}\right\} , \tag{2.817}$$

and after measurement the system is in the state

$$\hat{\rho}' = \frac{1}{p_m(3)}\hat{P}_{mm}(3)\hat{\rho}\hat{P}_{mm}(3) . \tag{2.818}$$

Applying the analogue of (2.696) to the density matrix (2.772) with $n_1 = n_2 = n_3 = n$, we obtain

$$
\begin{aligned}
p_m(3)&\hat{\rho}' \\
= &\frac{1}{n^3}\hat{1}(1) \otimes \hat{1}(2) \otimes \hat{P}_{mm}(3) + \\
&\frac{1}{2n^2}\sum_{j=1}^{s}\lambda_j(1)\left[\hat{\lambda}_j(1) \otimes \hat{1}(2) \otimes \hat{P}_{mm}(3)\right] + \\
&\frac{1}{2n^2}\sum_{k=1}^{s}\lambda_k(2)\left[\hat{1}(1) \otimes \hat{\lambda}_k(2) \otimes \hat{P}_{mm}(3)\right] + \\
&\frac{1}{2n^2}\sum_{l=1}^{n-1}w_l(3)w_l^{(m)}\left[\hat{1}(1) \otimes \hat{1}(2) \otimes \hat{P}_{mm}(3)\right] + \\
&\frac{1}{4n}\sum_{j,k=1}^{s}K_{jk}(1,2)\left[\hat{\lambda}_j(1) \otimes \hat{\lambda}_k(2) \otimes \hat{P}_{mm}(3)\right] + \\
&\frac{1}{4n}\sum_{j=1}^{s}\sum_{l=1}^{n-1}K_{js'+l}(1,3)w_l^{(m)}\left[\hat{\lambda}_j(1) \otimes \hat{1}(2) \otimes \hat{P}_{mm}(3)\right] + \\
&\frac{1}{4n}\sum_{k=1}^{s}\sum_{l=1}^{n-1}K_{ks'+l}(2,3)w_l^{(m)}\left[\hat{1}(1) \otimes \hat{\lambda}_k(2) \otimes \hat{P}_{mm}(3)\right] + \\
&\frac{1}{8}\sum_{j,k=1}^{s}\sum_{l=1}^{n-1}K_{jks'+l}(1,2,3)w_l^{(m)}\left[\hat{\lambda}_j(1) \otimes \hat{\lambda}_k(2) \otimes \hat{P}_{mm}(3)\right] . \tag{2.819}
\end{aligned}
$$

As before, $s' = n^2 - n$. Replacing $\hat{P}_{mm}(3)$ by its representation (cf. (2.695))

$$\hat{P}_{mm}(3) = \frac{1}{n}\hat{1}(3) + \frac{1}{2}\sum_{l=1}^{n-1}w_l^{(j)}\hat{w}_l(3) , \tag{2.820}$$

and taking into account that (cf. (2.698))

$$p_m(3) = \frac{1}{n} + \frac{1}{2}\sum_{l=1}^{n-1}w_l(3)w_l^{(m)} , \tag{2.821}$$

we finally obtain, assuming $p_m(3) \neq 0$:

$$\hat{\rho}' = \frac{1}{n^3}\hat{1}(1) \otimes \hat{1}(2) \otimes \hat{1}(3) +$$

$$\frac{1}{2n^2}\sum_{l'=1}^{n-1}w_{l'}^{(m)}\left[\hat{1}(1) \otimes \hat{1}(2) \otimes \hat{w}_{l'}(3)\right] +$$

$$\frac{1}{2n^2p_m(3)}\sum_{j=1}^{s}K_j^{(1)}(1,3)\left[\hat{\lambda}_j(1) \otimes \hat{1}(2) \otimes \hat{1}(3)\right] +$$

$$\frac{1}{2n^2p_m(3)}\sum_{k=1}^{s}K_k^{(1)}(2,3)\left[\hat{1}(1) \otimes \hat{\lambda}_k(2) \otimes \hat{1}(3)\right] +$$

$$\frac{1}{4np_m(3)}\sum_{j=1}^{s}\sum_{l=1}^{n-1}K_j^{(1)}(1,3)w_l^{(m)}\left[\hat{\lambda}_j(1) \otimes \hat{1}(2) \otimes \hat{w}_l(3)\right] +$$

$$\frac{1}{4np_m(3)}\sum_{k=1}^{s}\sum_{l=1}^{n-1}K_k^{(1)}(2,3)w_l^{(m)}\left[\hat{1}(1) \otimes \hat{\lambda}_k(2) \otimes \hat{w}_l(3)\right] +$$

$$\frac{1}{4np_m(3)}\sum_{j,k=1}^{s}K_{jk}^{(2)}(1,2,3)\left[\hat{\lambda}_j(1) \otimes \hat{\lambda}_k(2) \otimes \hat{1}(3)\right] +$$

$$\frac{1}{8p_m(3)}\sum_{j,k=1}^{s}\sum_{l=1}^{n-1}K_{jkl}^{(2)}(1,2,3)w_l^{(m)}\left[\hat{\lambda}_j(1) \otimes \hat{\lambda}_k(2) \otimes \hat{w}_l(3)\right] \quad (2.822)$$

with

$$K_j^{(1)}(1,3) = \frac{1}{n}\lambda_j(1) + \frac{1}{2}\sum_{l'=1}^{n-1}K_{js'+l'}(1,3)w_{l'}^{(m)} , \quad (2.823)$$

$$K_k^{(1)}(2,3) = \frac{1}{n}\lambda_k(2) + \frac{1}{2}\sum_{l'=1}^{n-1}K_{ks'+l'}(2,3)w_{l'}^{(m)} , \quad (2.824)$$

$$K_{jk}^{(2)}(1,2,3) = \frac{1}{n}K_{jk}(1,2) + \frac{1}{2}\sum_{l'=1}^{n-1}K_{jks'+l'}(1,2,3)w_{l'}^{(m)} . \quad (2.825)$$

Comparing (2.822) with $\hat{\rho}'$ in the form (2.772), we can identify the coherence vectors and correlation tensors after measurement by

$$\lambda'_{s'+l}(3) = w_l^{(m)} \, ,$$

$$K'_{js'+l'}(1,3) = \frac{w_l^{(m)}}{p_m(3)} \left[ \frac{1}{n}\lambda_j(1) + \frac{1}{2}\sum_{l'=1}^{n-1} K_{js'+l'}(1,3)w_{l'}^{(m)} \right] \, ,$$

$$K'_{ks'+l'}(2,3) = \frac{w_l^{(m)}}{p_m(3)} \left[ \frac{1}{n}\lambda_k(2) + \frac{1}{2}\sum_{l'=1}^{n-1} K_{ks'+l'}(2,3)w_{l'}^{(m)} \right] \, ,$$

$$K'_{jks'+l'}(1,2,3) = \frac{w_l^{(m)}}{p_m(3)} \left[ \frac{1}{n}K_{jk}(1,2) + \frac{1}{2}\sum_{l'=1}^{n-1} K_{jks'+l'}w_{l'}^{(m)} \right] \, ,$$

$$\lambda'_j(1) = \frac{1}{p_m(3)} \left[ \frac{1}{n}\lambda_j(1) + \frac{1}{2}\sum_{l=1}^{n-1} K_{js'+l}(1,3)w_l^{(m)} \right] \, ,$$

$$\lambda'_k(2) = \frac{1}{p_m(3)} \left[ \frac{1}{n}\lambda_k(2) + \frac{1}{2}\sum_{l=1}^{n-1} K_{ks'+l}(2,3)w_l^{(m)} \right] \, ,$$

$$K'_{jk}(1,2) = \frac{1}{p_m(3)} \left[ \frac{1}{n}K_{jk}(1,2) + \frac{1}{2}\sum_{l=1}^{n-1} K_{jks'+l}w_l^{(m)} \right]$$

(2.826)

with

$$j,k = 1,2,\ldots,s \, , \quad l,l' = 1,2,\ldots,n-1 \, , \quad s' = n^2 - n \, . \tag{2.827}$$

These equations constitute a generalization of the result found for 2 nodes (compare (2.702)). There are two sets: the vectors (tensors) involving the measured node 3, and the vectors (tensors) not involving node 3 (the last three equations). The former connect to a tensor of lower rank, the latter to a tensor of the next higher rank. This systematic form is easily extended to 4 and more nodes.

Again, transforming to the correlation tensors proper,

$$M_{ij}(\mu,\nu) = K_{ij}(\mu,\nu) - \lambda_i(\mu)\lambda_j(\nu) \, , \tag{2.828}$$

$$\begin{aligned} M_{ijk}(\mu,\nu,\zeta) = {} & K_{ijk}(\mu,\nu,\zeta) - \lambda_i(\mu)\lambda_j(\nu)\lambda_k(\zeta) - \\ & \lambda_i(\mu)M_{jk}(\nu,\zeta) - \lambda_j(\nu)M_{ik}(\mu,\zeta) - \\ & \lambda_k(\zeta)M_{ij}(\mu,\nu) \, , \end{aligned} \tag{2.829}$$

the first set of equations reduces to (cf. (2.705))

$$\lambda'_{s'+l}(3) = w_l^{(m)} \, ,$$
$$M'_{ik}(1,3) = M'_{jk}(2,3) = 0 \, ,$$
$$M'_{ijk}(1,2,3) = 0 \, ,$$

(2.830)

and those not involving the measured node 3 read

$$\lambda_i'(1) = \lambda_i(1) + \frac{1}{2p_m(3)} \sum_{l=1}^{n-1} M_{is'+l}(1,3)w_l^{(m)} ,$$

$$\lambda_j'(2) = \lambda_j(2) + \frac{1}{2p_m(3)} \sum_{l=1}^{n-1} M_{js'+l}(2,3)w_l^{(m)} ,$$

$$M_{ij}'(1,2) = M_{ij}(1,2) + \frac{1}{2p_m(3)} \sum_{l=1}^{n-1} M_{ijs'+l}(1,2,3)w_l^{(m)} -$$

$$\frac{1}{[2p_m(3)]^2} \sum_{l,l'=1}^{n-1} M_{is'+l}(1,3)M_{js'+l'}(2,3)w_l^{(m)}w_{l'}^{(m)} . \tag{2.831}$$

Only the first set does not contain any memory of the initial state. The second set, describing the "indirect" effects of the subsystem measurement, still contains parts of the previous state, though in a new "blend". The index $m$ refers to the assumed measurement outcome of subsystem 3, $a_m(3)$.

The equation for $M_{ij}'(1,2)$ signals that a measurement on subsystem 3 can transfer 2-node entanglement between subsystems 1, 3 and 2, 3 to the pair 1, 2, even if the latter do not interact physically! Entanglement can thus be generated by local dissipation (measurements), see Fig. 2.42. In the same way, single-node coherence, expressed by $\lambda_i'(1)$, $\lambda_j'(2)$, can be generated via 2-node coherence in 1, 3 and 2, 3 of the initial state.

*Example 2.5.3.* We apply the projection formulae to the Frenkel state (cf. (2.782)). Suppose we perform a measurement on particle 3 and obtain the result

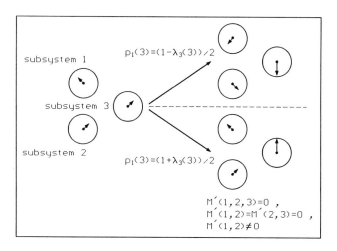

**Fig. 2.42.** Measurement of subsystem 3 in a 3-node network ($SU(2)$). After measurement only subsystem 3 is (in general) in a pure local state. The pair 1, 2 may be in a partly entangled state with local coherence vectors, length $< 1$

$$\lambda_3'(3) = w_1^{(1)} = -1 \tag{2.832}$$

(which occurs with probability $p_1(3) = 2/3$). Then

$$M_{ik}'(1,3) = M_{ik}'(2,3) = 0 , \tag{2.833}$$

$$M_{ijk}'(1,2,3) = 0 , \tag{2.834}$$

$$\lambda_i'(\nu) = \lambda_i(\nu) - \frac{1}{2p_1(3)} M_{i3}(\nu,3) \quad (\nu = 1,2) \tag{2.835}$$

so that

$$\lambda_i'(\nu) = 0 . \tag{2.836}$$

The measurement "transfers" coherence (entanglement) to the node pair not directly measured:

$$\begin{aligned} M_{ij}'(1,2) &= M_{ij}(1,2) - \frac{1}{2p_1(3)} M_{ij3}(1,2,3) - \\ &\quad \frac{1}{[2p_1(3)]^2} M_{i3}(1,3) M_{j3}(2,3) , \end{aligned} \tag{2.837}$$

$$M_{11}'(1,2) = M_{22}(1,2) = 1 , \quad M_{33}(1,2) = -1 . \tag{2.838}$$

## 2.6 *N*-Node States

Contents: $N$-node states, dipole–dipole interaction, quantum-dot array, Ising limit.

### 2.6.1 *N*-Node Hamilton Operator

For $N$ nodes of Hilbert space dimension $n$ each, the Hamilton operator

$$\hat{H} = \sum_\mu \hat{H}(\mu) + \sum_{\mu<\nu} \hat{H}(\mu,\nu) \tag{2.839}$$

has the representation

$$\begin{aligned} \hat{H} &= \frac{1}{n^N} \mathrm{Tr}\left\{\hat{H}\right\} \hat{1} + \frac{1}{2n^{N-1}} \sum_{\mu=1}^{N} \sum_{i=1}^{s} \mathcal{H}_i(\mu) \hat{\lambda}_i(\mu) + \\ &\quad \frac{1}{4n^{N-2}} \sum_{\mu<\nu}^{N} \sum_{i,j=1}^{s} \mathcal{H}_{ij}(\mu,\nu) \hat{\lambda}_i(\mu) \otimes \hat{\lambda}_j(\nu) \end{aligned} \tag{2.840}$$

with

$$\boxed{\begin{aligned}\mathcal{H}_i(\mu) &= \mathrm{Tr}\left\{\hat{H}\hat{\lambda}_i(\mu)\right\}\\ &= n^{N-1}\mathrm{Tr}_\mu\left\{\hat{H}(\mu)\hat{\lambda}_i(\mu)\right\} + \sum_{\nu\neq\mu}\mathrm{Tr}\left\{\hat{H}(\mu,\nu)\hat{\lambda}_i(\mu)\right\},\end{aligned}}\tag{2.841}$$

$$\boxed{\begin{aligned}\mathcal{H}_{i,j}(\mu,\nu) &= \mathrm{Tr}\left\{\hat{H}\hat{\lambda}_i(\mu)\otimes\hat{\lambda}_j(\nu)\right\}\\ &= n^{N-2}\mathrm{Tr}_\mu\mathrm{Tr}_\nu\left\{\hat{H}(\mu,\nu)\hat{\lambda}_i(\mu)\otimes\hat{\lambda}_j(\nu)\right\}.\end{aligned}}\tag{2.842}$$

Here we have applied (2.515); the $\hat{1}$-operators have been suppressed as factors $n^\alpha$. We stress the $SU(n)$ vector $(\mathcal{H}_i(\mu))$ and the tensors $(\mathcal{H}_{ij}(\mu,\nu))$ specify the set of model parameters (as far as the Hamiltonian part is concerned). Additional parameters enter in the description of the coupling to the environment. These parameters are classical and must be given as input. Of course, in typical situations they may be known only statistically (inhomogeneous ensemble, cf. Sect. 3.3.5.8) or even fluctuate in time (cf. Sect. 4.1.1).

*Example 2.6.1.* Consider the off-resonant interaction introduced in Sect. 2.4.2.1. For $n = 2$ we find

$$\hat{H} = \text{constant} + \frac{1}{2}\sum_\mu \Delta E(\mu)\hat{\sigma}_z(\mu)$$

$$- \frac{\hbar}{4}\sum_{\mu<\nu} 2C_{\mathrm{R}}^{\mu\nu}\hat{\sigma}_z(\mu)\otimes\hat{\sigma}_z(\nu).\tag{2.843}$$

A possible realization might consist of a quantum-dot array (see Fig. 2.43) of local charge transfer excitations. The individual node is sketched in Fig. 2.44. The 3 relevant levels are part of the total electronic spectrum, tailored by nano-design. Here, the lowest state (3) in the *conduction band* is delocalized over the whole quantum dot. The Fermi level has to be pinned between level (1) and the higher level (2) to make possible an optically induced charge transfer between the states (1) and (2) via the transient state (3) (cf. Fig. 1.17). The Hamiltonian for the single node $\nu$,

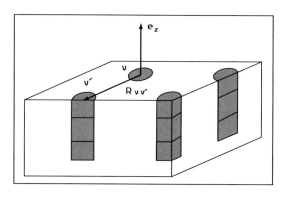

**Fig. 2.43.** Schematic structure of a quantum-dot array

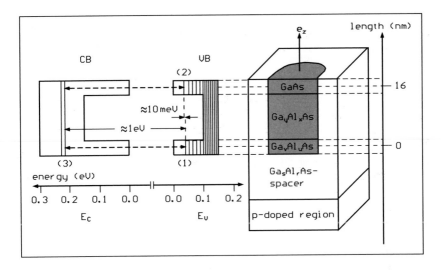

**Fig. 2.44.** Charge transfer node: schematic representation of a coupled pair of quantum dots and its spectrum. An asymmetric double well potential with valence band (VB) and conduction band (CB). Parameters: $y = 1 - x$, $v = 1 - u$, $s = 1 - r$, $x = 0.2$, $u = 0.01$, $r = 0.4$

$$\hat{H}(\nu) = \sum_{i=1}^{3} E_i(\nu)\hat{P}_{ii}(\nu) , \qquad (2.844)$$

is supplemented by

$$\hat{H}(\nu, \nu') = \sum_{l,l'=1}^{3} \hat{P}_{ll} \otimes \hat{P}_{l'l'}(\nu') h_{ll'}^{\nu\nu'} . \qquad (2.845)$$

The specific spatial localization of the three states of dot $\nu$ corresponds to different (static) dipole moments: we assume

$$d_1(\nu) = -d_2(\nu) = -de_z(\nu) , \quad d_3(\nu) = 0 . \qquad (2.846)$$

These dipole moments enter the dipole–dipole interaction term between the dots (cf. (2.560)):

$$h_{ll'}^{\nu\nu'} = \frac{1}{4\pi\epsilon_r\epsilon_0 |\boldsymbol{R}_{\nu\nu'}|^3} \left\{ d_{l_\nu}(\nu)d_{l_{\nu'}}(\nu') - \right.$$
$$\left. \frac{3}{|\boldsymbol{R}_{\nu\nu'}|^2} \left[\boldsymbol{R}_{\nu\nu'}d_{l_\nu}(\nu)\right]\left[\boldsymbol{R}_{\nu\nu'}d_{l_{\nu'}}(\nu')\right] \right\} , \quad (2.847)$$

where

$$\boldsymbol{R}_{\nu\nu'} = \boldsymbol{R}_\nu - \boldsymbol{R}_{\nu'} \qquad (2.848)$$

is the vector pointing from dot $\nu'$ to $\nu$ (compare Fig. 2.43). Even if the model (2.843) captures the most important features of this array, control will be constrained by the precision to which the parameters $\Delta E(\mu)$, $h^{\nu\nu'}_{ll'}$ can be specified.

While the representation (2.843) ends with 2-node terms, (there are typically only 2-node interactions), the corresponding representation for the density operator $\hat{\rho}$ requires the full expansion up to $N$-node correlations:

$$
\hat{\rho} = \frac{1}{n^N}\hat{1} + \frac{1}{2n^{N-1}} \sum_{\mu=1}^{N} \sum_{i=1}^{s} \lambda_i(\mu)\hat{\lambda}_i(\mu)
$$

$$
+ \frac{1}{4n^{N-2}} \sum_{\mu<\nu}^{N} \sum_{i,j=1}^{s} K_{ij}(\mu,\nu)\hat{\lambda}_i(\mu) \otimes \hat{\lambda}_j(\nu)
$$

$$
+ \frac{1}{8n^{N-3}} \sum_{\mu<\nu<\sigma}^{N} \sum_{i,j,k=1}^{s} K_{ijk}(\mu,\nu,\sigma)\hat{\lambda}_i(\mu) \otimes \hat{\lambda}_j(\nu) \otimes \hat{\lambda}_k(\sigma)
$$

$$
+ \dots . \tag{2.849}
$$

### 2.6.2 Physics of Entanglement

The state of a "classical" system can conveniently be described by a single vector (point) in the respective $\Gamma$ space. For an $N$-particle system in 3 dimensions this $\Gamma$ space is $6N$-dimensional (3 position and 3 momentum coordinates for each particle). Alternatively, the so-called 6-dimensional $\mu$ space can be applied, in which the system would be characterized by $N$ individual vectors for each subsystem. Both representations are completely equivalent (compare Fig. 2.45):

$$
r_c = \frac{N \dim\{\mu\text{space}\}}{\dim\{\Gamma\text{space}\}} = 1 . \tag{2.850}
$$

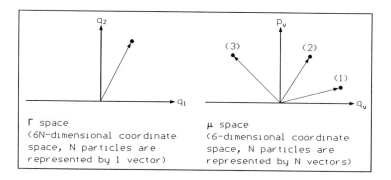

**Fig. 2.45.** Classical composite system. Only 2 coordinates ($q_1$, $q_2$ or $q_\nu$, $p_\nu$, respectively) and 3 particles ($\nu = 1, 2, 3$) are shown

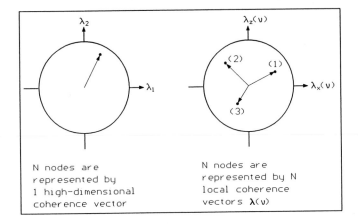

**Fig. 2.46.** Quantum mechanical composite system. Only 2 coherence vector components ($\lambda_1$, $\lambda_2$ or $\lambda_x(\nu)$, $\lambda_z(\nu)$, respectively) and 3 nodes ($\nu = 1, 2, 3$) are shown

For a quantum mechanical system the corresponding representations would be a $(n_{\text{tot}}^2 - 1)$-dimensional coherence vector characterizing the total system and the decomposition into $N$ local coherence vectors $\boldsymbol{\lambda}(\nu)$ of dimension $n^2 - 1$, i.e. $n_{\text{tot}} = n^N$. (Compare Fig. 2.46.)

This *local picture* is incomplete: this is indicated by the fact that the local entropies $S(\nu)$ are, in general, no longer zero even if the total entropy $S_{\text{tot}} = 0$. Furthermore, the local dimension is only $3N$ (in the above $SU(2)$ example), i.e. only a tiny fraction $r_q$ of the original state space dimension:

$$r_{\text{q}}(n, N) = \frac{N \dim \{SU(n)\}}{\dim \{SU(n_{\text{tot}})\}} = \frac{(n^2 - 1)\, N}{n^{2N} - 1} \ll 1 \text{ for } N \gg 1 . \qquad (2.851)$$

For example, for a network of four 2-level nodes $r_q \approx 1/84$. (Classically the total space is the direct sum, quantum mechanically the direct product of subspaces.) The local picture does not contain the notorious *quantum entanglements*, which are non-local in character. They carefully have to be distinguished from classical correlations.

What do we mean by "non-classicality"? Classical limits are often related to the mere size of an object, i.e. to a macroscopic limit. Macroscopic systems are, unavoidably, composite systems with a number of parts, defined in an appropriate way, becoming very large: $N \to \infty$. However, without further constraints there is no reason to expect this system to reside in a classical state: just to the contrary, the larger the system the more ways of entanglement exist. Classical properties are not intrinsic (cf., however, Sect. 4.1.2).

### 2.6.3 Ising Limit

For systems based on the harmonic oscillator model one usually considers the Glauber states $|\alpha\rangle$ as "classical" reference states (see Sect. 2.3.3.4):

$$\hat{\rho}(t) = |\alpha\rangle \langle\alpha| \ . \tag{2.852}$$

Amplitude expectation values in such states indeed coincide with the dynamics of the classical oscillator (cf. (3.18)). In particular, deviations from the corresponding Poisson distribution of number states $|j\rangle$ are then termed *non-classical*. This conception cannot be used in the present case. Here we introduce a *classical* (or *Ising*) *limit* of the $N$-node state, defined by

$$\hat{\rho} = \frac{1}{n^N}\hat{1} + \frac{1}{2n^{N-1}}\sum_{\mu=1}^{N}\sum_{i=1}^{n-1}\hat{\lambda}_{s'+i}(\mu)\hat{\lambda}_{s'+i}(\mu) \tag{2.853}$$

with

$$s' = n^2 - n \ . \tag{2.854}$$

This limit means that any coherence (single-node, 2-node, ..., etc.) is lost, which amounts to a tremendous reduction of the state space available. Also in this classical limit, large networks can be treated without severe numerical problems: the dynamics is governed by simple rate equations (cf. Sect. 3.3.7.7).

The *classical limit*, as far as the state space is concerned, may most intuitively be visualized for an $N$- (*pseudo*) *spin system*. If the local eigenbasis of $\hat{\sigma}_z(\nu)$, $\nu = 2, 3, \ldots, N$ is the basis of decoherence, the classical state space consists of the $2^N$ discrete product states

$$\begin{aligned} |\psi\rangle &= |\uparrow\downarrow\uparrow\uparrow\ldots\downarrow\rangle \\ &= |\uparrow(1)\rangle \otimes |\downarrow(2)\rangle \otimes |\uparrow(3)\rangle \otimes |\uparrow(4)\rangle \otimes \ldots \otimes |\downarrow(N)\rangle \ . \end{aligned} \tag{2.855}$$

Specific Hamilton models restricted to such states are known as *Ising models*. (There are extensions to local multi-level variants.) These Ising models represent the natural classical limit of the quantum networks studied in this book. The full quantum mechanical state space is exponentially larger: it contains all possible superpositions. However, having in mind the above local classical limit, it is useful to group those superpositions systematically into local 1-node, 2-node, ... $n$-node *coherence tensors* (i.e. a kind of *cluster expansion*). In the absence of damping these correlations will, in general, dominate the state and the dynamics of the system to the extent that local properties become virtually irrelevant. The local picture gains significance again due to the embedding into the environment: measurement and damping models are more often than not local. As a consequence, coherence in the respective basis becomes fragile, in fact, higher-order (i.e. "multi-node") coherence is more effectively damped out than lower-order coherence. Non-classical behaviour can thus survive in a classical environment only to the extent that those *dissipative channels* are (or can be made) weak enough.

## 2.7 Summary

In this chapter static properties of small *quantum networks* have been studied. The networks have been mainly restricted to $N = 1, 2, 3$ subsystems ("nodes"). Extensions to larger networks are straightforward though somewhat involved, unless one focusses on the classical limit. Furthermore, an intuitive understanding of larger networks will be severely limited until one reaches a statistical treatment in the very large $N$ limit.

Any concrete system is defined by a *Hamiltonian*, which, in the present case, decomposes into local node-Hamiltonians and node–node interactions. A fundamental interaction type is the *Coulomb coupling*, which has been specified here in parametrized form. Explicit examples include *spin networks*, correlated *photon pairs*, and *assemblies of few-level molecules* or few level *quantum dots*.

The state description has been in terms of the *density operator*. This operator, as well as any other observable or Hamiltonian, has been represented in terms of operators generating *local $SU(n)$ algebras*, where $n$ is the dimension of the local Hilbert subspace. Any local (Hermitian) operator is then represented by one scalar (its trace) and an $(n^2 - 1)$-dimensional vector. General non-local operators come in the form of a *cluster expansion* including $2, 3, 4 \ldots$ local nodes. They are specified by corresponding matrices of order $2, 3, 4 \ldots$ . In this way we were led to a natural description of *one-node* and *more-node coherence* (*entanglement*) as characteristic features of the density operator. These quantum coherences are non-classical features. They are relevant for deducing measurement results (here introduced via the von *Neumann projection postulate*).

Direct and indirect measurements have been discussed. Even though a more detailed account for measurement has to be postponed until Chap. 4, it should be clear that the reconstruction of a quantum state from actual measurement is severely restricted. These rules controlling information retrieval can be exploited in quantum communication tasks.

Incomplete knowledge about a state can be measured in terms of the corresponding *entropy*. Contrary to the classical domain, though, complete knowledge of the total system does not imply complete knowledge of its parts. In fact, non-local correlations (specified, for example, by a correlation index) imply non-zero subsystem entropy. The entropy, then, is no longer an additive quantity.

# 3. Quantum Dynamics

## 3.1 Introduction

In this chapter, *dynamic properties* of quantum networks will be addressed. We start with a discussion of *closed systems*. Principles of the dynamics of state vectors, density matrices, coherence vectors, and expectation values are studied, and the fundamental equations of motion (Schrödinger's equation, Liouville's equation, Heisenberg's equation of motion) are considered. As a fundamental example, the *optically driven 2-level system* is treated in detail leading to *Bloch equations without damping*. Coupled sets of Bloch equations describing 2- and 3-node functions are then introduced. Such coupled equations allow the discussion of systems composed of a number of subsystems ("nodes"). Finally, dynamic aspects of *open systems* under the influence of their respective surrounding are considered. A special *master equation*, the *Markovian master equation*, will be introduced. This will enable us to discuss *damping models* for 2- and 3-level systems as well as to introduce *generalized network equations* with local damping.

## 3.2 Unitary Dynamics

Contents: equations of motion, constants of motion, H models periodic in time, motion of coherence vectors (nutation, precession), Bloch equations without damping, network dynamics.

### 3.2.1 Liouville Equation

In this section, the dynamics of *closed systems* is discussed. In particular, the *Schrödinger equation* is considered, and the fundamental operator equation, the *Liouville equation*, is introduced.

**3.2.1.1 Unitary Dynamics of Pure States.** The time-dependence of state vectors $|\psi(t)\rangle$ is generated by the Hamiltonian of the system according to

$$i\hbar \frac{\partial}{\partial t} |\psi(t)\rangle = \hat{H}(t) |\psi(t)\rangle \ . \tag{3.1}$$

This *Schrödinger equation* is the fundamental equation specifying the so-called *Schrödinger picture*: in this picture the dynamics is carried by the state vectors, while the basic operators are time-independent. The Hamiltonian $\hat{H}$ may include time-dependent potential terms describing the influence of an (autonomous) environment.

Let the initial state $|\psi(0)\rangle$ and the state at a time $t$ be connected by the time-evolution operator $\hat{U}(t)$,

$$|\psi(t)\rangle = \hat{U}(t)\,|\psi(0)\rangle \ , \langle\psi(t)| = \langle\psi(0)|\,\hat{U}^{\dagger}(t) \ , \tag{3.2}$$

with

$$\hat{U}(0) = \hat{1} \ . \tag{3.3}$$

($\hat{U}^{\dagger}(t)$ defines the adjoint operator.) Inserting (3.2) into Schrödinger's equation, one obtains

$$\boxed{\mathrm{i}\hbar\frac{\partial}{\partial t}\hat{U}(t) = \hat{H}(t)\hat{U}(t) \ ,} \tag{3.4}$$

where

$$\boxed{-\mathrm{i}\hbar\frac{\partial}{\partial t}\hat{U}^{\dagger}(t) = \left[\hat{H}(t)\hat{U}(t)\right]^{\dagger} = \hat{U}^{\dagger}(t)\hat{H}(t)} \tag{3.5}$$

represents the adjoint equation. If the first equation is multiplied by the adjoint operator $\hat{U}^{\dagger}(t)$ (multiplication from the l.h.s.), and the second equation by $\hat{U}(t)$ (multiplication from the r.h.s.), then summing up both equations implies

$$\mathrm{i}\hbar\left[\hat{U}^{\dagger}(t)\frac{\partial\hat{U}(t)}{\partial t} + \frac{\partial\hat{U}^{\dagger}(t)}{\partial t}\hat{U}(t)\right] = \mathrm{i}\hbar\frac{\partial}{\partial t}\left[\hat{U}^{\dagger}(t)\hat{U}(t)\right] = 0 \tag{3.6}$$

so that $\hat{U}^{\dagger}(t)\hat{U}(t) = \hat{c} = $ constant. Since at time $t = 0$ the equation (3.3) holds, $\hat{c}$ has to be equal $\hat{1}$ so that

$$\hat{U}^{\dagger}(0)\hat{U}(0) = \hat{1} = \hat{U}^{\dagger}(t)\hat{U}(t) \ , \tag{3.7}$$

i.e. $\hat{U}(t)$ is a *unitary operator*. Solving (3.4) for $\partial\hat{H}/\partial t = 0$, one obtains the explicit solution

$$\boxed{\hat{U}(t) = \mathrm{e}^{-\mathrm{i}\hat{H}t/\hbar} \ .} \tag{3.8}$$

In the case of unitary operators $\hat{U}$ the scalar product $\left\langle\hat{U}f\,\middle|\,\hat{U}g\right\rangle$ $(f, g \in \mathbb{H})$ is identical with the scalar product $\langle f\,|\,g\rangle$, i.e. independent of time. $\hat{U}(t)$ defines a one-parameter unitary group.

*Example 3.2.1.* As a simple example consider in $SU(2)$:

$$|\psi(0)\rangle = \frac{1}{\sqrt{2}}\left[|1\rangle + |2\rangle\right] \tag{3.9}$$

with

$$\hat{H} = \frac{1}{2}\left(E_1 + E_2\right)\hat{1} + \frac{1}{2}\left(E_2 - E_1\right)\hat{w}_1 .$$  (3.10)

This implies

$$|\psi(t)\rangle = e^{-i\hat{H}t/\hbar}|\psi(0)\rangle = \frac{1}{\sqrt{2}}\left(|1\rangle e^{-iE_1t/\hbar} + |2\rangle e^{-iE_2t/\hbar}\right)$$  (3.11)

or

$$(\rho_{ij}(t)) = \frac{1}{2}\left(\begin{array}{cc} 1 & \exp\left(-i\omega_{12}t\right) \\ \exp\left(i\omega_{12}t\right) & 1 \end{array}\right)$$  (3.12)

so that the respective coherence vector reads

$$\lambda = [\cos\left(\omega_{12}t\right), \sin\left(\omega_{12}t\right), 0] .$$  (3.13)

*Example 3.2.2.* Glauber states (cf. Sect. 2.3.3.4) are very special non-stationary states. With

$$\hat{H} = \Delta E\hat{a}^+\hat{a} = \hbar\omega_0\hat{a}^+\hat{a}$$  (3.14)

we find

$$|\alpha(t)\rangle = e^{-i\hat{H}t/\hbar}|\alpha(0)\rangle$$
$$= e^{-|\alpha|^2/2}\sum_{j=1}^{\infty}\frac{\alpha(t)^{j-1}}{\sqrt{(j-1)!}}|j\rangle ,$$  (3.15)

where

$$\alpha(t) = \alpha e^{-i\omega_0 t} .$$  (3.16)

If the operator

$$\hat{x} = \sqrt{\frac{\hbar}{2m\omega_0}}\left(\hat{a}^+ + \hat{a}\right)$$  (3.17)

is interpreted as the elongation of the harmonic oscillator, the expectation value

$$d(t) = \langle\alpha(t)|\hat{x}|\alpha(t)\rangle$$
$$= |\alpha|\sqrt{\frac{2\hbar}{m\omega_0}}\cos\left(\omega_0 t - \varphi\right)$$  (3.18)

behaves like its classical counterpart. Schrödinger expected that such superpositions might also be found for other elementary models, i.e. coherent states behaving like "Kepler particles" in the $1/r$ potential. This seems not to be the case: in general, coherent states are thus not able to bridge the gap between quantum and classical behaviour.

**3.2.1.2 Density Matrix and Unitary Operator.** Consider a density operator defined by (2.251). Taking state vectors $|\psi(t)\rangle$ determined by Schrödinger's equation, this time-dependent density operator reads

$$\hat{\rho}(t) = \sum_n \rho_n |\psi_n(t)\rangle \langle \psi_n(t)| \; . \tag{3.19}$$

At time $t = 0$ the density operator is given by

$$\hat{\rho}(0) = \sum_n \rho_n |\psi_n(0)\rangle \langle \psi_n(0)| \tag{3.20}$$

so that, inserting (3.2), the density operator reads

$$\boxed{\hat{\rho}(t) = \hat{U}(t)\hat{\rho}(0)\hat{U}^\dagger(t) \; .} \tag{3.21}$$

Here, $\hat{U}(t)$ must be derived from (3.4). However, it is convenient to have an evolution equation directly for the density operator.

*Liouville Equation.* The derivative of (3.21) with respect to time is given by

$$\begin{aligned}
i\hbar \frac{\partial \hat{\rho}(t)}{\partial t} &= i\hbar \frac{\partial \hat{U}(t)}{\partial t} \hat{\rho}(0)\hat{U}^\dagger(t) + i\hbar \hat{U}(t)\hat{\rho}(0)\frac{\partial \hat{U}^\dagger(t)}{\partial t} \\
&= \hat{H}\hat{U}(t)\hat{\rho}(0)\hat{U}^\dagger(t) - \hat{U}(t)\hat{\rho}(0)\hat{U}^\dagger(t)\hat{H} \\
&= \hat{H}\hat{\rho}(t) - \hat{\rho}(t)\hat{H} \; ,
\end{aligned} \tag{3.22}$$

where we have used (3.4) and (3.5), so that

$$\boxed{i\hbar \frac{\partial}{\partial t}\hat{\rho}(t) = \left[\hat{H}, \hat{\rho}(t)\right]_- \; .} \tag{3.23}$$

This differential equation is called the *Liouville equation* or the *v. Neumann equation*. This equation is also valid if the Hamiltonian $\hat{H}$ is time-dependent.

In order to derive the corresponding matrix equation, a system of eigenvectors has to be specified.

*Matrix Equation.* Consider a system specified by the Hamiltonian

$$\hat{H}(t) = \hat{H}_0 + \hat{V}(t) \; , \tag{3.24}$$

and a set of eigenvectors defined by

$$\hat{H}_0 |E_n\rangle = E_n |E_n\rangle \; , \tag{3.25}$$

wherein $\hat{V}(t)$ defines the potential of external forces. Using the basis represented by the eigenvectors $|E_n\rangle$, the relation

$$\langle E_n| \hat{H}(t) |E_{n'}\rangle = E_n \delta_{nn'} + \langle E_n| \hat{V}(t) |E_{n'}\rangle \tag{3.26}$$

defines matrix elements of the Hamiltonian $\hat{H}(t)$.

Multiplying the Liouville equation (3.23) with $\langle E_n|$ from left, and $|E_{n'}\rangle$ from right, one obtains the differential equation

$$i\hbar\frac{\partial}{\partial t}\rho_{nn'}(t) = \left\langle E_n \left| \left[\hat{H}(t),\hat{\rho}(t)\right]_- \right| E_{n'} \right\rangle$$

$$= \left\langle E_n \left| \left[\hat{H}_0,\hat{\rho}(t)\right]_- \right| E_{n'} \right\rangle + \left\langle E_n \left| \left[\hat{V}(t),\hat{\rho}(t)\right]_- \right| E_{n'} \right\rangle$$

$$= (E_n - E_{n'})\,\rho_{nn'} + \left\langle E_n \left| \left[\hat{V}(t),\hat{\rho}(t)\right]_- \right| E_{n'} \right\rangle$$

$$= (E_n - E_{n'})\,\rho_{nn'} + \sum_{\nu}\langle E_n| \hat{V}(t)|E_\nu\rangle\,\rho_{\nu n'} -$$

$$\sum_{\mu}\rho_{n\mu}\,\langle E_\mu| \hat{V}(t)|E_{n'}\rangle \ . \tag{3.27}$$

Introducing the abbreviation

$$D_{nn'mm'}(t) = (E_n - E_{n'})\,\delta_{mn}\delta_{m'n'} + \langle E_n| \hat{V}(t)|E_m\rangle\,\delta_{m'n'} -$$
$$\langle E_{m'}| \hat{V}(t)|E_{n'}\rangle\,\delta_{mn} \ , \tag{3.28}$$

the matrix equation (3.27) can be rewritten:

$$i\hbar\frac{\partial}{\partial t}\rho_{nn'}(t) = \sum_{m,m'} D_{nn'mm'}(t)\rho_{mm'}(t) \ . \tag{3.29}$$

This equation determines the evolution of the matrix elements $\rho_{nn'}(t) = \langle E_n| \hat{\rho}(t)|E_{n'}\rangle$. A similar evolution equation can be derived in order to determine the behaviour of expectation values $\overline{A}(t)$.

*Evolution Equation for Expectation Values.* Consider

$$\overline{A}(t) = \mathrm{Tr}\left\{\hat{\rho}(t)\hat{A}\right\} \tag{3.30}$$

(cf. (2.293)) so that the derivative with respect to time is given by

$$i\hbar\frac{\partial}{\partial t}\overline{A}(t) = i\hbar\mathrm{Tr}\left\{\hat{\rho}(t)\frac{\partial \hat{A}}{\partial t}\right\} + i\hbar\mathrm{Tr}\left\{\frac{\partial \hat{\rho}(t)}{\partial t}\hat{A}\right\} \ . \tag{3.31}$$

Making use of the Liouville equation, and defining $\overline{\frac{\partial A(t)}{\partial t}} = \mathrm{Tr}\left\{\hat{\rho}(t)\frac{\partial \hat{A}}{\partial t}\right\}$, (3.31) can be written as

$$i\hbar\frac{\partial}{\partial t}\overline{A}(t) = i\hbar\overline{\frac{\partial A(t)}{\partial t}} + \mathrm{Tr}\left\{\left[\hat{H}(t),\hat{\rho}(t)\right]_-\hat{A}\right\} \ . \tag{3.32}$$

This differential equation determines the time-evolution of expectation values $\overline{A}(t)$ if the density matrix $\hat{\rho}(t)$ and the Hamiltonian $\hat{H}(t)$ are given.

**3.2.1.3 Entropy as a Constant of Motion.** According to (2.252), (2.340), the equation

$$S(t) = -k\mathrm{Tr}\left\{\hat{\rho}(t)\ln\hat{\rho}(t)\right\} \tag{3.33}$$

with

$$\text{Tr}\left\{\hat{\rho}(t)\right\} = 1 \tag{3.34}$$

determines a measure for the uncertainty of a state. Now consider the derivative

$$\frac{dS(t)}{dt} = -k\text{Tr}\left\{\frac{\partial\hat{\rho}(t)}{\partial t}\ln\hat{\rho}(t)\right\} - k\text{Tr}\left\{\hat{\rho}(t)\frac{\partial\ln\hat{\rho}(t)}{\partial t}\right\} . \tag{3.35}$$

With the help of Liouville's equation, this differential equation can be written as

$$\frac{dS(t)}{dt} = \frac{i}{\hbar}k\text{Tr}\left\{\left[\hat{H}(t),\hat{\rho}(t)\right]_{-}\ln\hat{\rho}(t)\right\} - k\text{Tr}\left\{\hat{\rho}(t)\frac{\partial\ln\hat{\rho}(t)}{\partial t}\right\} . \tag{3.36}$$

Due to the fact that a trace is invariant with respect to cyclic permutations, this expression is equivalent to

$$\frac{dS(t)}{dt} = \frac{i}{\hbar}k\text{Tr}\left\{\left[\ln\hat{\rho}(t),\hat{H}(t)\right]_{-}\hat{\rho}(t)\right\} - k\text{Tr}\left\{\hat{\rho}(t)\frac{\partial\ln\hat{\rho}(t)}{\partial t}\right\} . \tag{3.37}$$

Making use of the equation

$$i\hbar\frac{\partial}{\partial t}\ln\hat{\rho}(t) = \left[\hat{H}(t),\ln\hat{\rho}(t)\right]_{-} , \tag{3.38}$$

the time-derivative of $S$ reads

$$\frac{dS(t)}{dt} = k\text{Tr}\left\{\frac{\partial\ln\hat{\rho}(t)}{\partial t}\hat{\rho}(t)\right\} - k\text{Tr}\left\{\hat{\rho}(t)\frac{\partial\ln\hat{\rho}(t)}{\partial t}\right\} \tag{3.39}$$

so that, with $\text{Tr}\left\{\hat{\rho}(t)\frac{\partial\ln\hat{\rho}(t)}{\partial t}\right\} = \text{Tr}\left\{\frac{\partial\ln\hat{\rho}(t)}{\partial t}\hat{\rho}(t)\right\}$, one obtains

$$\boxed{\frac{dS(t)}{dt} = 0 .} \tag{3.40}$$

$S$ is thus a constant of motion: a pure state remains a pure state, and a mixed state remains a mixed state. This no longer holds for open systems, though.

### 3.2.2 Dynamics of the Coherence Vector

The motion of the coherence vector follows as a special case of the dynamics of expectation values. This motion will now be investigated.

**3.2.2.1 The Evolution Equation of the Coherence Vector.** An element $\lambda_i(t)$ of a time-dependent coherence vector is defined by $\lambda_i(t) = \text{Tr}\left\{\hat{\lambda}_i\hat{\rho}(t)\right\}$ (compare with (2.272)) so that the derivative with respect to time $t$ reads

$$\frac{\partial}{\partial t}\lambda_i(t) = \frac{\partial}{\partial t}\text{Tr}\left\{\hat{\lambda}_i\hat{\rho}(t)\right\} = \text{Tr}\left\{\hat{\lambda}_i\frac{\partial\hat{\rho}(t)}{\partial t}\right\} . \tag{3.41}$$

Then, using Liouville's equation again, the equation (3.41) can be recast into

$$\frac{\partial}{\partial t}\lambda_i(t) = \frac{1}{i\hbar}\mathrm{Tr}\left\{\hat{\lambda}_i\left[\hat{H}(t),\hat{\rho}(t)\right]_-\right\} . \tag{3.42}$$

Additionally, using the general relation

$$\mathrm{Tr}\left\{\hat{A}\left[\hat{B},\hat{C}\right]_-\right\} \overset{\text{cyclic permutation}}{=} \mathrm{Tr}\left\{\hat{B}\left[\hat{C},\hat{A}\right]_-\right\}$$

$$\overset{\text{exchange }\hat{C},\,\hat{A}}{=} -\mathrm{Tr}\left\{\hat{B}\left[\hat{A},\hat{C}\right]_-\right\} , \tag{3.43}$$

one obtains

$$\frac{\partial}{\partial t}\lambda_i(t) = -\frac{1}{i\hbar}\mathrm{Tr}\left\{\hat{H}(t)\left[\hat{\lambda}_i,\hat{\rho}(t)\right]_-\right\} . \tag{3.44}$$

Replacing $\hat{\rho}$ by the representation

$$\hat{\rho}(t) = \frac{1}{n}\hat{1} + \frac{1}{2}\sum_{k=1}^{s}\lambda_k(t)\hat{\lambda}_k \tag{3.45}$$

(compare with (2.271)), the commutator

$$\left[\hat{\lambda}_i,\hat{\rho}(t)\right]_- = \frac{1}{2}\sum_{k=1}^{s}\lambda_k(t)\left[\hat{\lambda}_i,\hat{\lambda}_k\right]_- \tag{3.46}$$

can easily be shown to follow so that the evolution equation (3.44) turns into

$$\boxed{\frac{\partial}{\partial t}\lambda_i(t) = \sum_{k=1}^{s}\Omega_{ik}\lambda_k(t)} \tag{3.47}$$

with

$$\boxed{\Omega_{ik} = -\frac{1}{2i\hbar}\mathrm{Tr}\left\{\hat{H}(t)\left[\hat{\lambda}_i,\hat{\lambda}_k\right]_-\right\} = -\Omega_{ki}} . \tag{3.48}$$

$\Omega_{ik}$ is antisymmetric and thus defines rotations in the $(n^2 - 1)$-dimensional space of the coherence vector. In the case of $\left[\hat{\lambda}_i,\hat{\lambda}_k\right]_- = 0$ (which holds, for example, in the $\hat{w}_l$ subspace), the change of $\lambda_i(t)$ with respect to time is independent of the change of $\lambda_k(t)$, and vice versa.

More explicitly, using

$$\hat{H}(t) = \frac{1}{2}\mathrm{Tr}\left\{\hat{H}(t)\right\}\hat{1} + \frac{1}{2}\sum_j\mathcal{H}_j\hat{\lambda}_j , \quad \mathrm{Tr}\left\{\hat{H}(t)\right\} = \text{constant} \tag{3.49}$$

(e.g. compare (2.178)) with

$$\mathcal{H}_j = \mathrm{Tr}\left\{\hat{H}(t)\hat{\lambda}_j\right\} \tag{3.50}$$

(compare with (2.180)), considering (2.71), and using the representation (2.70), the matrix elements $\Omega_{ik}$ defined by (3.48) can be written as

$$\Omega_{ik} = -\sum_j f_{ikj}\Gamma_j = \sum_j f_{ijk}\Gamma_j \ , \quad \Gamma_j = \frac{\mathcal{H}_j}{\hbar} \ . \tag{3.51}$$

The *rotation matrix* $\Omega_{ik}$ is thus specified by the Hamilton model and the structure constants of the underlying $SU(n)$ algebra. Inserting this representation into the evolution equation (3.47), one obtains

$$\frac{\partial}{\partial t}\lambda_i(t) = \sum_{j,k} f_{ijk}\Gamma_j\lambda_k(t) \ . \tag{3.52}$$

For $SU(2)$ the structure constants $f_{ijk}$ are identical with the elements of the totally antisymmetric $\varepsilon$ tensor (cf. (2.2)), and the r.h.s. of (3.52) is just the vector product between $\boldsymbol{\Gamma}$ and $\boldsymbol{\lambda}$, often written as

$$\frac{\partial}{\partial t}\boldsymbol{\lambda} = \boldsymbol{\Gamma} \times \boldsymbol{\lambda} \ . \tag{3.53}$$

This notation may be kept also for $SU(n)$, $n \geq 2$.

*Example 3.2.3.* Consider a 2-level system with transfer coupling (compare Sect. 2.1.3.4). Such a system is described by the vector

$$\hbar\boldsymbol{\Gamma} = (\operatorname{Re}\{V_{12}\}, \operatorname{Im}\{V_{12}\}, E_2 - E_1) \tag{3.54}$$

(compare with (2.218)). The elements $f_{ijk}$ unequal zero are $f_{123} = f_{231} = f_{312} = 1$, $f_{132} = f_{213} = f_{321} = -1$ so that, introducing the abbreviation

$$\hbar\omega_{21} = E_2 - E_1 \ , \tag{3.55}$$

the rotation matrix is given by

$$(\Omega_{ij}) = \begin{pmatrix} 0 & -\omega_{12} & -\operatorname{Im}\{V_{12}\}/\hbar \\ \omega_{12} & 0 & -\operatorname{Re}\{V_{12}\}/\hbar \\ \operatorname{Im}\{V_{12}\}/\hbar & \operatorname{Re}\{V_{12}\}/\hbar & 0 \end{pmatrix} \ . \tag{3.56}$$

The equation for the coherence vector

$$\boldsymbol{\lambda} = \{u_{12}(t), v_{12}(t), w_1(t)\} \tag{3.57}$$

then reads

$$\begin{aligned}
i = 1: \quad & \dot{u}_{12}(t) = \frac{1}{\hbar}\operatorname{Im}\{V_{12}\}w_1(t) - \omega_{21}v_{12}(t) \ , \\
i = 2: \quad & \dot{v}_{12}(t) = \omega_{21}u_{12}(t) - \frac{1}{\hbar}\operatorname{Re}\{V_{12}\}w_1(t) \ , \\
i = 3: \quad & \dot{w}_1(t) = \frac{1}{\hbar}\operatorname{Re}\{V_{12}\}v_{12}(t) - \frac{1}{\hbar}\operatorname{Im}\{V_{12}\}u_{12}(t) \ .
\end{aligned} \tag{3.58}$$

During this motion, the length of the time-dependent coherence vectors remains constant. This is now shown for the general case.

**3.2.2.2 Length of Time-Dependent Coherence Vectors.** Let us consider

$$\frac{\partial}{\partial t}|\lambda(t)|^2 = 2\left[\dot{\lambda}(t)\lambda(t)\right] = 2\sum_i \dot{\lambda}_i(t)\lambda_i(t) . \tag{3.59}$$

Making use of (3.52), this equation can be replaced by

$$\frac{\partial}{\partial t}|\lambda(t)|^2 = 2\sum_j \Gamma_j \sum_{ik} f_{ijk}\lambda_k(t)\lambda_i(t) . \tag{3.60}$$

However, due to $f_{ijk} = -f_{kji}$, the sum $\sum_{ik} f_{ijk}\lambda_k(t)\lambda_i(t)$ is identically zero, so that the expression (3.60) is equivalent to

$$\boxed{\frac{\partial}{\partial t}|\lambda(t)|^2 = 0 .} \tag{3.61}$$

Therefore, the length of a time-dependent coherence vector is a constant of motion:

$$|\lambda(t)|^2 = \text{constant} . \tag{3.62}$$

Using the expression (3.62), the trace defined by (2.276) can be reformulated:

$$\boxed{C(n,2) = \text{Tr}\left\{\hat{\rho}^2(t)\right\} = \frac{1}{n} + \frac{1}{2}|\lambda(t)|^2 = \text{constant} ,} \tag{3.63}$$

i.e. also the trace $\text{Tr}\left\{\hat{\rho}^2(t)\right\}$ is a constant of motion. In the same way, it can be shown that $C(n, q > 2)$ are also constants of motion.

**3.2.2.3 Dynamics of Expectation Values.** Applying the differential operator $\partial/\partial t$ to

$$\boxed{\left\langle\hat{A}\right\rangle = \text{Tr}\left\{\hat{\rho}\hat{A}\right\} = \frac{1}{n}\text{Tr}\left\{\hat{A}\right\} + \frac{1}{2}\sum_{j=1}^s A_j\lambda_j(t)} \tag{3.64}$$

(cf. (2.296)) with

$$A_j = \text{Tr}\left\{\hat{A}\hat{\lambda}_j\right\} := \text{constant} \tag{3.65}$$

(cf. (2.295)), one obtains the differential equation

$$\frac{\partial}{\partial t}\left\langle\hat{A}\right\rangle = \frac{1}{2}\sum_{j=1}^s A_j\dot{\lambda}_j(t) = \frac{1}{2}\sum_{j,j',k=1}^s f_{jj'k}\Gamma_{j'}\lambda_k(t)A_j . \tag{3.66}$$

This equation determines the evolution of an expectation value $\left\langle\hat{A}\right\rangle$.

**3.2.2.4 Constant of Precession.** If the Hamiltonian $\hat{H}$ is time-independent, i.e. if

$$\frac{\partial \boldsymbol{\Gamma}}{\partial t} = 0 , \tag{3.67}$$

one finds that the scalar product $\boldsymbol{\lambda}(t)\boldsymbol{\Gamma}$ is a constant of motion.

In order to verify this assertion, we consider the derivative

$$
\begin{aligned}
\frac{\partial}{\partial t}[\boldsymbol{\lambda}(t)\boldsymbol{\Gamma}] &= \frac{\partial}{\partial t}\sum_i \lambda_i(t)\Gamma_i \\
&= \sum_i \Gamma_i \frac{\partial}{\partial t}\lambda_i(t) \\
&= \sum_{i,j,k} f_{ijk}\Gamma_i\Gamma_j\lambda_k(t) \\
&= \sum_k \lambda_k(t)\sum_{i,j} f_{ijk}\Gamma_i\Gamma_j \ .
\end{aligned} \tag{3.68}
$$

Taking into account that the structure constants obey the relation $f_{ijk} = -f_{jik}$, the sum $\sum_{i,j} f_{ijk}\Gamma_i\Gamma_j$ vanishes for each $k$. Therefore, (3.68) reduces to

$$\boxed{\frac{\partial}{\partial t}[\boldsymbol{\lambda}(t)\boldsymbol{\Gamma}] = 0 \ .} \tag{3.69}$$

This means that the directional cosine

$$\boldsymbol{\lambda}(t)\boldsymbol{\Gamma} = \cos\Theta\,|\boldsymbol{\lambda}(t)|\,|\boldsymbol{\Gamma}| \tag{3.70}$$

remains constant during the motion. In particular, any state $\boldsymbol{\lambda} \sim \boldsymbol{\Gamma}$ is thus stationary.

As a simple example in $SU(2)$ consider the 2-level system of Sect. 2.2.5.1 with $\hbar\boldsymbol{\Gamma} = (0,0,E_2 - E_1)$. Then $\lambda_z$ is a constant of motion: so, not only the energy eigenstates $\boldsymbol{\lambda} = (0,0,\pm1)$, but also mixed states are stationary (cf. also Sect. 3.2.3.7).

### 3.2.3 Hamilton Model with Periodic Time-Dependence

In this section special time-dependent Hamiltonians are considered. In particular, *Hamiltonians periodic in time* are studied, the *rotating wave approximation (RWA)* is introduced, and unitary transformations are considered, which allow the removal of the time-dependence. Finally, *Bloch equations* for 2- and 3-level systems are studied.

**3.2.3.1 Optically Driven 2-Level System.** Consider a time-dependent Hamiltonian

$$\boxed{\hat{H}(t) = \hat{H}_0 + \hat{H}_{\mathrm{L}}(t) \ ,} \tag{3.71}$$

wherein the Hamiltonian without external forces, $\hat{H}_0$, is defined by

$$\hat{H}_0 = E_1 \hat{P}_{11} + E_2 \hat{P}_{22} \, . \tag{3.72}$$

The projection operators $\hat{P}_{11}$, $\hat{P}_{22}$ define two stationary states with the energies $E_1$, $E_2$, respectively.

Assuming *optical interactions* described by the Hamiltonian

$$\hat{H}_{\mathrm{L}}(t) = -\widehat{d\epsilon} = -d \left( \hat{P}_{12} + \hat{P}_{21} \right) \epsilon \tag{3.73}$$

with the electric field

$$\epsilon = \epsilon_0 \cos \omega t = \frac{1}{2} \epsilon_0 \left( \mathrm{e}^{\mathrm{i}\omega t} + \mathrm{e}^{-\mathrm{i}\omega t} \right) \, , \tag{3.74}$$

the total Hamiltonian $\hat{H}(t)$ represents a system with external forces periodic in time. Inserting $\epsilon$ into $\hat{H}_{\mathrm{L}}$, and using the abbreviation

$$g = g^* = -\frac{d\epsilon_0}{\hbar} \, , \tag{3.75}$$

these external forces are described by

$$\hat{H}_{\mathrm{L}}(t) = \frac{1}{2} g \hbar \left( \hat{P}_{12} \mathrm{e}^{-\mathrm{i}\omega t} + \hat{P}_{21} \mathrm{e}^{\mathrm{i}\omega t} \right) + \hat{H}_{\mathrm{L}}^{(2)}(t) \tag{3.76}$$

with

$$\hat{H}_{\mathrm{L}}^{(2)}(t) = \frac{1}{2} g \hbar \left( \hat{P}_{12} \mathrm{e}^{\mathrm{i}\omega t} + \hat{P}_{21} \mathrm{e}^{-\mathrm{i}\omega t} \right) \, . \tag{3.77}$$

$\hat{H}_{\mathrm{L}}^{(2)}(t)$ describes the so-called *energy-non-conserving terms* responsible, for example, for the *Lamb shift* or the *Bloch-Siegert shift*. Neglecting the term $\hat{H}_{\mathrm{L}}^{(2)}(t)$ (the so-called *rotating wave approximation* (RWA)), one obtains the interaction Hamiltonian

$$\hat{H}_{\mathrm{L}}(t) = \frac{1}{2} g \hbar \left( \hat{P}_{12} \mathrm{e}^{-\mathrm{i}\omega t} + \hat{P}_{21} \mathrm{e}^{\mathrm{i}\omega t} \right) \, . \tag{3.78}$$

The explicit time-dependence can be removed by a time-dependent unitary transformation. Applying $\hat{U}(t) = \hat{U}_{11}(\alpha_1)$ with $\alpha_1 = -\omega t$ and $\hat{U}_{11}$ (as introduced in Sect. 2.2.3.3) to the Hamiltonian (3.71) given by

$$\hat{H} = E_1 \hat{P}_{11} + E_2 \hat{P}_{22} + \frac{1}{2} g \hbar \left( \hat{P}_{12} \mathrm{e}^{-\mathrm{i}\omega t} + \hat{P}_{21} \mathrm{e}^{\mathrm{i}\omega t} \right) \, , \tag{3.79}$$

we obtain

$$\hat{H}' = \hat{U}_{11}^\dagger \hat{H} \hat{U}_{11} = E_1 \hat{P}_{11} + E_2 \hat{P}_{22} + \frac{1}{2} g \hbar \left( \hat{P}_{12} + \hat{P}_{21} \right) \, . \tag{3.80}$$

In this *rotating reference frame* the Hamiltonian is time-independent! This transformation is not unique: we could have used $\hat{U}_{22}$ with $\alpha_2 = \omega t$ as well, or a combination of the two.

We note that the inclusion of $\hat{H}_{\mathrm{L}}^{(2)}(t)$ would imply in $\hat{H}'$ the additional term $\frac{1}{2}g\hbar\left(\hat{P}_{12}e^{2i\omega t} + \hat{P}_{21}e^{-2i\omega t}\right)$ thus spoiling the intended time-independence.

*Effective Time-independent Schrödinger Equation.* (See [117].) Due to the fact that the operator $\hat{U}^\dagger(t)\hat{U}(t) = \hat{U}(t)\hat{U}^\dagger(t)$ represents a unit operator $\hat{1}$ (cf. (3.7)), the Schrödinger equation (3.1) can be recast into the form

$$i\hbar\frac{\partial}{\partial t}\hat{U}(t)\hat{U}^\dagger(t)\,|\psi(t)\rangle = \hat{H}(t)\hat{U}(t)\hat{U}^\dagger(t)\,|\psi(t)\rangle \ . \tag{3.81}$$

Multiplying this partial differential equation with $\hat{U}^\dagger$, this equation yields

$$i\hbar\hat{U}^\dagger(t)\frac{\partial}{\partial t}\hat{U}(t)\hat{U}^\dagger(t)\,|\psi(t)\rangle = \hat{U}^\dagger(t)\hat{H}(t)\hat{U}(t)\hat{U}^\dagger(t)\,|\psi(t)\rangle \ . \tag{3.82}$$

Using

$$\begin{aligned}
\hat{U}^\dagger(t)\frac{\partial}{\partial t}\hat{U}(t) &= \hat{U}^\dagger(t)\frac{\partial}{\partial t}\left[\hat{U}(t)\right] + \hat{U}^\dagger(t)\hat{U}(t)\frac{\partial}{\partial t} \\
&= \hat{U}^\dagger(t)\frac{\partial}{\partial t}\left[\hat{U}(t)\right] + \frac{\partial}{\partial t} \ ,
\end{aligned} \tag{3.83}$$

and taking (for the present choice)

$$\hat{U}^\dagger(t)\frac{\partial}{\partial t}\left[\hat{U}(t)\right] = -i\omega\hat{P}_{11} = \text{const.} \tag{3.84}$$

into account, the Schrödinger equation finally results in

$$\boxed{i\hbar\frac{\partial}{\partial t}\,|\psi(t)\rangle' = \left(\hbar\omega\hat{P}_{11} + \hat{H}'\right)|\psi(t)\rangle' \ ,} \tag{3.85}$$

wherein

$$\boxed{|\psi(t)\rangle' = \hat{U}^\dagger(t)\,|\psi(t)\rangle} \tag{3.86}$$

defines the state vectors in the rotating reference frame, and

$$\hat{\rho}' = \hat{U}^\dagger\hat{\rho}\hat{U} \tag{3.87}$$

the corresponding density operator.

Introducing the *effective Hamiltonian*

$$\begin{aligned}
\hat{H}_{\mathrm{eff}} &= \hbar\omega\hat{P}_{11} + \hat{H}' \\
&= (E_1 + \hbar\omega)\,\hat{P}_{11} + E_2\hat{P}_{22} + \frac{1}{2}g\hbar\left(\hat{P}_{12} + \hat{P}_{21}\right) \ ,
\end{aligned} \tag{3.88}$$

the Schrödinger equation (3.85) reduces to

$$\boxed{i\hbar\frac{\partial}{\partial t}\,|\psi(t)\rangle' = \hat{H}_{\mathrm{eff}}\,|\psi(t)\rangle' \ .} \tag{3.89}$$

From a formal point of view, this equation is equivalent to the original Schrödinger equation (3.1). This treatment of time-dependent Hamiltonians is related to the concept of *quasienergies* and the *Floquet theorem* (cf. [71]).

Using the representation (2.74) and

$$\sum_i \hat{P}_{ii} = 1 \, , \tag{3.90}$$

the projection operators occurring in the effective Hamiltonian can be expressed by generating operators:

$$\boxed{\hat{H}_{\text{eff}} = \frac{1}{2} \left( E_1 + E_2 + \hbar\omega \right) \hat{1} + \frac{1}{2} \left( E_2 - E_1 - \hbar\omega \right) \hat{w}_1 + \frac{g\hbar}{2} \hat{u}_{12} \, .} \tag{3.91}$$

This Hamiltonian has the form of a 2-level system with transfer coupling as discussed in Sect. 2.2.5.5.

*Control Parameters.* Comparing (3.91) in $SU(2)$ with the general representation

$$\hat{H}(t) = \frac{1}{2} \text{Tr} \left\{ \hat{H}(t) \right\} \hat{1} + \frac{\hbar}{2} \sum_j \Gamma_j \hat{\lambda}_j \, , \quad \Gamma_j = \frac{\mathcal{H}_j}{\hbar} \tag{3.92}$$

(cf. (3.49)), the identification

$$\boxed{\text{Tr} \left\{ \hat{H}(t) \right\} = E_1 + E_2 \, ,} \tag{3.93}$$

$$\boxed{\boldsymbol{\Gamma} = (g, 0, \delta)} \tag{3.94}$$

follows, where we have introduced the *detuning parameter*

$$\delta = \frac{E_2 - E_1}{\hbar} - \omega = \omega_{21} - \omega \, . \tag{3.95}$$

The Hamiltonian $\hat{H}(t)$ is thus specified by its trace and by the vector $\boldsymbol{\Gamma}$, where this $\boldsymbol{\Gamma}$ vector contains adjustable parameters $g$, $\delta$: the system can be controlled. The applicability of the rotating wave approximation requires the relation $\delta, g \ll \omega$ to be fulfilled. Such unitary transformations to rotating frames can be used also for time-independent Hamiltonians ($g = 0$): it is sometimes convenient to *remove* part of the coherent oscillations by choosing $\delta = 0$.

**3.2.3.2 Bloch Equations.** The Schrödinger equation (3.89) determines state vectors in rotating wave approximation. The motion of the corresponding coherence vector $\boldsymbol{\lambda} = (u_{12}, v_{12}, w_1)$ is defined by the evolution equation (3.52). Identifying the structure constants $f_{ijk}$ of the evolution equation (3.52) with elements of the $\varepsilon$ tensor, and identifying the coherence vector $\boldsymbol{\lambda}$ with the polarization vector $\boldsymbol{P} = (P_x, P_y, P_z)$, one obtains the differential equation

$$\boxed{\frac{\partial}{\partial t} P_i = \sum_{j,k} \varepsilon_{ijk} \Gamma_j P_k = (\boldsymbol{\Gamma} \times \boldsymbol{P})_i \, ,} \tag{3.96}$$

where the matrix $(\Omega_{ij})$, which is defined by (3.51), is here given by

$$(\Omega_{ij}) = \begin{pmatrix} 0 & -\delta & 0 \\ +\delta & 0 & -g \\ 0 & +g & 0 \end{pmatrix} . \tag{3.97}$$

This evolution equation represents the famous *Bloch equations without damping*. $\boldsymbol{P}$ is called *Bloch vector*. Inserting the $\boldsymbol{\Gamma}$ vector (3.94) into this system of equations, one finds

$$\begin{aligned} \dot{P}_x + \delta P_y &= 0 , \\ \dot{P}_y - \delta P_x + g P_z &= 0 , \\ \dot{P}_z - g P_y &= 0 . \end{aligned} \tag{3.98}$$

*Stationary Solutions.* In the case

$$\frac{\partial}{\partial t} \boldsymbol{P} = 0 , \tag{3.99}$$

(3.98) reduces to a system of equations which determines stationary solutions:

$$\begin{aligned} \delta P_y &= 0 , \\ -\delta P_x + g P_z &= 0 , \\ g P_y &= 0 . \end{aligned} \tag{3.100}$$

In the case $\delta = g = 0$, $\boldsymbol{P}$ is undefined. However, if at least one parameter of the 2-dimensional set of parameters $\{\delta, g\}$ is unequal zero, (3.100) determines non-trivial solutions. Taking this assumption as a basis, the first and the third equation of (3.100) determine the polarization in $y$ direction:

$$P_y = 0 . \tag{3.101}$$

Additionally, assuming

$$P_x^2 + P_z^2 = P_0^2 , \tag{3.102}$$

the second equation of (3.100) results in

$$\delta^2 P_x^2 = g^2 P_z^2 = g^2 \left( P_0^2 - P_x^2 \right) . \tag{3.103}$$

Solving this equation, one obtains relations which determine the polarizations in $x$ and in $z$ direction:

$$\begin{aligned} P_x^2 &= P_0^2 \frac{g^2}{\delta^2 + g^2} , \\ P_z^2 &= P_0^2 \frac{\delta^2}{\delta^2 + g^2} . \end{aligned} \tag{3.104}$$

$P_0^2 = 1$ would hold for a pure state. Due to the fact that in the stationary case the absolute value of the vector product (3.96) can be written as

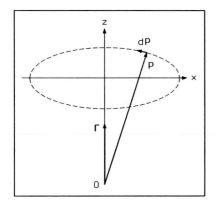

**Fig. 3.1.** Precession of the polarization vector in the $x, y$ plane

$$|\boldsymbol{\Gamma} \times \boldsymbol{P}| = |\boldsymbol{P}||\boldsymbol{\Gamma}| \sin \angle(\boldsymbol{\Gamma}, \boldsymbol{P}) = 0 \, , \tag{3.105}$$

the vectors $\boldsymbol{P}$ and $\boldsymbol{\Gamma}$ are aligned in a parallel or antiparallel way.

*Non-Stationary Solution.* Let us first study the case $g = 0$ for which the transformation into the rotating reference frame is obsolete, and $\delta$ may be identified with $(E_2 - E_1)/\hbar$. Then $P_z(t) = P_z(0) = $ constant, and the remaining equations

$$\begin{aligned} \dot{P}_x &= -\delta P_y \, , \\ \dot{P}_y &= \delta P_x \end{aligned} \tag{3.106}$$

imply

$$\ddot{P}_x = -\delta \dot{P}_y = \delta^2 P_x \, , \tag{3.107}$$

which is solved for $P_x(0) = 1$ by

$$\begin{aligned} P_x &= \cos(\delta t) \, , \\ \dot{P}_x &= -\delta \sin(\delta t) = -\delta P_y \, . \end{aligned} \tag{3.108}$$

The vector $\boldsymbol{P}$ thus performs a precession (so-called *spin precession*) around the $z$ axis, i.e. around $\boldsymbol{\Gamma} = (0, 0, \delta)$. This behaviour is shown in Fig. 3.1.

In the general case, using initial conditions

$$\begin{aligned} P_x(t)|_{t=0} &= P_x(0) \, , \\ P_y(t)|_{t=0} &= P_y(0) \, , \\ P_z(t)|_{t=0} &= P_z(0) \, , \end{aligned} \tag{3.109}$$

and introducing the *Rabi frequency*

$$\sqrt{\delta^2 + g^2} := \Omega_{\mathrm{R}} \, , \tag{3.110}$$

the solution of the Bloch equations is given by

$$P_x(t) = P_x(0) - \frac{\delta}{\Omega_R} P_y(0) \sin \Omega_R t +$$
$$\left[ \frac{\delta^2}{\Omega_R^2} P_x(0) - \frac{\delta g}{\Omega_R^2} P_z(0) \right] (\cos \Omega_R t - 1) \ ,$$
$$P_y(t) = P_y(0) \cos \Omega_R t +$$
$$\left[ \frac{\delta}{\Omega_R} P_x(0) - \frac{g}{\Omega_R} P_z(0) \right] \sin \Omega_R t \ ,$$
$$P_z(t) = P_z(0) + \frac{g}{\Omega_R} P_y(0) \sin \Omega_R t +$$
$$\left[ -\frac{g\delta}{\Omega_R^2} P_x(0) + \frac{g^2}{\Omega_R^2} P_z(0) \right] (\cos \Omega_R t - 1) \ .$$

$$(3.111)$$

Inserting this solution into the original equations (3.98), its validity can easily be shown. A direct derivation of (3.111) is possible based on the method of *Laplace transforms*. (3.111) represents a periodic motion with frequency $\Omega_R$.

Considering the special case

$$P_x(0) = P_y(0) = 0 \ , \tag{3.112}$$

(3.111) reduces to

$$P_x(t) = -\frac{\delta g}{\Omega_R^2} P_z(0) (\cos \Omega_R t - 1) \ ,$$
$$P_y(t) = -\frac{g}{\Omega_R} P_z(0) \sin \Omega_R t \ , \tag{3.113}$$
$$P_z(t) = P_z(0) + \frac{g^2}{\Omega_R^2} P_z(0) (\cos \Omega_R t - 1) \ .$$

For time-independent Hamiltonians there is, according to Sect. 3.2.2.4, a constant of motion given by

$$\Gamma \lambda = g P_x + \delta P_z = \text{constant} \ . \tag{3.114}$$

We easily convince ourselves that this relation is fulfilled by (3.111) and (3.113). However, this holds here only in the rotating reference frame and by using the rotating wave approximation.

The systems of equations (3.113) and (3.111) include the possibility of *nutation*. Due to the fact that optical driving forces are responsible, the nutation may be characterized as an *optical nutation*. For $\delta = 0$ and $P_z(0) = 1$, a rotation in the $y, z$ plane, perpendicular to the vector $\Gamma = (g, 0, 0)$, results:

$$\boxed{P(t) = (0, -\sin gt, \cos gt) \ .} \tag{3.115}$$

For $\delta = 0$ the Rabi frequency $\Omega_R$ is identical with $g$ and defines the angular velocity of the rotation. Figure 3.2 shows this kind of rotation. The general situation is depicted in Fig. 3.3.

### 3.2.3.3 Transformation Between Rotating and Laboratory Frame.

The Bloch equations considered so far are valid within a rotating reference frame; the effective Hamiltonian (3.91) and the corresponding Schrödinger equation (3.89) are also defined in this frame. Due to (2.259), the density operator in the *laboratory frame* is given by

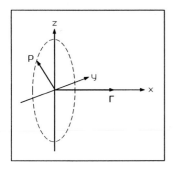

**Fig. 3.2.** Nutation of the polarization vector in the $y, z$ plane

$$\hat{\rho} = \sum_{i,j} \rho_{ij} \left| i \right\rangle \left\langle j \right| = \sum_{i,j} \rho_{ij} \hat{P}_{ij} \ . \tag{3.116}$$

Applying the unitary transformation defined by (2.145), and considering a 2-level system $(i, j = 1, 2)$, the projection operators $\hat{P}'_{ij}$ in the rotating frame and the original projection operators $\hat{P}_{ij}$ are connected by the scheme of transformations (2.152). Inserting (2.152) into (3.116), and using

$$\boxed{\rho_{ij} f_{ij}(t) = \rho'_{ij} \ (i, j = 1, 2)} \tag{3.117}$$

with

$$\boxed{f_{11}(t) = f_{22}(t) = 1 \ , \ \ f_{12}(t) = e^{i\omega t} \ , \ \ f_{21}(t) = e^{-i\omega t} \ ,} \tag{3.118}$$

the density operator (3.116) turns into

$$\hat{\rho} = \sum_{i,j=1}^{2} \rho'_{ij} \hat{P}_{ij} \ . \tag{3.119}$$

This formulation is nothing but the density operator $\hat{\rho}$ expressed by matrix elements $\rho'_{ij}$ of the density matrix $(\rho'_{ij})$ of the rotating frame (and by projection operators of the laboratory frame), where (3.117) defines the transformation rule of these matrix elements.

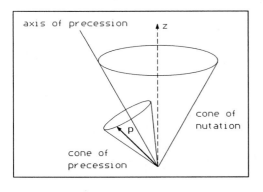

**Fig. 3.3.** Precession and nutation

Using (2.283),

$$\rho'_{12} = \frac{1}{2}\left(u'_{12} + iv'_{12}\right) \ , \quad \rho'_{21} = \frac{1}{2}\left(u'_{12} - iv'_{12}\right) \ ,$$

$$\rho'_{11} = \frac{1}{2}\left(1 - w'_1\right) \ , \quad \rho'_{22} = \frac{1}{2}\left(1 + w'_1\right) \ ,$$

$$(3.120)$$

and identifying $u'_{12} = P_x$, $v'_{12} = P_y$, $w'_1 = P_z$, the matrix elements in the rotating frame are

$$\rho'_{12} = \frac{1}{2}\left(P_x + iP_y\right) \ , \quad \rho'_{21} = \frac{1}{2}\left(P_x - iP_y\right) \ ,$$

$$\rho'_{11} = \frac{1}{2}\left(1 - P_z\right) \ , \quad \rho'_{22} = \frac{1}{2}\left(1 + P_z\right) \ .$$

$$(3.121)$$

Inserting these relations into the transformation rule (3.117), the matrix elements $\rho_{ij}$ are

$$\boxed{\begin{aligned} \rho_{12} &= \frac{1}{2}\left(P_x + iP_y\right)e^{-i\omega t} \ , \quad \rho_{21} = \frac{1}{2}\left(P_x - iP_y\right)e^{i\omega t} \ , \\ \rho_{11} &= \frac{1}{2}\left(1 - P_z\right) \ , \quad \rho_{22} = \frac{1}{2}\left(1 + P_z\right) \ . \end{aligned}}$$

$$(3.122)$$

These equations define the connection of the matrix elements $\rho_{ij}$ in the laboratory frame with the elements $P_i$ of the coherence vector in the rotating frame, as calculated in the previous sections.

Using Euler's relation $e^{\pm i\omega t} = \cos\omega t \pm i\sin\omega t$, and taking $u_{12}(t) = \rho_{12} + \rho_{21}$, $v_{12}(t) = i(\rho_{12} - \rho_{21})$, $w_1(t) = -\rho_{11} + \rho_{22}$ into account, the elements $u_{12}(t)$, $v_{12}(t)$, $w_1(t)$ of the coherence vector in the laboratory frame can be expressed by $P_x$, $P_y$, $P_z$ in the rotating frame as

$$\boxed{\begin{aligned} u_{12}(t) &= P_x \cos\omega t + P_y \sin\omega t \ , \\ v_{12}(t) &= P_x \sin\omega t - P_y \cos\omega t \ , \\ w_1(t) &= P_z \ . \end{aligned}}$$

$$(3.123)$$

**3.2.3.4 Dressed States.** (See [62].) The matrix representation of the effective Hamiltonian (3.91) (without the constant part $\frac{1}{2}\left(E_1 + E_2 + \hbar\omega\right)\hat{1}$) is – with respect to the eigenstates of $\hat{\sigma}_z$ ($|-1\rangle$, $|+1\rangle$) – given by

$$\boxed{\mathsf{H}^{(\mathrm{eff})} = \frac{\hbar}{2}\begin{pmatrix} -\delta & g \\ g & +\delta \end{pmatrix} \ .}$$

$$(3.124)$$

This matrix defines the effective time-independent Schrödinger equation:

$$\boxed{\mathsf{H}^{(\mathrm{eff})}\boldsymbol{\Psi} = E'\boldsymbol{\Psi} \ , \quad \boldsymbol{\Psi} = \begin{pmatrix} a \\ b \end{pmatrix} \ , \quad a^2 + b^2 = 1 \ .}$$

$$(3.125)$$

The eigenvalues $E'$ have to be determined by solving the *characteristic determinant*

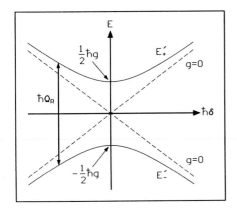

**Fig. 3.4.** Energy spectrum of the effective Hamiltonian as a function of detuning $\delta$

$$\begin{vmatrix} -\frac{\hbar\delta}{2} - E' & \frac{\hbar g}{2} \\ \frac{\hbar g}{2} & +\frac{\hbar\delta}{2} - E' \end{vmatrix} = E'^2 - \frac{\hbar^2\delta^2 + \hbar^2 g^2}{4} = 0 \ . \tag{3.126}$$

One obtains

$$\boxed{E'_\pm = \pm\frac{\hbar}{2}\sqrt{g^2 + \delta^2}} \tag{3.127}$$

so that the energy difference $\Delta E'$ of the two eigenstates is given by

$$\Delta E' = \hbar\sqrt{g^2 + \delta^2} = \hbar\Omega_R \ , \tag{3.128}$$

wherein $\Omega_R$ again is the Rabi frequency. (See Fig. 3.4.)

Inserting the result (3.127) into the linear system of equations defined by (3.125), and using $a_\pm^2 = 1 - b_\pm^2$, the components of the column matrix $\boldsymbol{\Psi}$ can be calculated as follows

$$b_\pm^2 = \frac{1}{2}\frac{\sqrt{g^2 + \delta^2} \mp \delta}{\sqrt{g^2 + \delta^2}} \ , \quad a_\pm^2 = \frac{1}{2}\frac{\sqrt{g^2 + \delta^2} \pm \delta}{\sqrt{g^2 + \delta^2}} \ . \tag{3.129}$$

Specializing in the case $\delta = 0$, the components $a, b$ are defined by

$$a^2 = b^2 = \frac{1}{2} \tag{3.130}$$

for both states. Multiplying the first equation of (3.125) with $a$, the second equation with $b$, and summing up both, one obtains the equation $E'_\pm = \hbar g a b$ so that

$$\begin{aligned} a = b = \frac{1}{\sqrt{2}} \ , \quad & E_+ = \frac{1}{2}\hbar|g| > 0 \ , \\ a = -b = \frac{1}{\sqrt{2}} \ , \quad & E_- = -\frac{1}{2}\hbar|g| < 0 \ . \end{aligned} \tag{3.131}$$

$a, b$ represent the coefficients in a superposition of two basis vectors. In the case of (3.131) the two superpositions are

$$\left|\Psi^+\right\rangle = \frac{1}{\sqrt{2}}\left(\left|-1\right\rangle + \left|+1\right\rangle\right) ,$$
$$\left|\Psi^-\right\rangle = \frac{1}{\sqrt{2}}\left(\left|-1\right\rangle - \left|+1\right\rangle\right) . \tag{3.132}$$

As we see from (2.31), these state vectors are eigenvectors of the operator $\hat{\sigma}_x$. The states defined by (3.132) as well as the general states determined by $a_\pm$, $b_\pm$ according to (3.129) are called *dressed states*.

*Polarization.* For the dressed states described by

$$\boldsymbol{\Psi}_\pm = \begin{pmatrix} a_\pm \\ b_\pm \end{pmatrix} , \tag{3.133}$$

the components $P_i$ of the polarization vector

$$\boldsymbol{P} = (P_x, P_y, P_z) = \left\langle \Psi^\pm | \hat{\boldsymbol{\sigma}} | \Psi^\pm \right\rangle , \quad \hat{\boldsymbol{\sigma}} = (\hat{\sigma}_x, \hat{\sigma}_y, \hat{\sigma}_z) \tag{3.134}$$

are defined by

$$P_i = \begin{pmatrix} a_\pm \\ b_\pm \end{pmatrix} \left(\sigma^i\right) \begin{pmatrix} a_\pm \\ b_\pm \end{pmatrix} \quad (i = 1, 2, 3, x, y, z) , \tag{3.135}$$

where $\left(\sigma^i\right)$ denotes Pauli matrices. Inserting (3.129) into the polarizations $P_i$, the result is

$$P_z^2 = \frac{\delta^2}{\delta^2 + g^2} ,$$
$$P_x^2 = \frac{g^2}{\delta^2 + g^2} , \tag{3.136}$$
$$P_y^2 = 0 .$$

This result is identical with the $P_0 = 1$ (pure state) case obtained already for the stationary Bloch equations (compare with (3.104), (3.101)).

### 3.2.3.5 Optically Driven 3-Level System. The Hamiltonian for a 3-level system is specified by

$$\hat{H} = E_1 \hat{P}_{11} + E_2 \hat{P}_{22} + E_3 \hat{P}_{33} . \tag{3.137}$$

The RWA can also be applied if several frequencies $\omega_j$ $(j = 1, 2, \ldots)$ are involved provided

$$\boxed{\delta_j, g_j \ll \omega_j , \quad |\omega_j - \omega_k| .} \tag{3.138}$$

Generalizing the results of Sect. 3.2.3.1, and assuming the RWA to be applicable, the interaction with 2 monochromatic light fields may be specified within 2 different scenarios, the so-called $\nu$ and $\Lambda$ *scenario*.

$\nu$ *Scenario.* The $\nu$ scenario (see Fig. 3.5) is defined by the coupling Hamiltonian

$$\hat{H}_{\mathrm{L}}(t) = \hat{H}_{\mathrm{L}}^{(21)}(t) + \hat{H}_{\mathrm{L}}^{(31)}(t) ,$$
(3.139)

where

$$\hat{H}_{\mathrm{L}}^{(ij)}(t) = \frac{1}{2} g_{ij} \hbar \left( \hat{P}_{ij} e^{-i\omega^{(ij)}t} + \hat{P}_{ji} e^{i\omega^{(ij)}t} \right) .$$
(3.140)

$g_{ij}$ are the respective coupling constants, and $\omega^{(ij)}$ represent the laser frequencies. $\hbar\omega^{(ij)}$ need not coincide with the transition energy $\omega_{ij} = E_i - E_j$. It is assumed here that this laser field is definitely far off resonance (with respect to any other transition).

We can now generalize the transformation method outlined in Sect. 3.2.3.1 to the present case: with

$$\hat{U}_{22} = e^{i\omega^{(21)}t\hat{P}_{22}} , \quad \hat{U}_{33} = e^{i\omega^{(31)}t\hat{P}_{33}}$$
(3.141)

the time-dependence can be removed by

$$\hat{H}' = \hat{U}_{22}^\dagger \hat{U}_{33}^\dagger \hat{H} \hat{U}_{33} \hat{U}_{22} .$$
(3.142)

The effective Hamiltonian then reads (cf. Sect. 2.2.3.3)

$$\hat{H}_{\mathrm{eff}} = E_1 \hat{P}_{11} + \left( E_2 - \hbar\omega^{(21)} \right) \hat{P}_{22} + \left( E_3 - \hbar\omega^{(31)} \right) \hat{P}_{33} +$$
$$\frac{1}{2} g_{21} \hbar \left( \hat{P}_{12} + \hat{P}_{21} \right) + \frac{1}{2} g_{31} \hbar \left( \hat{P}_{13} + \hat{P}_{31} \right) ,$$
(3.143)

and the $SU(3)$ form is given by (cf. (2.188))

$$\hat{H}_{\mathrm{eff}} = \frac{1}{3} \left( E_1 + E_2 + E_3 - \hbar\omega^{(21)} - \hbar\omega^{(31)} \right) \hat{1} +$$
$$\frac{1}{2} \left( E_2 - E_1 - \hbar\omega^{(21)} \right) \hat{w}_1 -$$
$$\frac{1}{2\sqrt{3}} \left( E_1 + E_2 - 2E_3 + 2\hbar\omega^{(31)} - \hbar\omega^{(21)} \right) \hat{w}_2 +$$
$$\frac{\hbar}{2} g_{21} \hat{u}_{12} + \frac{\hbar}{2} g_{31} \hat{u}_{13} .$$
(3.144)

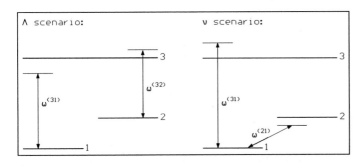

**Fig. 3.5.** $\Lambda$ and $\nu$ scenario

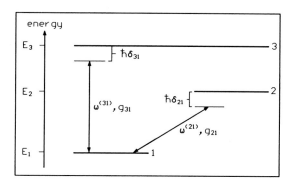

**Fig. 3.6.** $\nu$ scenario with laser frequencies, and coupling constants. In this example, detuning parameters $\delta_{31}, \delta_{21} > 0$ are considered

Introducing the detuning parameters (cf. Fig. 3.6)

$$\hbar\delta_{21} = E_2 - E_1 - \hbar\omega^{(21)} \,, \quad \hbar\delta_{31} = E_3 - E_1 - \hbar\omega^{(31)} \,, \tag{3.145}$$

one finally gets

$$\hat{H}_{\text{eff}} = \text{const.}\hat{1} + \frac{\hbar}{2}\left[\delta_{21}\hat{w}_1 + \frac{1}{\sqrt{3}}\left(2\delta_{31} - \delta_{21}\right)\hat{w}_2 + g_{21}\hat{u}_{12} + g_{31}\hat{u}_{13}\right] \tag{3.146}$$

so that the non-zero components of the characterizing $SU(3)$ vector are given by

$$
\begin{aligned}
\mathcal{H}_1 &= \text{Tr}\left\{\hat{H}_{\text{eff}}\hat{u}_{12}\right\} = \hbar g_{21} \,, \\
\mathcal{H}_2 &= \text{Tr}\left\{\hat{H}_{\text{eff}}\hat{u}_{13}\right\} = \hbar g_{31} \,, \\
\mathcal{H}_7 &= \text{Tr}\left\{\hat{H}_{\text{eff}}\hat{w}_1\right\} = \hbar\delta_{21} \,, \\
\mathcal{H}_8 &= \text{Tr}\left\{\hat{H}_{\text{eff}}\hat{w}_2\right\} = \frac{\hbar}{\sqrt{3}}\left(2\delta_{31} - \delta_{21}\right) \,.
\end{aligned}
\tag{3.147}
$$

The corresponding trace $\mathcal{H}_0$ is given by

$$\mathcal{H}_0 = \text{Tr}\left\{\hat{H}_{\text{eff}}\right\} = E_1 + E_2 + E_3 - \hbar\omega^{(21)} - \hbar\omega^{(31)} \,. \tag{3.148}$$

*$\Lambda$ Scenario.* In the $\Lambda$ scenario we replace (3.139) by (see Fig. 3.5)

$$\hat{H}_{\text{L}}(t) = \hat{H}_{\text{L}}^{(31)}(t) + \hat{H}_{\text{L}}^{(32)}(t) \,. \tag{3.149}$$

The unitary transformation with

$$\hat{U}_{22} = e^{-i\omega^{(32)}t\hat{P}_{22}} \,, \quad \hat{U}_{33} = e^{-i\omega^{(31)}t\hat{P}_{11}} \tag{3.150}$$

and

$$\hat{H}' = \hat{U}_{22}^\dagger\hat{U}_{11}^\dagger\hat{H}\hat{U}_{11}\hat{U}_{22} \tag{3.151}$$

then leads to the effective Hamiltonian

$$\hat{H}_{\text{eff}} = \left(E_1 + \hbar\omega^{(31)}\right)\hat{P}_{11} + \left(E_2 + \hbar\omega^{(32)}\right)\hat{P}_{22} + E_3\hat{P}_{33} +$$
$$\frac{1}{2}g_{31}\hbar\left(\hat{P}_{13} + \hat{P}_{31}\right) + \frac{1}{2}g_{32}\hbar\left(\hat{P}_{23} + \hat{P}_{32}\right) . \tag{3.152}$$

The $SU(3)$ form is given by

$$\boxed{\mathcal{H}_0 = \text{Tr}\left\{\hat{H}_{\text{eff}}\right\} = E_1 + E_2 + E_3 + \hbar\omega^{(31)} + \hbar\omega^{(32)}} \tag{3.153}$$

and

$$\boxed{\begin{aligned}
\mathcal{H}_2 &= \text{Tr}\left\{\hat{H}_{\text{eff}}\hat{u}_{13}\right\} = \hbar g_{31} , \\
\mathcal{H}_3 &= \text{Tr}\left\{\hat{H}_{\text{eff}}\hat{u}_{23}\right\} = \hbar g_{32} , \\
\mathcal{H}_7 &= \text{Tr}\left\{\hat{H}_{\text{eff}}\hat{w}_1\right\} = \hbar\left(\delta_{31} - \delta_{32}\right) , \\
\mathcal{H}_8 &= \text{Tr}\left\{\hat{H}_{\text{eff}}\hat{w}_2\right\} = \frac{\hbar}{\sqrt{3}}\left(\delta_{31} + \delta_{32}\right) .
\end{aligned}} \tag{3.154}$$

All other terms are zero. The detuning parameters are defined by

$$\hbar\delta_{32} = E_3 - E_2 - \hbar\omega^{(32)} , \quad \hbar\delta_{31} = E_3 - E_1 - \hbar\omega^{(31)} . \tag{3.155}$$

The unitary transformation for eliminating the time-dependence works at most for an open chain: $n-1$ laser fields connecting $n$ levels.

**3.2.3.6 Generalized Rotating Frames.** We have by now encountered some examples for which an oscillatory time-dependence of a Hamiltonian could be removed by means of a unitary transformation. Obviously, this is not always possible: counterrotating terms beyond the RWA are examples. We discuss a generalized scenario.

Consider the Hamiltonian

$$\hat{H}(t) = \hat{H}_0 + \hat{H}' + \hat{H}_{\text{L}}(t) , \tag{3.156}$$

where

$$\hat{H}_0 = \sum_{j=1}^{n} E_j^0 \hat{P}_{jj} , \tag{3.157}$$

$$\hat{1} = \sum_{j=1}^{n} \hat{P}_{jj} \tag{3.158}$$

is supposed to define a non-degenerate spectrum, and

$$\hat{H}' = \frac{1}{2}\sum_{i>j}\left(V_{ij}\hat{P}_{ij} + V_{ji}\hat{P}_{ji}\right) \tag{3.159}$$

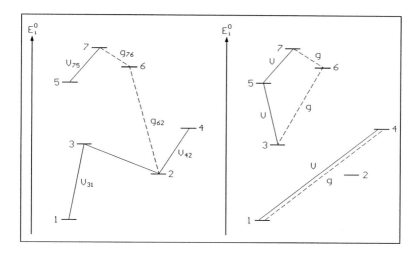

**Fig. 3.7.** No closed loop (part a)) and closed loop (part b))

introduces transfer coupling (cf. Sect. 2.2.5.5). $\hat{H}_L(t)$ describes time-dependent couplings of the RWA-type:

$$\hat{H}_L(t) = \frac{\hbar}{2} \sum_{i<j} \left( g_{ij}\hat{P}_{ij}e^{-i\omega^{(ij)}t} + g_{ji}\hat{P}_{ji}e^{i\omega^{(ij)}t} \right) . \tag{3.160}$$

We have to assume the applicability of the RWA, and $V_{ij} = V_{ji}^*$, $g_{ij} = g_{ji}^*$. Obviously, the $V_{ij}$ terms are merely special cases of the $g_{ij}$ terms with $\omega^{(ij)} = 0$. One can then compose a unitary transformation (cf. Sect. 2.2.3.3)

$$\hat{U} = \prod_{k=1}^{m} \hat{U}_{kk}(\alpha_k) \tag{3.161}$$

which removes the time-dependence, provided the set of interactions $g_{ij}$, $V_{ij}$ does not form any closed loops: for the case shown in Fig. 3.7, part a), we may choose

$$\hat{U} = \hat{U}_{55}(\alpha_5)\,\hat{U}_{77}(\alpha_7)\,\hat{U}_{66}(\alpha_6) \tag{3.162}$$

with

$$\alpha_5 = \alpha_7 = \left(-\omega^{(62)} + \omega^{(76)}\right)t , \quad \alpha_6 = \omega^{(62)}t . \tag{3.163}$$

(Note that the transformation $\hat{U}_{77}$ makes the nominally time-independent interaction $V_{57}$ time-dependent, which then has to be removed by the last transformation $\hat{U}_{55}$.) Closed loops spoil this procedure: in Fig. 3.7, part b) the step-wise removal of time-dependence is not possible. This also holds for transitions driven by 2 different time-dependent forces. This method can be generalized also to composite systems.

**3.2.3.7 Stationary States and Adiabatic Following.** For a time-independent Hamilton system, specified by the vector $\boldsymbol{\Gamma}$, a stationary state is defined by

$$\dot{\boldsymbol{\lambda}} = 0 . \qquad (3.164)$$

According to (3.52), such a coherence vector is constrained by the $s$ conditions

$$\sum_{j,k} f_{ijk} \Gamma_j \lambda_k = (\boldsymbol{\Gamma} \times \boldsymbol{\lambda})_i = 0 , \quad i = 1, 2, \ldots, s . \qquad (3.165)$$

Special stationary states are the eigenstates $\boldsymbol{\lambda}^{(m)}$ (pure states).

*Example 3.2.4.* Consider the 3-level Hamiltonian (2.188) with

$$\mathcal{H}_7 = \hbar \Gamma_7 = \hbar \omega_{21} \qquad (3.166)$$

and

$$\mathcal{H}_8 = \hbar \Gamma_8 = \frac{\hbar}{\sqrt{3}} (\omega_{31} + \omega_{32}) \qquad (3.167)$$

so that

$$|\boldsymbol{\Gamma}|^2 = \omega_{21}^2 + \frac{1}{3} (\omega_{31} + \omega_{32})^2 . \qquad (3.168)$$

The three eigenstates (cf. Fig. 2.10) are

$$\lambda_7^{(1)} = w_1^{(1)} = -1 , \quad \lambda_8^{(1)} = w_2^{(1)} = -\frac{1}{\sqrt{3}} , \qquad (3.169)$$

$$\lambda_7^{(2)} = w_1^{(2)} = 1 , \quad \lambda_8^{(2)} = w_2^{(2)} = -\frac{1}{\sqrt{3}} , \qquad (3.170)$$

$$\lambda_7^{(3)} = w_1^{(3)} = 0 , \quad \lambda_8^{(3)} = w_2^{(3)} = \frac{2}{\sqrt{3}} \qquad (3.171)$$

with

$$|\boldsymbol{\lambda}|^2 = \frac{4}{3} . \qquad (3.172)$$

Taking into account the properties of $f_{ijk}$ (cf. (2.78)), we immediately confirm equation (3.165). The directional cosines (cf. (3.70))

$$\boxed{\cos \Theta^{(m)} = \frac{\sum_j \Gamma_j \lambda_j^{(m)}}{|\boldsymbol{\Gamma}| |\boldsymbol{\lambda}|}} \qquad (3.173)$$

are here

$$\cos \Theta^{(1)} = -\frac{\sqrt{3}}{2} \frac{1}{|\boldsymbol{\Gamma}|} \left[ \omega_{21} + \frac{1}{\sqrt{3}} (\omega_{31} + \omega_{32}) \right] ,$$

$$\cos \Theta^{(2)} = \frac{\sqrt{3}}{2} \frac{1}{|\boldsymbol{\Gamma}|} \left[ \omega_{21} - \frac{1}{3} (\omega_{31} + \omega_{32}) \right] , \qquad (3.174)$$

$$\cos \Theta^{(3)} = \frac{1}{\sqrt{3}} \frac{1}{|\boldsymbol{\Gamma}|} (\omega_{31} + \omega_{32}) .$$

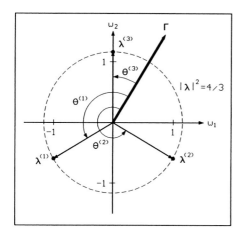

**Fig. 3.8.** Angles between the $\Gamma$ vector and the eigenstates in $SU(3)$

In Fig. 3.8 the angles between $\boldsymbol{\Gamma}$ (Hamilton model) and the corresponding eigenstates $\boldsymbol{\lambda}^{(m)}$, $m = 1, 2, 3$ are depicted.

The length of the vectors, the directional cosines as well as (3.165) are invariant under rotations in the $s$-dimensional vector space. Such rotations are induced by unitary transformations of the underlying basis functions $|m\rangle$ spanning the $n$-dimensional Hilbert space. Unitary transformations also result from unitary dynamics. These invariance properties are readily exploited for a direct diagonalization procedure: suppose that $\hat{H}$ (i.e. $\boldsymbol{\Gamma}$) is given with respect to any complete basis $|\alpha\rangle$. We then calculate from (3.165) the stationary eigenstates $\boldsymbol{\lambda}^{(m)}$ with respect to this original basis, and from (3.173) the respective directional cosines. If the eigenbasis of $\hat{H}$ is used for the generating operators $\hat{\lambda}_i$, i.e.

$$\left[\hat{H}, \hat{w}_l\right]_{-} = 0 \text{ for } l = 1, 2, \ldots, n , \tag{3.175}$$

we know that the $n$ eigenvectors of $\hat{H}$ are $(m = 1, 2, 3 \ldots, n)$

$$\boldsymbol{\lambda}^{(m)} = \left\{0, 0, \ldots, w_1^{(m)}, w_2^{(m)}, \ldots, w_{n-1}^{(m)}\right\} . \tag{3.176}$$

Knowing $|\boldsymbol{\Gamma}|$ and the directional cosines, we can thus calculate $\boldsymbol{\Gamma}$ with respect to this new eigenbasis. From $\boldsymbol{\Gamma}$ the eigenvalues are found according to (2.110):

$$E^{(\nu)}/\hbar = \frac{1}{n}\Gamma_0 + \frac{1}{2}\sum_{l=1}^{n-1}\Gamma_l w_l^{(\nu)} . \tag{3.177}$$

*Example 3.2.5.* Consider the 2-level system represented by (2.218) with

$$\mathcal{H}_1 = \hbar\Gamma_1 = \hbar g , \tag{3.178}$$

$$\mathcal{H}_2 = \hbar\Gamma_2 = 0 , \tag{3.179}$$

$$\mathcal{H}_3 = \hbar\Gamma_3 = \hbar\omega_{21} \qquad (3.180)$$

and

$$|\boldsymbol{\Gamma}| = \sqrt{g^2 + \omega_{21}^2} \; . \qquad (3.181)$$

$\boldsymbol{\Gamma}$ is thus not within the $w_l$ subspace. Identifying $f_{ijk}$ with $\varepsilon_{ijk}$ according to (2.2), we find with (3.165) the system of equations

$$0 = -\Gamma_3\lambda_2 \; ,$$
$$0 = \Gamma_3\lambda_1 - \Gamma_1\lambda_3 \; , \qquad (3.182)$$
$$0 = \Gamma_1\lambda_2 \; ,$$

from which we conclude

$$\lambda_2 = 0 \qquad (3.183)$$

and

$$\lambda_1 = \frac{g}{\omega_{21}}\lambda_3 \; . \qquad (3.184)$$

For a pure state we require

$$\lambda_1^2 + \lambda_3^2 = 1 \qquad (3.185)$$

so that

$$\lambda_1^2 = \frac{g^2}{\omega_{21}^2 + g^2} \; , \quad \lambda_3^2 = \frac{\omega_{21}^2}{\omega_{21}^2 + g^2} \; . \qquad (3.186)$$

With $\omega_{21}, g > 0$, $\lambda_1$ and $\lambda_3$ must have the same sign:

$$\boxed{\boldsymbol{\lambda}^\pm = \pm\frac{1}{\sqrt{\omega_{21}^2 + g^2}} \{g, 0, \omega_{21}\} \; .} \qquad (3.187)$$

This vector is parallel (antiparallel) to $\boldsymbol{\Gamma}$, i.e. the directional cosines are

$$\cos\Theta^{(\pm)} = \sum_j \Gamma_j\lambda_j^{(\pm)} \frac{1}{\sqrt{\omega_{21}^2 + g^2}} = \pm 1 \; . \qquad (3.188)$$

The scalar product is invariant with respect to any orthogonal transformation leading from $\boldsymbol{\Gamma}, \boldsymbol{\lambda}$ to $\tilde{\boldsymbol{\Gamma}}, \tilde{\boldsymbol{\lambda}}$:

$$\sum_j \tilde{\Gamma}_j\tilde{\lambda}_j^{(\pm)} = \pm\sqrt{\omega_{21}^2 + g^2} \; . \qquad (3.189)$$

For

$$\tilde{\boldsymbol{\lambda}} = (0, 0, \pm 1) \qquad (3.190)$$

we thus find

$$\tilde{\Gamma}_3 = \sqrt{\omega_{21}^2 + g^2} > 0 \qquad (3.191)$$

and

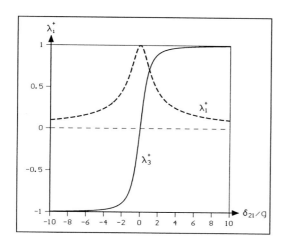

**Fig. 3.9.** Adiabatic following: the stationary state represented by $\left\{\lambda_1^+, 0, \lambda_3^+\right\}$ "follows" the changing vector $(g, 0, \delta_{21})$

$$E_\pm/\hbar = \Gamma_0 \pm \sqrt{\omega_{21}^2 + g^2} \ . \tag{3.192}$$

This is an alternative derivation of the dressed states, introduced in Sect. 3.2.3.4: $g$ is then the coupling to the light field, and $\omega_{21}$ is (in RWA) replaced by the detuning parameter $\delta_{21}$, which can now be positive or negative and controlled from the outside:

$$\boldsymbol{\lambda}^\pm = \pm \frac{1}{\sqrt{\delta_{21}^2 + g^2}} \{g, 0, \delta_{21}\} \ . \tag{3.193}$$

The state $\boldsymbol{\lambda}^\pm$ corresponds to the energy $E'_\pm$ as of (3.127). For given $g > 0$ we can thus *adiabatically invert* the population by slowly sweeping the detuning from $\delta/g \ll -1$ to $\delta/g \gg 1$, where *adiabatically*, here, means that (cf. [107])

$$\left|\dot{\boldsymbol{\lambda}}\right| / |\boldsymbol{\lambda}| \ll |\boldsymbol{\Gamma}| \ . \tag{3.194}$$

In Fig. 3.9 the *adiabatic following* of the coherence vector

$$\boldsymbol{\lambda}^+ = \left\{\lambda_1^+, 0, \lambda_3^+\right\} \tag{3.195}$$

is shown: $\boldsymbol{\lambda}^+$ "follows" the changing vector

$$\boldsymbol{\Gamma} = (g, 0, \delta_{21}) \ . \tag{3.196}$$

(In a more realistic model, spontaneous decay would have to be included.) These considerations can be generalized to driven $n$-level systems, like the $\nu$ scenario in $SU(3)$.

**3.2.3.8 Dressed States of Driven 3-Level System.** We consider the $\Lambda$ scenario specified by Fig. 3.10. We conclude from (3.152) that the Hamilton matrix is given by

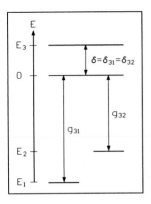

**Fig. 3.10.** Energy spectrum and parameters for a special $\Lambda$ scenario

$$H^{(\mathrm{eff})} = \begin{pmatrix} E_1 + \hbar\omega^{(31)} & 0 & g_{31}/2 \\ 0 & E_2 + \hbar\omega^{(32)} & g_{32}/2 \\ g_{31}/2 & g_{32}/2 & E_3 \end{pmatrix}$$

$$= \frac{1}{2} \begin{pmatrix} 0 & 0 & g_{31} \\ 0 & 0 & g_{32} \\ g_{31} & g_{32} & 2\delta \end{pmatrix} \tag{3.197}$$

so that the eigenvalue equation reads

$$H^{(\mathrm{eff})}\boldsymbol{\Psi} = E'\boldsymbol{\Psi} . \tag{3.198}$$

Thus, the eigenvalues follow from

$$\det \begin{pmatrix} -E' & 0 & g_{31}/2 \\ 0 & -E' & g_{32}/2 \\ g_{31}/2 & g_{32}/2 & \delta - E' \end{pmatrix}$$

$$= -E' \left[ (\delta - E') - \frac{1}{4}|g_{32}|^2 \right] + \frac{1}{4}|g_{31}|^2 E'$$

$$= 0 , \tag{3.199}$$

i.e.

$$E_1' = 0 , \quad E_{2,3}' = \frac{1}{2}(\delta \pm \Omega) \tag{3.200}$$

with

$$\Omega^2 = \delta^2 + g^2 , \tag{3.201}$$

and

$$g^2 = |g_{31}|^2 + |g_{32}|^2 . \tag{3.202}$$

Let the corresponding eigenvalues be denoted by

$$\boldsymbol{\Psi} = \begin{pmatrix} a \\ b \\ c \end{pmatrix} \tag{3.203}$$

with

$$a^2 + b^2 + c^2 = 1 \ . \tag{3.204}$$

From the eigenvalue equation above then follows for $E' = 0$:

$$0 = g_{31}c \ ,$$
$$0 = g_{32}c \ , \tag{3.205}$$
$$0 = g_{31}a + g_{32}b + 2\delta c \ .$$

Thus, with $c = 0$, $a^2 + b^2 = 1$, and $\tan \Theta = g_{31}/g_{32}$, one obtains

$$\Psi = \begin{pmatrix} \cos \Theta \\ -\sin \Theta \\ 0 \end{pmatrix} \tag{3.206}$$

or

$$\boxed{|\psi_1\rangle = \cos \Theta \, |1\rangle - \sin \Theta \, |2\rangle \ .} \tag{3.207}$$

This is a coherent state, not involving the transient state $|3\rangle$ and independent of detuning $\delta$. It can be adjusted by the laser field intensities. The population is said to be "trapped" in $|\psi_1\rangle$ (as there is no decay via state $|3\rangle$).

The two states with energy eigenvalues $E'_{2,3}$ are determined by

$$0 = -(\delta \pm \Omega)\, a + g_{31}c \ ,$$
$$0 = -(\delta \pm \Omega)\, b + g_{32}c \ , \tag{3.208}$$
$$0 = g_{31}a + g_{32}b - (\delta \mp \Omega)\, c$$

with

$$a = \frac{g_{31}}{\delta \pm \Omega}c \ ,$$
$$b = \frac{g_{32}}{\delta \pm \Omega}c \ . \tag{3.209}$$

Observing $a^2 + b^2 + c^2 = 1$ one obtains

$$a = \frac{g_{31}\,(\delta \pm \Omega)}{g^2 + (\delta \pm \Omega)^2} \ ,$$
$$b = \frac{g_{32}\,(\delta \pm \Omega)}{g^2 + (\delta \pm \Omega)^2} \ , \tag{3.210}$$
$$c = \frac{(\delta \pm \Omega)^2}{g^2 + (\delta \pm \Omega)^2} \ .$$

These eigenstates invlove the transient state $|3\rangle$ and depend on all three control parameters $g_{31}$, $g_{21}$, $\delta$.

### 3.2.4 Heisenberg Picture

In this section the *Heisenberg picture* is introduced. *Heisenberg operators* are studied, related *equations of motion* are discussed, and *time-dependent correlation functions* are considered.

**3.2.4.1 Heisenberg's Equation of Motion.** According to (3.21) the density operator is (in the Schrödinger picture) given by

$$\hat{\rho}(t) = \hat{U}(t)\hat{\rho}(0)\hat{U}^{\dagger}(t) \tag{3.211}$$

with

$$\hat{U}^{\dagger}(t)\hat{U}(t) = \hat{U}(t)\hat{U}^{\dagger}(t) = \hat{1} , \tag{3.212}$$

while the basic observables $\hat{A}$ are explicitly time-independent. All the operators in the Heisenberg picture are related to those in the Schrödinger picture by the unitary transformation

$$\hat{\rho}^{(\mathrm{H})} = \hat{U}^{\dagger}(t)\hat{\rho}(t)\hat{U}(t) = \hat{\rho}(0) = \text{constant} \tag{3.213}$$

$$\hat{A}^{(\mathrm{H})}(t) = \hat{U}^{\dagger}(t)\hat{A}\hat{U}(t) . \tag{3.214}$$

Using this picture, the dynamics is represented by time-dependent operators and time-independent density matrices (or time-independent state vectors). Therefore, also the (transformed) coherence vector is time-independent:

$$\boldsymbol{\lambda}^{(\mathrm{H})}(t) = \boldsymbol{\lambda}(0) = \text{constant} . \tag{3.215}$$

Equations of motion are now to be derived for the observables $\hat{A}$. Differentiating the expression (3.214) with respect to time, assuming *Schrödinger operators* $\hat{A}$ with

$$\frac{\partial \hat{A}}{\partial t} = 0 , \tag{3.216}$$

and using

$$\hat{U}^{\dagger}(t)\hat{H}\hat{U}(t) := \hat{H}^{(\mathrm{H})} = \hat{U}^{\dagger}(t)\hat{U}(t)\hat{H} = \hat{H} , \tag{3.217}$$

one obtains the evolution equation

$$\boxed{\; i\hbar\frac{\partial}{\partial t}\hat{A}^{(\mathrm{H})}(t) = \left[\hat{A}^{(\mathrm{H})}(t), \hat{H}\right]_{-} = -\left[\hat{H}, \hat{A}^{(\mathrm{H})}(t)\right]_{-} . \;} \tag{3.218}$$

This equation of motion is called *Heisenberg equation*. This equation determines the motion of time-dependent *Heisenberg operators* $\hat{A}^{(\mathrm{H})}(t)$. In this equation the Schrödinger operator $\hat{H}$ occurs, because – due to (3.217) – a Hamiltonian $\hat{H}$ in the Schrödinger picture is equivalent to a Hamiltonian $\hat{H}^{(\mathrm{H})}$ in the Heisenberg picture.

**3.2.4.2 Casimir Operators.** Inserting the representation

$$\hat{H} = \frac{1}{2}\mathrm{Tr}\left\{\hat{H}\right\}\hat{1} + \frac{1}{2}\sum_{j=1}^{s}\mathcal{H}_j\hat{\lambda}_j \ , \ \ \mathcal{H}_j = \mathrm{Tr}\left\{\hat{H}\hat{\lambda}_j\right\} \tag{3.219}$$

(use (3.49) for a time-independent Hamiltonian $\hat{H}$) into the Heisenberg equation (3.218), the operator equation turns into

$$\boxed{i\hbar\frac{\partial}{\partial t}\hat{A}^{(\mathrm{H})}(t) = -\frac{1}{2}\sum_{j=1}^{s}\mathcal{H}_j\left[\hat{\lambda}_j, \hat{A}^{(\mathrm{H})}(t)\right]_-}\ . \tag{3.220}$$

Equation (3.220) is an equation of motion expressed in terms of generating operators of the special unitary group $SU(n)$ ($s = n^2 - 1$). If now

$$\boxed{\left[\hat{\lambda}_j, \hat{A}^{(\mathrm{H})}(t)\right]_- = \left[\hat{\lambda}_j, \hat{A}(t)\right]_- = 0 \ \ (j = 1, 2, \ldots, s)} \tag{3.221}$$

(the commutator relations are invariant under unitary transformations, i.e. they are the same in the Schrödinger and the Heisenberg picture), the r.h.s. of (3.220) is equal zero so that $\hat{A}^{(\mathrm{H})}(t)$ is a constant of motion:

$$i\hbar\frac{\partial}{\partial t}\hat{A}^{(\mathrm{H})}(t) = 0 \ . \tag{3.222}$$

Operators which commute with all generating operators of a group $SU(n)$ are called *Casimir operators* (compare with Racah's theorem after equation (2.73)), which are thus specified by the $s$ conditions (3.221). According to Racah's theorem, there are $r = n - 1$ independent Casimir operators.

**3.2.4.3 Time-Evolution of $\mathcal{A}_j^{(\mathrm{H})}$.** Applying the representation (2.85) to the Heisenberg operator, we obtain

$$\hat{A}^{(\mathrm{H})}(t) = \frac{1}{n}\mathrm{Tr}\left\{\hat{A}^{(\mathrm{H})}(t)\right\}\hat{1} + \frac{1}{2}\sum_{j=1}^{s}\mathcal{A}_j^{(\mathrm{H})}(t)\hat{\lambda}_j \ , \tag{3.223}$$

$$\mathcal{A}_j^{(\mathrm{H})}(t) = \mathrm{Tr}\left\{\hat{A}^{(\mathrm{H})}(t)\hat{\lambda}_j\right\} \ . \tag{3.224}$$

We first note that the trace of $\hat{A}^{(\mathrm{H})}(t)$ is constant so that its dynamics does not require further analysis:

$$\mathcal{A}_0 = \mathrm{Tr}\left\{\hat{A}^{(\mathrm{H})}(t)\right\} = \mathrm{Tr}\left\{\hat{U}^\dagger(t)\hat{A}\hat{U}(t)\right\} = \mathrm{Tr}\left\{\hat{A}\right\} = \mathrm{constant} \ . \tag{3.225}$$

Inserting the Heisenberg equation (3.218) into

$$\frac{\partial}{\partial t}\mathcal{A}_j^{(\mathrm{H})}(t) = \mathrm{Tr}\left\{\hat{\lambda}_j\frac{\partial}{\partial t}\hat{A}^{(\mathrm{H})}(t)\right\} \ , \tag{3.226}$$

this derivative can be replaced by

$$\frac{\partial}{\partial t}\mathcal{A}_j^{(\mathrm{H})}(t) = -\frac{1}{i\hbar}\mathrm{Tr}\left\{\hat{\lambda}_j\left[\hat{H},\hat{A}^{(\mathrm{H})}(t)\right]_-\right\}$$

$$= \frac{1}{i\hbar}\mathrm{Tr}\left\{\hat{H}\left[\hat{\lambda}_j,\hat{A}^{(\mathrm{H})}(t)\right]_-\right\} . \tag{3.227}$$

Making use of the commutator relation

$$\left[\hat{\lambda}_j,\hat{A}^{(\mathrm{H})}(t)\right]_- = \frac{1}{2}\sum_{k=1}^{s}\mathcal{A}_k^{(\mathrm{H})}(t)\left[\hat{\lambda}_j,\hat{\lambda}_k\right]_- \tag{3.228}$$

(insert (3.223) into the commutator $\left[\hat{\lambda}_j,\hat{A}^{(\mathrm{H})}(t)\right]_- = \hat{\lambda}_j\hat{A}^{(\mathrm{H})}(t) - \hat{A}^{(\mathrm{H})}(t)\hat{\lambda}_j$),
the derivative (3.226) finally reads

$$\boxed{\frac{\partial}{\partial t}\mathcal{A}_j^{(\mathrm{H})}(t) = \frac{1}{2i\hbar}\sum_{k=1}^{s}\mathrm{Tr}\left\{\hat{H}\left[\hat{\lambda}_j,\hat{\lambda}_k\right]_-\right\}\mathcal{A}_k^{(\mathrm{H})}(t) .} \tag{3.229}$$

This equation determines the motion of $\mathcal{A}_j^{(\mathrm{H})}(t)$.

Using the matrix elements

$$-\frac{1}{2i\hbar}\mathrm{Tr}\left\{\hat{H}(t)\left[\hat{\lambda}_i,\hat{\lambda}_k\right]_-\right\} = \Omega_{ik} = \sum_j f_{ijk}\Gamma_j , \quad \Gamma_j = \frac{\mathcal{H}_j}{\hbar} \tag{3.230}$$

(cf. (3.48), (3.51)), the equation of motion (3.229) can finally be expressed
by

$$\frac{\partial}{\partial t}\mathcal{A}_j^{(\mathrm{H})}(t) = -\sum_k \Omega_{jk}\mathcal{A}_k^{(\mathrm{H})}(t) \tag{3.231}$$

or

$$\boxed{\frac{\partial}{\partial t}\mathcal{A}_i^{(\mathrm{H})}(t) = -\sum_{j,k} f_{ijk}\Gamma_j\mathcal{A}_k^{(\mathrm{H})}(t) .} \tag{3.232}$$

These equations describe rotations (see (3.48)ff.) in the $n^2 - 1$-dimensional
space.

Comparing the evolution equation (3.232) (which determines $\mathcal{A}_k^{(\mathrm{H})}(t)$ in
the Heisenberg picture) with the equation (3.52) (which determines the time-
dependent coherence vector $\boldsymbol{\lambda}(t)$ in the Schrödinger picture), it is obvious
that the vectors $\boldsymbol{\lambda}(t)$ and $\left(\mathcal{A}_k^{(\mathrm{H})}(t)\right)$ are just counter-rotating. Figure 3.11
illustrates this situation in the $u_{12}, w_1$ plane.

**3.2.4.4 Dynamics of Expectation Values (Heisenberg Picture).** In
the Heisenberg picture, expectation values are defined by

$$\left\langle \hat{A}^{(\mathrm{H})}(t)\right\rangle = \mathrm{Tr}\left\{\hat{\rho}(0)\hat{A}^{(\mathrm{H})}(t)\right\} . \tag{3.233}$$

It is obvious that this expectation value is the same as in the Schrödinger
picture:

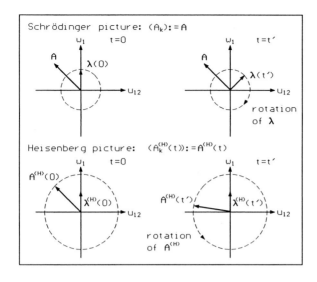

**Fig. 3.11.** Heisenberg picture and Schrödinger picture. The motion of the coherence vector and of the vector $\mathcal{A}_k^{(\mathrm{H})}(t)$

$$\mathrm{Tr}\left\{\hat{\rho}(0)\hat{A}^{(\mathrm{H})}(t)\right\} = \mathrm{Tr}\left\{\hat{\rho}(0)\hat{U}^{\dagger}(t)\hat{A}\hat{U}(t)\right\} = \mathrm{Tr}\left\{\hat{U}^{\dagger}(t)\rho(0)\hat{U}(t)\hat{A}\right\}$$

$$= \mathrm{Tr}\left\{\hat{\rho}(t)\hat{A}\right\} . \tag{3.234}$$

Differentiating (3.234) with respect to time, one obtains

$$\frac{\partial}{\partial t}\left\langle \hat{A}^{(\mathrm{H})}(t)\right\rangle = \mathrm{Tr}\left\{\hat{\rho}(0)\frac{\partial}{\partial t}\hat{A}^{(\mathrm{H})}(t)\right\} . \tag{3.235}$$

Inserting the Heisenberg operator (3.223), and taking into account that the trace $\mathrm{Tr}\left\{\hat{A}^{(\mathrm{H})}\right\}$ is a constant of motion, the derivative (3.235) yields

$$\frac{\partial}{\partial t}\left\langle \hat{A}^{(\mathrm{H})}(t)\right\rangle = \frac{1}{2}\sum_{j=1}^{s}\mathrm{Tr}\left\{\hat{\rho}(0)\hat{\lambda}_j\frac{\partial}{\partial t}\mathcal{A}_j^{(\mathrm{H})}(t)\right\} . \tag{3.236}$$

Using formula (3.45), the density operator $\hat{\rho}(0)$ is given by

$$\hat{\rho}(0) = \frac{1}{n}\hat{1} + \frac{1}{2}\sum_{k=1}^{s}\lambda_k(0)\hat{\lambda}_k . \tag{3.237}$$

Inserting this operator into (3.236), and using the trace relations

$$\mathrm{Tr}\left\{\hat{\lambda}_k\hat{\lambda}_j\right\} = 2\delta_{kj} \tag{3.238}$$

(cf. (2.275)), the derivative (3.236) reduces to

$$\frac{\partial}{\partial t}\left\langle \hat{A}^{(\mathrm{H})}(t)\right\rangle = \frac{1}{2}\sum_{k=1}^{s}\lambda_k(0)\frac{\partial}{\partial t}\mathcal{A}_k^{(\mathrm{H})}(t) . \tag{3.239}$$

Then, replacing $\frac{\partial}{\partial t}\mathcal{A}_k^{(\mathrm{H})}(t)$ by (3.232), one obtains the evolution equation

$$\boxed{\frac{\partial}{\partial t}\left\langle \hat{A}^{(\mathrm{H})}(t)\right\rangle = \frac{1}{2}\sum_{j,j',k=1}^{s} f_{jj'k}\Gamma'_j \mathcal{A}_j^{(\mathrm{H})}(t)\lambda_k(0)\,.}$$

(3.240)

For the special unitary group $SU(2)$ ($f_{ijk} = \varepsilon_{ijk}$), the r.h.s. of this formula can be interpreted as the volume of a parallelepiped spanned by the 3 vectors $\boldsymbol{\Gamma}$, $\boldsymbol{A}^{(\mathrm{H})}$, and $\boldsymbol{\lambda}(0)$. This volume is the same in both pictures, cf. Fig. 3.11: equation (3.240) coincides with (3.66).

### 3.2.4.5 2-Time-Correlation Functions.
Consider the two Heisenberg operators in $SU(2)$, $\hat{A}^{(\mathrm{H})}$, $\hat{B}^{(\mathrm{H})}$, with

$$\mathrm{Tr}\left\{\hat{A}^{(\mathrm{H})}(t_1)\right\} = 0\,,\quad \mathrm{Tr}\left\{\hat{B}^{(\mathrm{H})}(t_2)\right\} = 0\,.$$

(3.241)

Using the representation (3.223), these two Heisenberg operators can be written as

$$\hat{A}^{(\mathrm{H})}(t_1) = \frac{1}{2}\sum_{j=1}^{3} \mathcal{A}_j^{(\mathrm{H})}(t_1)\hat{\sigma}_j\,,$$

(3.242)

$$\hat{B}^{(\mathrm{H})}(t_2) = \frac{1}{2}\sum_{j=1}^{3} \mathcal{B}_j^{(\mathrm{H})}(t_2)\hat{\sigma}_j\,,$$

(3.243)

where $\hat{\lambda}_j = \hat{\sigma}_j$ was used. Thus, the operator product $\hat{A}^{(\mathrm{H})}(t_1)\hat{B}^{(\mathrm{H})}(t_2)$ reads

$$\hat{A}^{(\mathrm{H})}(t_1)\hat{B}^{(\mathrm{H})}(t_2) = \frac{1}{4}\sum_{i,j=1}^{3} \mathcal{A}_i^{(\mathrm{H})}(t_1)\mathcal{B}_j^{(\mathrm{H})}(t_2)\hat{\sigma}_i\hat{\sigma}_j\,.$$

(3.244)

Correlation functions have already been introduced in Sect. 2.4. The 2-time-correlation function $C(t_1,t_2)$ can be defined by

$$C(t_1,t_2) = \left\langle \hat{A}^{(\mathrm{H})}(t_1)\hat{B}^{(\mathrm{H})}(t_2)\right\rangle\,.$$

(3.245)

We thus find

$$C(t_1,t_2) = \mathrm{Tr}\left\{\hat{\rho}(0)\hat{A}^{(\mathrm{H})}(t_1)\hat{B}^{(\mathrm{H})}(t_2)\right\}$$

$$= \frac{1}{4}\sum_{i,j=1}^{3} \mathcal{A}_i^{(\mathrm{H})}(t_1)\mathcal{B}_j^{(\mathrm{H})}(t_2)\mathrm{Tr}\left\{\hat{\rho}(0)\hat{\sigma}_i\hat{\sigma}_j\right\}\,.$$

(3.246)

Inserting the density operator

$$\hat{\rho}(0) = \frac{1}{2}\hat{1} + \frac{1}{2}\sum_{k=1}^{3} P_k(0)\hat{\sigma}_k$$

(3.247)

(use (3.237) and replace $\hat{\lambda}_k$, $\lambda_k(0)$ by $\hat{\sigma}_k$, $P_k(0)$), (3.246) can be recast into

$$C(t_1, t_2) = \frac{1}{8} \sum_{i,j=1}^{3} \mathcal{A}_i^{(\mathrm{H})}(t_1) \mathcal{B}_j^{(\mathrm{H})}(t_2) \mathrm{Tr}\left\{\hat{\sigma}_i \hat{\sigma}_j\right\} +$$

$$\frac{1}{8} \sum_{i,j,k=1}^{3} \mathcal{A}_i^{(\mathrm{H})}(t_1) \mathcal{B}_j^{(\mathrm{H})}(t_2) \mathrm{Tr}\left\{\hat{\sigma}_i \hat{\sigma}_j \hat{\sigma}_k\right\} P_k(0) . \tag{3.248}$$

Making use of the trace relations (2.24) and (2.25), this equation reduces to

$$C(t_1, t_2) = \frac{1}{2} \sum_{i=1}^{3} \mathcal{A}_i^{(\mathrm{H})}(t_1) \mathcal{B}_i^{(\mathrm{H})}(t_2) +$$

$$\frac{\mathrm{i}}{4} \sum_{i,j,k=1}^{3} \varepsilon_{ijk} \mathcal{A}_i^{(\mathrm{H})}(t_1) \mathcal{B}_j^{(\mathrm{H})}(t_2) P_k(0) . \tag{3.249}$$

*Example 3.2.6.* Let

$$\hat{A}^{(\mathrm{H})}(t = 0) = \hat{B}^{(\mathrm{H})}(t = 0) = \hat{\sigma}_z \tag{3.250}$$

and the state $\boldsymbol{P}(0)$ represent complete polarization, $\boldsymbol{P} = (1, 0, 0)$, $\boldsymbol{P} = (0, 1, 0)$, or $\boldsymbol{P} = (0, 0, 1)$. Then the second sum of (3.249) vanishes, $C(t_1, t_2)$ remains real. Assuming

$$\boxed{\begin{aligned} \mathcal{A}_i^{(\mathrm{H})}(t_1) &= 2\left(0, \sin g t_1, \cos g t_1\right) , \\ \mathcal{B}_i^{(\mathrm{H})}(t_2) &= 2\left(0, \sin g t_2, \cos g t_2\right) , \end{aligned}} \tag{3.251}$$

the correlation term (3.249) reduces to

$$\boxed{C(t_1, t_2) = \sin g t_1 \sin g t_2 + \cos g t_1 \cos g t_2 = \cos g(t_1 - t_2) .} \tag{3.252}$$

This $C(t_1, t_2)$ represents a special example of a 2-time-correlation function. We note that this 2-time function has the same form as the 2-angle correlation function for a $SU(2) \otimes SU(2)$ system in an EPR state, as given in (2.731). One can show that, as a consequence, temporal Bell inequalities are violated: the respective "histories" are inconsistent (cf. [105]).

### 3.2.5 Network Dynamics

In this section, coupled sets of equations for the *dynamics of quantum networks* are investigated. In particular, the Bloch equations are extended to composite systems. 1-node and 2-node interactions are considered. We start with basic commutator relations, the density operator, and the basic Hamiltonian.

**3.2.5.1 Commutators and Anticommutators in Product Space.** Consider composite systems defined in a product space composed of various subspaces. Let $\hat{A}(\nu)$, $\hat{C}(\nu)$ be operators in subspace $\nu$, and $\hat{B}(\nu')$, $\hat{D}(\nu')$ operators in subspace $\nu'$.

*Commutator Relations.* Between these operators the commutator is defined, as usual, by

$$
\left[ \hat{A}(\nu) \otimes \hat{B}(\nu'), \hat{C}(\nu) \otimes \hat{D}(\nu') \right]_-
$$

$$
= \hat{A}(\nu)\hat{C}(\nu) \otimes \hat{B}(\nu')\hat{D}(\nu') - \hat{C}(\nu)\hat{A}(\nu) \otimes \hat{D}(\nu')\hat{B}(\nu')
$$

$$
= \frac{1}{2}\hat{A}(\nu)\hat{C}(\nu) \otimes \hat{B}(\nu')\hat{D}(\nu') - \frac{1}{2}\hat{C}(\nu)\hat{A}(\nu) \otimes \hat{B}(\nu')\hat{D}(\nu') +
$$

$$
\frac{1}{2}\hat{A}(\nu)\hat{C}(\nu) \otimes \hat{D}(\nu')\hat{B}(\nu') - \frac{1}{2}\hat{C}(\nu)\hat{A}(\nu) \otimes \hat{D}(\nu')\hat{B}(\nu') +
$$

$$
\frac{1}{2}\hat{A}(\nu)\hat{C}(\nu) \otimes \hat{B}(\nu')\hat{D}(\nu') - \frac{1}{2}\hat{A}(\nu)\hat{C}(\nu) \otimes \hat{D}(\nu')\hat{B}(\nu') +
$$

$$
\frac{1}{2}\hat{C}(\nu)\hat{A}(\nu) \otimes \hat{B}(\nu')\hat{D}(\nu') - \frac{1}{2}\hat{C}(\nu)\hat{A}(\nu) \otimes \hat{D}(\nu')\hat{B}(\nu')
$$

$$
= \frac{1}{2}\left[ \hat{A}(\nu)\hat{C}(\nu) - \hat{C}(\nu)\hat{A}(\nu) \right] \otimes \hat{B}(\nu')\hat{D}(\nu') +
$$

$$
\frac{1}{2}\left[ \hat{A}(\nu)\hat{C}(\nu) - \hat{C}(\nu)\hat{A}(\nu) \right] \otimes \hat{D}(\nu')\hat{B}(\nu') +
$$

$$
\frac{1}{2}\hat{A}(\nu)\hat{C}(\nu) \otimes \left[ \hat{B}(\nu')\hat{D}(\nu') - \hat{D}(\nu')\hat{B}(\nu') \right] +
$$

$$
\frac{1}{2}\hat{C}(\nu)\hat{A}(\nu) \otimes \left[ \hat{B}(\nu')\hat{D}(\nu') - \hat{D}(\nu')\hat{B}(\nu') \right] \tag{3.253}
$$

(cf. (2.518)). Introducing the *anticommutator*

$$
\left[ \hat{E}(\nu), \hat{F}(\nu) \right]_+ = \hat{E}(\nu)\hat{F}(\nu) + \hat{F}(\nu)\hat{E}(\nu) , \tag{3.254}
$$

(3.253) can be decomposed as

$$
\boxed{
\begin{aligned}
&\left[ \hat{A}(\nu) \otimes \hat{B}(\nu'), \hat{C}(\nu) \otimes \hat{D}(\nu') \right]_- \\
&= \frac{1}{2}\left[ \hat{A}(\nu), \hat{C}(\nu) \right]_- \otimes \left[ \hat{B}(\nu'), \hat{D}(\nu') \right]_+ + \\
&\quad \frac{1}{2}\left[ \hat{A}(\nu), \hat{C}(\nu) \right]_+ \otimes \left[ \hat{B}(\nu'), \hat{D}(\nu') \right]_- .
\end{aligned}}
\tag{3.255}
$$

The commutator relation (3.255) is valid for any set of operators $\hat{A}$, $\hat{B}$, $\hat{C}$, $\hat{D}$. The following special cases are easily shown to follow:

$$
\hat{B}(\nu') = \hat{D}(\nu') = \hat{1}(\nu') :
$$

$$
\left[ \hat{A}(\nu) \otimes \hat{1}(\nu'), \hat{C}(\nu) \otimes \hat{1}(\nu') \right]_- = \left[ \hat{A}(\nu), \hat{C}(\nu) \right]_- \otimes \hat{1}(\nu') ,
$$

$$
\hat{A}(\nu) = \hat{C}(\nu) = \hat{1}(\nu) : \tag{3.256}
$$

$$
\left[ \hat{1}(\nu) \otimes \hat{B}(\nu'), \hat{1}(\nu) \otimes \hat{D}(\nu') \right]_- = \hat{1}(\nu) \otimes \left[ \hat{B}(\nu'), \hat{D}(\nu') \right]_- ,
$$

$\hat{B}(\nu') = \hat{1}(\nu')$ :
$$\left[\hat{A}(\nu) \otimes \hat{1}(\nu'), \hat{C}(\nu) \otimes \hat{D}(\nu')\right]_- = \left[\hat{A}(\nu), \hat{C}(\nu)\right]_- \otimes \hat{D}(\nu') ,$$

$\hat{A}(\nu) = \hat{1}(\nu)$ :
$$\left[\hat{1}(\nu) \otimes \hat{B}(\nu'), \hat{C}(\nu) \otimes \hat{D}(\nu')\right]_- = \hat{C}(\nu) \otimes \left[\hat{B}(\nu'), \hat{D}(\nu')\right]_- ,$$

(3.257)

$\hat{B}(\nu') = \hat{1}(\nu')$ ,  $\hat{C}(\nu) = \hat{1}(\nu)$ :
$$\left[\hat{A}(\nu) \otimes \hat{1}(\nu'), \hat{1}(\nu) \otimes \hat{D}(\nu')\right]_- = 0 .$$

(3.258)

*Commutator Relations of Generating Operators.* Identifying the operators $\hat{A}$, $\hat{B}$, $\hat{C}$, $\hat{D}$ with the generating operators $\hat{\lambda}_i$ in $SU(n)$, and considering a system composed of two subsystems, the general relation (3.255) results in

$$\left[\hat{\lambda}_j(1) \otimes \hat{\lambda}_k(2), \hat{\lambda}_{j'}(1) \otimes \hat{\lambda}_{k'}(2)\right]_-$$
$$= \frac{1}{2}\left[\hat{\lambda}_j(1), \hat{\lambda}_{j'}(1)\right]_- \otimes \left[\hat{\lambda}_k(2), \hat{\lambda}_{k'}(2)\right]_+ +$$
$$\frac{1}{2}\left[\hat{\lambda}_j(1), \hat{\lambda}_{j'}(1)\right]_+ \otimes \left[\hat{\lambda}_k(2), \hat{\lambda}_{k'}(2)\right]_- .$$

(3.259)

Making use of the commutator representation (2.70), the commutator (3.259) turns into

$$\left[\hat{\lambda}_j(1) \otimes \hat{\lambda}_k(2), \hat{\lambda}_{j'}(1) \otimes \hat{\lambda}_{k'}(2)\right]_-$$
$$= \mathrm{i} \sum_{l=1}^{s_1} f_{jj'l}(1)\hat{\lambda}_l(1) \otimes \left[\hat{\lambda}_k(2), \hat{\lambda}_{k'}(2)\right]_+ +$$
$$\mathrm{i} \left[\hat{\lambda}_j(1), \hat{\lambda}_{j'}(1)\right]_+ \otimes \sum_{l'=1}^{s_2} f_{kk'l}(2)\hat{\lambda}_{l'}(2)$$

(3.260)

with the special cases

$$\left[\hat{\lambda}_j(1) \otimes \hat{1}(2), \hat{\lambda}_{j'}(1) \otimes \hat{1}(2)\right]_- = 2\mathrm{i} \sum_{l=1}^{s_1} f_{jj'l}(1)\hat{\lambda}_l(1) \otimes \hat{1}(2) ,$$

(3.261)

$$\left[\hat{\lambda}_j(1) \otimes \hat{1}(2), \hat{\lambda}_{j'}(1) \otimes \hat{\lambda}_k(2)\right]_- = 2\mathrm{i} \sum_{l=1}^{s_1} f_{jj'l}(1)\hat{\lambda}_l(1) \otimes \hat{\lambda}_k(2) .$$

(3.262)

$f_{ijk}$ denotes the structure constants of the considered $SU(n)$ algebra.

For the anticommutators we may use (cf. Sect. 2.2.2.5)

$$\left[\hat{\lambda}_o(v), \hat{\lambda}_{o'}(v)\right]_+ = \frac{4}{n_v}\hat{1}(v)\delta_{oo'} + 2 \sum_{p=1}^{s_v} d_{oo'p}(v)\hat{\lambda}_p(v)$$

(3.263)

with

$$4d_{oo'p}(v) = \text{Tr}\left\{ \left[ \hat{\lambda}_o(v), \hat{\lambda}_{o'}(v) \right]_+ \hat{\lambda}_p(v) \right\} .$$ (3.264)

**3.2.5.2 2-Node System.** The respective *network Hamiltonian* is defined by

$$\hat{H} = \frac{1}{n_1 n_2} \text{Tr}\left\{ \hat{H} \right\} \hat{1} + \frac{1}{2n_1} \sum_{j=1}^{s_1} \mathcal{H}_j(1) \left[ \hat{\lambda}_j(1) \otimes \hat{1}(2) \right]$$

$$+ \frac{1}{2n_2} \sum_{k=1}^{s_2} \mathcal{H}_k(2) \left[ \hat{1}(1) \otimes \hat{\lambda}_k(2) \right]$$

$$+ \frac{1}{4} \sum_{j=1}^{s_1} \sum_{k=1}^{s_2} \mathcal{H}_{jk}(1,2) \left[ \hat{\lambda}_j(1) \otimes \hat{\lambda}_k(2) \right]$$ (3.265)

with

$$\mathcal{H}_j(v) = \text{Tr}\left\{ \hat{H} \hat{\lambda}_j(v) \right\} ,$$ (3.266)

$$\mathcal{H}_{jk}(1,2) = \text{Tr}\left\{ \hat{H} \left[ \hat{\lambda}_j(1) \otimes \hat{\lambda}_j(2) \right] \right\}$$ (3.267)

and

$$s_j = n_j^2 - 1 .$$ (3.268)

The state can be described by the density operator (2.599),

$$\hat{\rho} = \frac{1}{n_1 n_2} \hat{1} + \frac{1}{2n_2} \sum_{j=1}^{s_1} \lambda_j(1) \left[ \hat{\lambda}_j(1) \otimes \hat{1}(2) \right]$$

$$+ \frac{1}{2n_1} \sum_{k=1}^{s_2} \lambda_k(2) \left[ \hat{1}(1) \otimes \lambda_k(2) \right]$$

$$+ \frac{1}{4} \sum_{j=1}^{s_1} \sum_{k=1}^{s_2} K_{jk}(1,2) \left[ \hat{\lambda}_j(1) \otimes \hat{\lambda}_k(2) \right] .$$ (3.269)

**3.2.5.3 Network Equations.** We now insert the Hamiltonian (3.265) and the density operator (3.269) into the Liouville equation

$$i\hbar \frac{\partial}{\partial t} \hat{\rho} = \left[ \hat{H}, \hat{\rho} \right]_-$$ (3.270)

(cf. (3.23)). Taking into account the commutator relations introduced above, and comparing the coefficients of the occurring product operators, one obtains a coupled system of equations: the time-evolution of the components of the local coherence vectors, $\lambda_i(v)$, reads

$$\dot{\lambda}_i(1) = \frac{1}{n_2} \sum_{j,k} f_{ijk}(1) \Gamma_j(1) \lambda_k(1) +$$
$$\frac{1}{2} \sum_{j,k,l} f_{ijk}(1) \Gamma_{jl}(1,2) K_{kl}(1,2) ,$$
$$\dot{\lambda}_i(2) = \frac{1}{n_1} \sum_{j,k} f_{ijk}(2) \Gamma_j(2) \lambda_k(2) +$$
$$\frac{1}{2} \sum_{j,k,l} f_{ijk}(2) \Gamma_{lj}(1,2) K_{lk}(1,2) ,$$

(3.271)

while the correlation tensor is controlled by

$$\dot{K}_{ij}(1,2) = \frac{1}{n_2} \sum_{k,l} f_{kli}(1) \Gamma_{kj}(1,2) \lambda_l(1) +$$
$$\frac{1}{n_1} \sum_{k,l} f_{klj}(2) \Gamma_{ik}(1,2) \lambda_l(2) +$$
$$\frac{1}{n_2} \sum_{k,l} f_{lki}(1) \Gamma_l(1) K_{kj}(1,2) +$$
$$\frac{1}{n_1} \sum_{k,l} f_{klj}(2) \Gamma_k(2) K_{il}(1,2) +$$
$$\frac{1}{2} \sum_{k,l,k',l'} d_{ll'j}(2) f_{kk'i}(1) \Gamma_{kl}(1,2) K_{k'l'}(1,2) +$$
$$\frac{1}{2} \sum_{k,l,k',l'} d_{kk'i}(1) f_{ll'j}(2) \Gamma_{kl}(1,2) K_{k'l'}(1,2) .$$

(3.272)

Here, the abbreviations

$$\Gamma_i(v) = \frac{\mathcal{H}_i(v)}{\hbar} , \quad \Gamma_{ik}(1,2) = \frac{\mathcal{H}_{ik}(1,2)}{\hbar}$$

(3.273)

have been used. These equations determine the dynamics of the network in $SU(n_1) \otimes SU(n_2)$ in the absence of damping. They will thus be called *coherent network equations*.

*Networks with 1-Node Interactions.* If only 1-node interactions are present (i.e. if terms $\Gamma_{ij}(1,2)$ are assumed zero), these two systems of equations reduce to

$$\dot{\lambda}_i(1) = \frac{1}{n_2} \sum_{k=1}^{s_1} \Omega_{ik}^{(1)} \lambda_k(1) \,,$$

$$\dot{\lambda}_i(2) = \frac{1}{n_1} \sum_{k=1}^{s_2} \Omega_{ik}^{(2)} \lambda_k(2) \,, \tag{3.274}$$

$$\dot{K}_{ij}(1,2) = \frac{1}{n_2} \sum_{k=1}^{s_1} \Omega_{ik}^{(1)} K_{kj}(1,2) + \frac{1}{n_1} \sum_{k=1}^{s_2} K_{ik}(1,2) \Omega_{jk}^{(2)} \,,$$

with the matrix elements

$$\Omega_{ik}^{(1)} = \sum_{j=1}^{s_1} f_{ijk}(1) \Gamma_j(1) \tag{3.275}$$

and

$$\Omega_{ik}^{(2)} = \sum_{j=1}^{s_2} f_{ijk}(2) \Gamma_j(2) \,. \tag{3.276}$$

Using the notation

$$\Omega_{ik}^{(v)} := \Omega \begin{pmatrix} v & v \\ i & k \end{pmatrix} \,, \tag{3.277}$$

$$\lambda_i(v) := \lambda \begin{pmatrix} v \\ i \end{pmatrix} \,, \tag{3.278}$$

$$K_{ij}(1,2) := K \begin{pmatrix} 1 & 2 \\ i & j \end{pmatrix} \tag{3.279}$$

($v = 1, 2$), and the *sum convention* (i.e. summation with respect to equal indices is implied), the network equations (3.274) reduce to

$$\boxed{\begin{aligned} \dot{\lambda} \begin{pmatrix} 1 \\ i \end{pmatrix} &= \frac{1}{n_2} \Omega \begin{pmatrix} 1 & 1 \\ i & k \end{pmatrix} \lambda \begin{pmatrix} 1 \\ k \end{pmatrix} \,, \\ \dot{\lambda} \begin{pmatrix} 2 \\ j \end{pmatrix} &= \frac{1}{n_1} \Omega \begin{pmatrix} 2 & 2 \\ j & l \end{pmatrix} \lambda \begin{pmatrix} 2 \\ l \end{pmatrix} \,, \\ \dot{K} \begin{pmatrix} 1 & 2 \\ i & j \end{pmatrix} &= \frac{1}{n_2} \Omega \begin{pmatrix} 1 & 1 \\ i & k \end{pmatrix} K \begin{pmatrix} 1 & 2 \\ k & j \end{pmatrix} + \frac{1}{n_1} \Omega \begin{pmatrix} 2 & 2 \\ j & l \end{pmatrix} K \begin{pmatrix} 1 & 2 \\ i & l \end{pmatrix} \,. \end{aligned}} \tag{3.280}$$

The last line is seen to follow from the first two by replacing $\lambda \begin{pmatrix} 1 \\ i \end{pmatrix}$ and $\lambda \begin{pmatrix} 2 \\ j \end{pmatrix}$ by $K \begin{pmatrix} 1 & 2 \\ i & j \end{pmatrix}$ and summing over the 2 possibilities. One may then build up the dynamical equation for the next higher correlation tensor $K \begin{pmatrix} 1 & 2 & 3 \\ i & j & k \end{pmatrix}$ out of those for $\lambda \begin{pmatrix} 1 \\ i \end{pmatrix}, \lambda \begin{pmatrix} 2 \\ j \end{pmatrix}, \lambda \begin{pmatrix} 3 \\ k \end{pmatrix}$ according to the corresponding rule.

*Networks with 2-Node Interactions.* When, additionally, 2-node interactions
have to be taken into account (i.e. when terms $\Gamma_{ij}(1,2)$ are unequal zero),
the network equations (3.274) have to be replaced by

$$
\dot{\lambda}_i(1) = \frac{1}{n_2} \sum_{k=1}^{s_1} \Omega_{ik}^{(1)} \lambda_k(1) + \frac{1}{2} \sum_{k,l=1}^{s_1,s_2} Q_{ikl}^{(1)} K_{kl}(1,2) := \dot{\lambda}_i^c(1) \;,
$$

$$
\dot{\lambda}_i(2) = \frac{1}{n_1} \sum_{k=1}^{s_2} \Omega_{ik}^{(2)} \lambda_k(2) + \frac{1}{2} \sum_{k,l=1}^{s_2,s_1} Q_{lik}^{(2)} K_{lk}(1,2) := \dot{\lambda}_i^c(2) \;,
$$
(3.281)

$$
\dot{K}_{ij}(1,2) = \frac{1}{2} \sum_{l=1}^{s_1} Q_{ilj}^{(1)} \lambda_l(1) + \frac{1}{2} \sum_{l=1}^{s_2} Q_{ijl}^{(2)} \lambda_l(2) +
$$

$$
\frac{1}{n_2} \sum_{k=1}^{s_1} \Omega_{ik}^{(1)} K_{kj}(1,2) + \frac{1}{n_1} \sum_{l=1}^{s_2} \Omega_{jl}^{(2)} K_{il}(1,2) +
$$

$$
\frac{1}{2} \sum_{k',l'=1}^{s_1,s_2} D_{ik'jl'}^{(12)} K_{k'l'}(1,2) := \dot{K}_{ij}^c(1,2) \;.
$$
(3.282)

The notations $\dot{\lambda}_i^c(v)$ and $\dot{K}_{ij}^c(1,2)$ are introduced for later reference.

Using the notation (3.277)–(3.279), this set of equations can be written
as (sum convention)

$$
\boxed{
\begin{aligned}
\dot{\lambda}\begin{pmatrix}1\\i\end{pmatrix} &= \frac{1}{n_2}\Omega\begin{pmatrix}1\ 1\\i\ k\end{pmatrix}\lambda\begin{pmatrix}1\\k\end{pmatrix} + \frac{1}{2}Q\begin{pmatrix}1\ 1\ 2\\i\ k\ l\end{pmatrix}K\begin{pmatrix}1\ 2\\k\ l\end{pmatrix} , \\
\dot{\lambda}\begin{pmatrix}2\\j\end{pmatrix} &= \frac{1}{n_1}\Omega\begin{pmatrix}2\ 2\\j\ l\end{pmatrix}\lambda\begin{pmatrix}2\\l\end{pmatrix} + \frac{1}{2}Q\begin{pmatrix}1\ 2\ 2\\l\ j\ k\end{pmatrix}K\begin{pmatrix}1\ 2\\l\ k\end{pmatrix}
\end{aligned}
}
$$
(3.283)

and

$$
\boxed{
\begin{aligned}
\dot{K}\begin{pmatrix}1\ 2\\i\ j\end{pmatrix} &= \frac{1}{n_2}\Omega\begin{pmatrix}1\ 1\\i\ k\end{pmatrix}K\begin{pmatrix}1\ 2\\k\ j\end{pmatrix} + \frac{1}{n_1}\Omega\begin{pmatrix}2\ 2\\j\ l\end{pmatrix}K\begin{pmatrix}1\ 2\\i\ l\end{pmatrix} + \\
&\quad \frac{1}{n_2}Q\begin{pmatrix}1\ 1\ 2\\i\ m\ j\end{pmatrix}\lambda\begin{pmatrix}1\\m\end{pmatrix} + \frac{1}{n_1}Q\begin{pmatrix}1\ 2\ 2\\i\ j\ m\end{pmatrix}\lambda\begin{pmatrix}2\\m\end{pmatrix} + \\
&\quad \frac{1}{2}D\begin{pmatrix}1\ 1\ 2\ 2\\i\ k'\ j\ l'\end{pmatrix}K\begin{pmatrix}1\ 2\\k'\ l'\end{pmatrix} .
\end{aligned}
}
$$
(3.284)

The matrix elements

$$
\begin{aligned}
Q_{ijk}^{(1)} &= \sum_{l=1}^{s_1} f_{ilj}(1)\Gamma_{lk}(1,2) = \frac{1}{\hbar}\sum_{l=1}^{s_1} f_{ilj}(1)\mathcal{H}_{lk}(1,2) \\
&:= Q\begin{pmatrix}1\ 1\ 2\\i\ j\ k\end{pmatrix} ,
\end{aligned}
$$
(3.285)

$$Q_{ijk}^{(2)} = \sum_{l=1}^{s_2} f_{jlk}(2)\Gamma_{il}(1,2) = \frac{1}{\hbar}\sum_{l=1}^{s_2} f_{jlk}(2)\mathcal{H}_{il}(1,2)$$

$$:= Q\begin{pmatrix} 1 & 2 & 2 \\ i & j & k \end{pmatrix}, \tag{3.286}$$

and

$$D_{ik'jl'}^{(12)} = \sum_{k,l=1}^{s_1,s_2} \left[d_{kk'i}(1)f_{ll'j}(2) + d_{ll'j}(2)f_{kk'i}(1)\right]\Gamma_{kl}(1,2)$$

$$:= D\begin{pmatrix} 1 & 1 & 2 & 2 \\ i & k' & j & l' \end{pmatrix} \tag{3.287}$$

derive from the 2-node interactions. The evolution equation for the correlation tensor proper M, defined by

$$M_{ij}(1,2) = K_{ij}(1,2) - \lambda_i(1)\lambda_j(2)$$

$$= K\begin{pmatrix} 1 & 2 \\ i & j \end{pmatrix} - \lambda\begin{pmatrix} 1 \\ i \end{pmatrix}\lambda\begin{pmatrix} 2 \\ j \end{pmatrix}, \tag{3.288}$$

is then given by

$$\dot{M}\begin{pmatrix} 1 & 2 \\ i & j \end{pmatrix} = \dot{K}\begin{pmatrix} 1 & 2 \\ i & j \end{pmatrix} - \dot{\lambda}\begin{pmatrix} 1 \\ i \end{pmatrix}\lambda\begin{pmatrix} 2 \\ j \end{pmatrix} - \lambda\begin{pmatrix} 1 \\ i \end{pmatrix}\dot{\lambda}\begin{pmatrix} 2 \\ j \end{pmatrix}$$

$$= \frac{1}{n_2}\Omega\begin{pmatrix} 1 & 1 \\ i & k \end{pmatrix} M\begin{pmatrix} 1 & 2 \\ k & j \end{pmatrix} +$$

$$\frac{1}{n_1}\Omega\begin{pmatrix} 2 & 2 \\ j & l \end{pmatrix} M\begin{pmatrix} 1 & 2 \\ i & l \end{pmatrix} +$$

$$\frac{1}{n_2}Q\begin{pmatrix} 1 & 1 & 2 \\ i & m & j \end{pmatrix}\lambda\begin{pmatrix} 1 \\ m \end{pmatrix} +$$

$$\frac{1}{n_1}Q\begin{pmatrix} 1 & 2 & 2 \\ i & j & m \end{pmatrix}\lambda\begin{pmatrix} 2 \\ m \end{pmatrix} -$$

$$\frac{1}{2}Q\begin{pmatrix} 1 & 1 & 2 \\ i & k & l \end{pmatrix} M\begin{pmatrix} 1 & 2 \\ k & l \end{pmatrix}\lambda\begin{pmatrix} 2 \\ j \end{pmatrix} -$$

$$\frac{1}{2}Q\begin{pmatrix} 1 & 2 & 2 \\ l & j & k \end{pmatrix} M\begin{pmatrix} 1 & 2 \\ l & k \end{pmatrix}\lambda\begin{pmatrix} 1 \\ i \end{pmatrix} -$$

$$\frac{1}{2}Q\begin{pmatrix} 1 & 1 & 2 \\ i & k & l \end{pmatrix}\lambda\begin{pmatrix} 1 \\ k \end{pmatrix}\lambda\begin{pmatrix} 2 \\ l \end{pmatrix}\lambda\begin{pmatrix} 2 \\ j \end{pmatrix} -$$

$$\frac{1}{2}Q\begin{pmatrix} 1 & 2 & 2 \\ l & j & k \end{pmatrix}\lambda\begin{pmatrix} 1 \\ l \end{pmatrix}\lambda\begin{pmatrix} 2 \\ k \end{pmatrix}\lambda\begin{pmatrix} 1 \\ i \end{pmatrix} +$$

$$\frac{1}{2}D\begin{pmatrix} 1 & 1 & 2 & 2 \\ i & k' & j & l' \end{pmatrix} M\begin{pmatrix} 1 & 2 \\ k' & l' \end{pmatrix} +$$

$$\frac{1}{2}D\begin{pmatrix} 1 & 1 & 2 & 2 \\ i & k' & j & l' \end{pmatrix}\lambda\begin{pmatrix} 1 \\ k' \end{pmatrix}\lambda\begin{pmatrix} 2 \\ l' \end{pmatrix}. \tag{3.289}$$

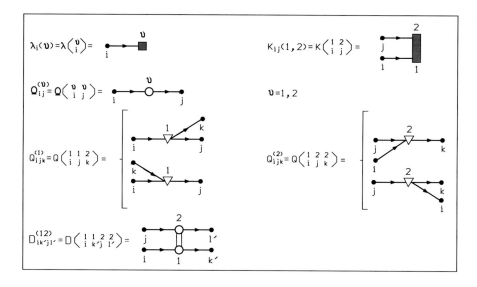

**Fig. 3.12.** Elements for a diagrammatic representation of network equations

This replaces (3.284). We immediately see that if $M\begin{pmatrix} 1 & 2 \\ i & j \end{pmatrix} = 0$ $(i, j = 1, 2, 3)$ at some initial time, it remains so unless appropriate 2-node interactions are present, i.e. $Q \neq 0$, $D \neq 0$.

These equations can be represented graphically, using a form similar to Feynman diagrams. In Fig. 3.12 the basic elements are sketched: the correlation tensors of order $q$ are symbolized by shaded rectangles with $q$ (cartesian) indices $i, j, \dots$. Local interactions $\Omega$ will be represented by open circles, non-local interactions $Q$ by open triangles. In Fig. 3.13 the resulting representation of the network equations (3.283)–(3.284) is shown. Each line stands for one specific node index $v = 1, 2$. We note that the $\Omega$ terms (circles) only act locally within one line. The $Q$ terms connect different lines in pairs.

**3.2.5.4 Example in $SU(2) \otimes SU(2)$.** We specialize in $n_1 = n_2 = 2$. Typical 2-node interactions in $SU(2) \otimes SU(2)$ appear in the form

$$(\mathcal{H}_{ij}) = \hbar (\Gamma_{ij}) = 2\hbar \begin{pmatrix} C_{\mathrm{F}} & 0 & 0 \\ 0 & C_{\mathrm{F}} & 0 \\ 0 & 0 & -C_{\mathrm{R}} \end{pmatrix} \qquad (3.290)$$

(compare the considerations presented in Sect. 2.4.2), where $(\Gamma_{ij})$ contains all matrix elements $\Gamma_{ij}(1, 2)$. These interactions may arise between different charge states from single electron transfer (see Fig. 3.14, a)), between local energy states from strain fields (within a solid-state matrix) or Coulomb interactions. In the latter case, $C_{\mathrm{F}}$ describes the exchange of excitation between a given pair of nodes (see Fig. 3.14, b)). $C_{\mathrm{R}}$ renormalizes the local transition

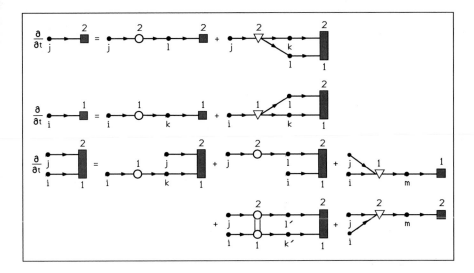

**Fig. 3.13.** Diagrammatic representation of network equations including 2-node interactions. First line: one-node functions $\dot\lambda_j(2) = F[\lambda_l(2), K_{lk}(1,2)]$. Middle: one-node functions $\dot\lambda_i(1) = F[\lambda_k(1), K_{kl}(1,2)]$. Bottom line: 2-node functions $\dot K_{ij}(1,2) = F[\lambda_m(1), \lambda_m(2), K_{kj}(1,2), K_{il}(1,2)]$

energy in node $\nu$ in response to a given local eigenstate of $\nu'$ (cf. Fig. 3.14, c)): the dipole–dipole coupling modifies the eigenspectrum of the network while leaving the product eigenstates unchanged. Even the most complicated networks will not go beyond the 2-node type of interaction if one restricts oneself to such fundamental interactions.

We now consider two 2-level systems, each driven by an electromagnetic field (RWA) and mutually coupled via Coulomb interactions. The Hamiltonian is specified by six parameters:

$$
\begin{aligned}
\mathcal{H}_x(1) &= \hbar\Gamma_x(1) = 2\hbar g^{(1)} , \\
\mathcal{H}_x(2) &= \hbar\Gamma_x(2) = 2\hbar g^{(2)} , \\
\mathcal{H}_z(1) &= \hbar\Gamma_z(1) = 2\hbar\delta^{(1)} , \\
\mathcal{H}_z(2) &= \hbar\Gamma_z(2) = 2\hbar\delta^{(2)} , \\
\mathcal{H}_{xx} &= \hbar\Gamma_{xx} = 2\hbar C_{\mathrm{F}} , \\
\mathcal{H}_{yy} &= \hbar\Gamma_{yy} = 2\hbar C_{\mathrm{F}} , \\
\mathcal{H}_{zz} &= \hbar\Gamma_{zz} = -2\hbar C_{\mathrm{R}} , \\
\text{other terms} &= 0 .
\end{aligned}
\tag{3.291}
$$

The single-node interaction based on the Hamiltonian vector components implies $(v = 1, 2)$

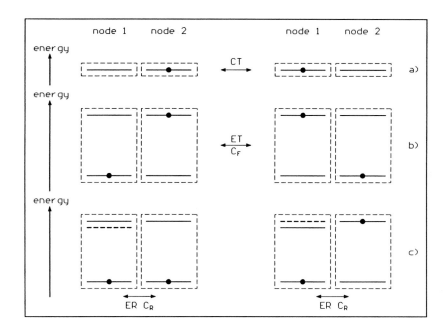

**Fig. 3.14.** Typical 2-node interactions. a) Charge transfer (CT). b) Excitation transfer (ET). c) Energy renormalization (ER), symmetric with respect to node 1 and 2

$$\Omega \begin{pmatrix} v\ v \\ 1\ 2 \end{pmatrix} = -\Omega \begin{pmatrix} v\ v \\ 2\ 1 \end{pmatrix} = -2\delta^{(v)}\ ,$$
$$\Omega \begin{pmatrix} v\ v \\ 2\ 3 \end{pmatrix} = -\Omega \begin{pmatrix} v\ v \\ 3\ 2 \end{pmatrix} = -2g^{(v)}$$

(3.292)

(compare (3.97)) and the 2-node interactions result in

$$Q \begin{pmatrix} 1\ 1\ 2 \\ i\ k\ 1 \end{pmatrix} = Q \begin{pmatrix} 1\ 2\ 2 \\ 1\ i\ k \end{pmatrix} = +f_{i1k}2C_{\mathrm{F}}\ ,$$
$$Q \begin{pmatrix} 1\ 1\ 2 \\ i\ k\ 2 \end{pmatrix} = Q \begin{pmatrix} 1\ 2\ 2 \\ i\ 2\ k \end{pmatrix} = +f_{i2k}2C_{\mathrm{F}}\ ,$$
$$Q \begin{pmatrix} 1\ 1\ 2 \\ i\ k\ 3 \end{pmatrix} = Q \begin{pmatrix} 1\ 2\ 2 \\ 3\ i\ k \end{pmatrix} = -f_{i3k}2C_{\mathrm{R}}$$

(3.293)

(compare (3.285)–(3.286)). Identifying the structure constants $f_{ijk}$ with the elements of the $\varepsilon$ tensor (cf. (2.2)), and using, again, the notation $Q_{ijk}^{(1)}$ and $Q_{ijk}^{(2)}$ (see (3.285)–(3.286)), this scheme reduces to

$$Q_{231}^{(1)} = Q_{312}^{(1)} = Q_{123}^{(2)} = Q_{231}^{(2)} = -2C_{\rm F} \ ,$$

$$Q_{321}^{(1)} = Q_{132}^{(1)} = Q_{132}^{(2)} = Q_{213}^{(2)} = +2C_{\rm F} \ ,$$

$$Q_{123}^{(1)} = Q_{312}^{(2)} = +2C_{\rm R} \ ,$$

$$Q_{213}^{(1)} = Q_{321}^{(2)} = -2C_{\rm R} \ .$$

(3.294)

All other terms (including the matrix elements $D_{ik'jl'}^{(12)}$) are zero. Identifying the indices $1,2,3$ with $x,y,z$, and inserting these matrix elements into the network equations (3.283)–(3.284), the motion of the 2 coherence vectors is determined by

$$
\begin{aligned}
\dot{\lambda}_x(1) &= -\delta^{(1)}\lambda_y(1) + C_{\rm F}K_{zy} + C_{\rm R}K_{yz} \ , \\
\dot{\lambda}_x(2) &= -\delta^{(2)}\lambda_y(2) + C_{\rm F}K_{yz} + C_{\rm R}K_{zy} \ , \\
\dot{\lambda}_y(1) &= +\delta^{(1)}\lambda_x(1) - g^{(1)}\lambda_z(1) - C_{\rm R}K_{zx} - C_{\rm F}K_{zx} \ , \\
\dot{\lambda}_y(2) &= +\delta^{(2)}\lambda_x(2) - g^{(2)}\lambda_z(2) - C_{\rm R}K_{zx} - C_{\rm F}K_{xz} \ , \\
\dot{\lambda}_z(1) &= +g^{(1)}\lambda_y(1) + C_{\rm F}\left(K_{yx} - K_{xy}\right) \ , \\
\dot{\lambda}_z(2) &= +g^{(2)}\lambda_y(2) + C_{\rm F}\left(K_{xy} - K_{yx}\right) \ ,
\end{aligned}
$$

(3.295)

and the motion of the correlation tensor by

$$
\begin{aligned}
\dot{K}_{xx} &= -\delta^{(1)}K_{yx} - \delta^{(2)}K_{xy} \ , \\
\dot{K}_{xy} &= +C_{\rm F}\lambda_z(1) - C_{\rm F}\lambda_z(2) - \delta^{(1)}K_{yy} + \delta^{(2)}K_{xx} - g^{(2)}K_{xz} \ , \\
\dot{K}_{xz} &= +C_{\rm R}\lambda_y(1) + C_{\rm F}\lambda_y(2) - \delta^{(1)}K_{yz} + g^{(2)}K_{xy} \ , \\
\dot{K}_{yy} &= -g^{(1)}K_{zy} + \delta^{(1)}K_{xy} - g^{(2)}K_{yz} + \delta^{(2)}K_{yx} \ , \\
\dot{K}_{yx} &= -C_{\rm F}\lambda_z(1) + C_{\rm F}\lambda_z(2) + \delta^{(1)}K_{xx} - g^{(1)}K_{zx} - \delta^{(1)}K_{yy} \ , \\
\dot{K}_{yz} &= -C_{\rm F}\lambda_x(2) - g^{(1)}K_{zz} + \delta^{(1)}K_{xz} + g^{(2)}K_{yy} \ , \\
\dot{K}_{zz} &= +g^{(1)}K_{yz} + g^{(2)}K_{zy} \ , \\
\dot{K}_{zx} &= +C_{\rm F}\lambda_y(1) + C_{\rm R}\lambda_y(2) + g^{(1)}K_{yx} - \delta^{(2)}K_{zy} \ , \\
\dot{K}_{zy} &= -C_{\rm F}\lambda_x(1) - C_{\rm R}\lambda_x(2) + g^{(1)}K_{yy} - g^{(2)}K_{zz} \ .
\end{aligned}
$$

(3.296)

In the same way, we can calculate (3.289) for this particular model.

### 3.2.5.5 Dynamics of Entanglement and Controlled NOT-operation.

Consider (3.283) and (3.284) for $n_1 = n_2 = 2$ and $C_{\rm F} = 0$. We then find in terms of $M_{ij}$ (see (3.288)–(3.289)):

$$
\begin{aligned}
\dot{\lambda}_1(1) &= -\left[\delta^{(1)} - C_{\rm R}\lambda_3(2)\right]\lambda_2(1) + C_{\rm R}M_{23} \ , \\
\dot{\lambda}_2(1) &= \left[\delta^{(1)} - C_{\rm R}\lambda_3(2)\right]\lambda_1(1) - g^{(1)}\lambda_3(1) - C_{\rm R}M_{13} \ , \\
\dot{\lambda}_3(1) &= g^{(1)}\lambda_2(1) \ .
\end{aligned}
$$

(3.297)

(The system of equations for $\lambda(2)$ follows replacing the node-index 1 by 2 and flipping $M_{ij} \to M_{ji}$.) The evolution of $M_{ij}$ is controlled by

$$\dot{M}_{11} = \delta^{(1)} M_{21} + \delta^{(2)} M_{12} - C_R M_{23} \lambda_1(2) - C_R M_{32} \lambda_1(1) -$$
$$C_R \lambda_2(1) \lambda_1(2) \lambda_3(2) - C_R \lambda_1(1) \lambda_3(1) \lambda_2(2) ,$$
$$\dot{M}_{12} = -\delta^{(2)} M_{11} - g^{(2)} M_{13} + \delta^{(1)} M_{22} - C_R M_{23} \lambda_2(2) +$$
$$C_R \lambda_1(1) \lambda_3(1) \lambda_1(2) - C_R \lambda_2(1) \lambda_2(2) \lambda_3(2) ,$$
$$\dot{M}_{13} = \delta^{(1)} M_{23} + g^{(2)} M_{12} + C_R \lambda_2(1) + C_R M_{23} \lambda_3(2) -$$
$$C_R \lambda_2(1) \lambda_3(2) \lambda_3(2) ,$$
$$\dot{M}_{21} = -\delta^{(1)} M_{11} - g^{(1)} M_{31} + \delta^{(2)} M_{22} - C_R M_{32} \lambda_2(1) +$$
$$C_R \lambda_1(1) \lambda_1(2) \lambda_3(2) - C_R \lambda_2(1) \lambda_3(1) \lambda_3(2) ,$$
$$\dot{M}_{22} = -g^{(1)} M_{32} - \delta^{(1)} M_{12} - g^{(2)} M_{23} - \delta^{(2)} M_{21} +$$
$$C_R M_{13} \lambda_2(2) + C_R M_{31} \lambda_2(1) + C_R \lambda_1(1) \lambda_2(2) \lambda_3(2) +$$
$$C_R \lambda_2(1) \lambda_3(1) \lambda_1(2) ,$$
$$\dot{M}_{23} = -g^{(1)} M_{33} - \delta^{(2)} M_{21} - g^{(2)} M_{23} - \delta^{(1)} M_{13} - C_R \lambda_1(1) +$$
$$C_R M_{13} \lambda_3(2) + C_R \lambda_1(1) \lambda_3(2) \lambda_3(2) ,$$
$$\dot{M}_{31} = \delta^{(2)} M_{32} + g^{(1)} M_{21} + C_R \lambda_2(2) + C_R M_{32} \lambda_3(1) -$$
$$C_R \lambda_3(1) \lambda_3(1) \lambda_2(2) ,$$
$$\dot{M}_{32} = -g^{(2)} M_{33} - \delta^{(1)} M_{12} - g^{(1)} M_{32} - \delta^{(2)} M_{31} - C_R \lambda_1(2) +$$
$$C_R M_{31} \lambda_3(1) + C_R \lambda_3(1) \lambda_3(1) \lambda_1(2) ,$$
$$\dot{M}_{33} = g^{(1)} M_{23} + g^{(2)} M_{32} . \tag{3.298}$$

*Stationary Solutions.* A trivial solution is

$$\lambda(1) = \lambda(2) = 0 , \quad M = 0 . \tag{3.299}$$

Also an entangled state with

$$M_{ij} = \delta_{ij} a_i \quad (|a_i| \le 1) \tag{3.300}$$

is stationary if

$$\lambda_1(j) = \lambda_2(j) = 0 \quad (j = 1, 2) , \tag{3.301}$$

$$g^{(1)} = g^{(2)} = 0 , \quad \delta^{(1)} = \delta^{(2)} = 0 . \tag{3.302}$$

(The last condition means that $M_{ij}$ is stationary only in the appropriate rotating frame.)

*Non-Stationary Solutions.* Let

$$\dot{M}(t = 0) = 0 . \tag{3.303}$$

This holds true for all times under certain conditions: only terms not containing any $M_{ij}$ will contribute to $\dot{M}$. The remaining terms also vanish in each equation if at all times

$$\lambda(1) = \lambda(2) = 0 , \quad \text{or} \quad \lambda_3(1) = \pm 1 , \quad \text{or} \quad \lambda_3(2) = \pm 1 . \tag{3.304}$$

Thus, for

$$g^{(2)} = 0 \tag{3.305}$$

and

$$\lambda(2) = (0, 0, \pm 1) = \text{const.} \tag{3.306}$$

we only have to consider the Bloch equations for subsystem 1 with the effective detuning parameter

$$\tilde{\delta}^{(1)} = \delta^{(1)} - C_R \lambda_3(2) . \tag{3.307}$$

Choosing

$$\delta^{(1)} = -C_R , \tag{3.308}$$

with a sufficiently large parameter $C_R$ implies that subsystem 1 is in resonance only if subsystem 2 is in state

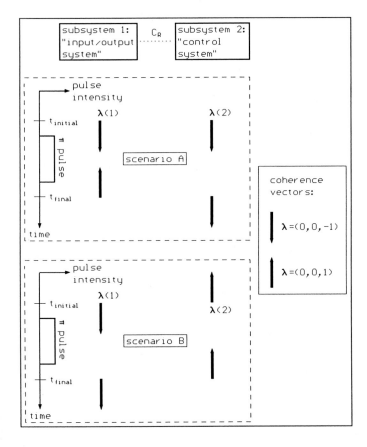

**Fig. 3.15.** Controlled NOT-operation: the position of the coherence vectors before and after the pulse $g^{(1)}\Delta t = \pi$

$$\lambda_3(2) = -1 \ . \tag{3.309}$$

This is a conditional dynamics: if we apply a $\pi$ *pulse* defined by

$$g^{(1)}\Delta t = \pi \tag{3.310}$$

to the initial state

$$\lambda_3(1) = \mp 1 \ , \tag{3.311}$$

it will flip to

$$\lambda_3'(1) = \pm 1 \ , \tag{3.312}$$

provided

$$\lambda_3(2) = -1 \tag{3.313}$$

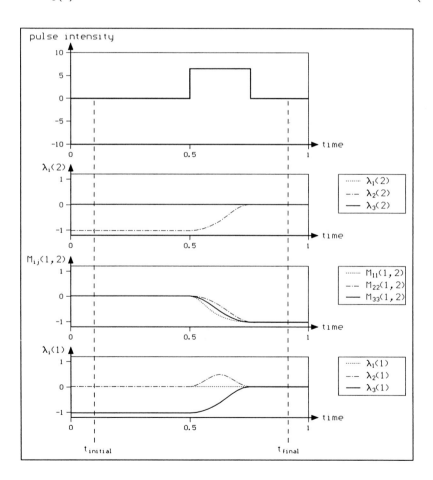

**Fig. 3.16.** Switch process with back action

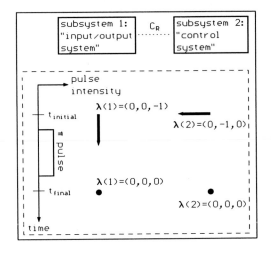

**Fig. 3.17.** Switch process with back action: the position of the coherence vectors in the corresponding Bloch sphere before and after the pulse

and no action takes place otherwise (in technical terms: this is a so-called *controlled NOT-operation* in which subsystem 2 represents the *control system*). In Fig. 3.15 the scenario with $\lambda(2) = (0,0,-1)$ as well as the scenario with $\lambda(2) = (0,0,1)$ is shown. The two subsystems are interacting in a classical sense: there is no back action on the control system 2 (even though it may detune subsystem 1) and both systems never get entangled. We can obtain the opposite action for

$$\delta^{(1)} = +C_{\mathrm{R}} \ . \tag{3.314}$$

What happens if we prepare subsystem 2 (the control system) in a superposition of states (something we could not do with a classical controlled NOT-element) (cf. [87])? Such a superposition is equivalent to putting (at time $t = 0$)

$$\lambda(2) = (0,1,0) \ , \tag{3.315}$$

say. Figure 3.16 shows the dynamical evolution. The driving force $g^{(1)}$ now also affects the control system: after time $\Delta t = \pi/g^{(1)}$ the 2 subsystems are perfectly entangled. The motion of the coherence vectors in the corresponding projected Bloch sphere is depicted in Fig. 3.17. This result should be compared with Fig. 3.15.

**3.2.5.6 Non-locality in Time.** Consider again two 2-level systems interacting via the non-resonant Coulomb interaction $C_{\mathrm{R}}$. Let system S be driven continuously by an electromagnetic field $g^{(S)}$ (detuning $\delta^{(S)} = 0$). The coherence vector $\lambda(S)$ exercises a rotation in the $\lambda_y, \lambda_z$ plane with Rabi frequency $\Omega_{\mathrm{R}}(S) = g^{(S)}$ (cf. (3.115)). System M is initially in its ground state $\lambda(M) = (0,0,-1)$. We then apply at time $t_1$ a $\pi$ pulse on system M ($g^{(M)}\Delta t_{\mathrm{M}} = \pi$, $\Delta t_{\mathrm{M}} \ll \Omega_{\mathrm{R}}^{-1}(S)$) with a laser frequency such that resonance results if system S is in the ground state. This establishes a "logic": the system M should be found in state $\lambda(M) = (0,0,1)$ if system S has been in the

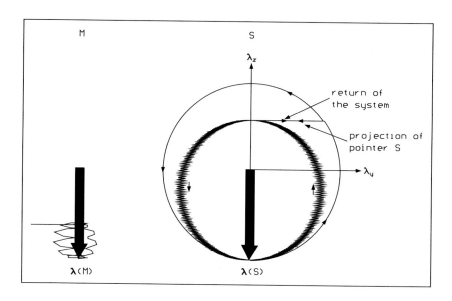

**Fig. 3.18.** Two 2-level systems are coupled by $C_R$. System S (r.h.s.) is driven by a light field $g^{(S)}$, system M (l.h.s.) is exposed to a short $\pi$ pulse at time $t_1$

ground state $\lambda(S) = (0, 0, -1)$ in the time window $t_1 \ldots t_1 + \Delta t_M$. Note that no dissipation is involved: this is not a measurement and the dynamics is completely reversible. This is shown in Figs. 3.18, 3.19: initially both systems are in the ground state (Bloch pointers of length 1 pointing down). Then the pointer of system 1 rotates, driven by the light field $g^{(S)}$, counter-clockwise until the pointer S is "projected" (during the short $\pi$ pulse on system M) on the $z$ axis (exactly as if an actual ensemble measurement had been performed). The pointer M also shrinks along the $z$ axis; it then oscillates in the $\lambda_x, \lambda_y$ plane while pointer S continues to rotate with the original Rabi frequency on a smaller, shifted circle. Both systems are entangled. A $2\pi$ pulse applied at time $t_1$ would leave the system (almost) unaffected. We can make the $\pi$ pulse a $2\pi$ pulse later on. A second $\pi$ pulse applied when pointer S is upward makes the system return (almost) to its original state. (In Fig. 3.18 this reversibility is only approximate; the up and down motion of pointer M is also counter-clockwise and shown by the broken wavy lines.) Instead of applying the second $\pi$ pulse to undo all the action we can perform an actual measurement on system M at any later time. This measurement (for a single system) will find system M either up or down, destroying the entanglement with its partner S (and reversibility). By this measurement also the state of S is changed: it is kicked into a state it would have reached if the measurement had been performed already at time $t_1$ with that very result. In this sense we can manipulate histories, a rather direct indication of the *time-non-locality* of

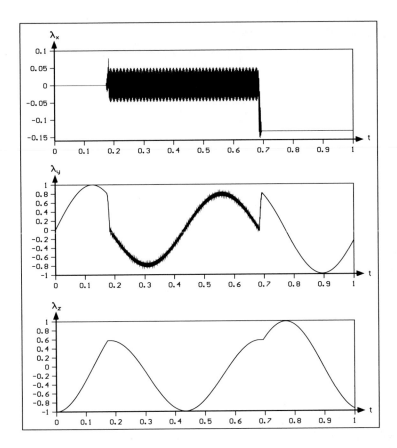

**Fig. 3.19.** The time-evolution of the components $\lambda_i(S)$ (cf. [76])

quantum networks. A formal consequence are the *temporal Bell inequalities* (cf. [109]).

**3.2.5.7 $N$-Node Dynamics.** We first generalize the result of (3.280), valid for $N = 2$ and local 1-node interactions, to a lattice of $N$ $SU(n)$ nodes. We note that the last line of (3.280) follows from the first two by formally increasing the order of the respective correlation tensor (here, a vector) by 1 and summing over all the node interactions involved. This recipe gives for $N = 3$ the third-order correlation:

$$
\dot{K} \begin{pmatrix} 1 & 2 & 3 \\ i & j & m \end{pmatrix} = \frac{1}{n^{N-1}} \Omega \begin{pmatrix} 1 & 1 \\ i & k \end{pmatrix} K \begin{pmatrix} 1 & 2 & 3 \\ k & j & m \end{pmatrix} +
$$
$$
\frac{1}{n^{N-1}} \Omega \begin{pmatrix} 2 & 2 \\ j & l \end{pmatrix} K \begin{pmatrix} 1 & 2 & 3 \\ i & l & m \end{pmatrix} +
$$
$$
\frac{1}{n^{N-1}} \Omega \begin{pmatrix} 3 & 3 \\ m & n \end{pmatrix} K \begin{pmatrix} 1 & 2 & 3 \\ i & j & n \end{pmatrix} . \tag{3.316}
$$

In general, one finds (up to third-order terms)

$$\dot{\lambda}\begin{pmatrix}\mu\\i\end{pmatrix} = \frac{1}{n^{N-1}}\Omega\begin{pmatrix}\mu\ \mu\\i\ \ k\end{pmatrix}\lambda\begin{pmatrix}\mu\\k\end{pmatrix},$$

(3.317)

$$\dot{K}\begin{pmatrix}\mu\ \nu\\i\ \ j\end{pmatrix} = \frac{1}{n^{N-1}}\Omega\begin{pmatrix}\mu\ \mu\\i\ \ k\end{pmatrix}K\begin{pmatrix}\mu\ \nu\\k\ \ j\end{pmatrix} + \frac{1}{n^{N-1}}\Omega\begin{pmatrix}\nu\ \nu\\j\ \ l\end{pmatrix}K\begin{pmatrix}\mu\ \nu\\i\ \ l\end{pmatrix},$$

(3.318)

$$\dot{K}\begin{pmatrix}\mu\ \nu\ \sigma\\i\ \ j\ \ m\end{pmatrix} = \frac{1}{n^{N-1}}\Omega\begin{pmatrix}\mu\ \mu\\i\ \ k\end{pmatrix}K\begin{pmatrix}\mu\ \nu\ \sigma\\k\ \ j\ \ m\end{pmatrix} +$$
$$\frac{1}{n^{N-1}}\Omega\begin{pmatrix}\nu\ \nu\\j\ \ l\end{pmatrix}K\begin{pmatrix}\mu\ \nu\ \sigma\\i\ \ l\ \ m\end{pmatrix} +$$
$$\frac{1}{n^{N-1}}\Omega\begin{pmatrix}\sigma\ \sigma\\m\ \ n\end{pmatrix}K\begin{pmatrix}\mu\ \nu\ \sigma\\i\ \ j\ \ n\end{pmatrix}.$$

(3.319)

Finally, including 2-node interactions, one generalizes (3.283) and (3.284) to yield

$$\dot{\lambda}\begin{pmatrix}\mu\\i\end{pmatrix} = \frac{1}{n^{N-1}}\Omega\begin{pmatrix}\mu\ \mu\\i\ \ k\end{pmatrix}\lambda\begin{pmatrix}\mu\\k\end{pmatrix} + \frac{1}{2n^{N-2}}Q\begin{pmatrix}\mu\ \mu\ \nu\\i\ \ k\ \ l\end{pmatrix}K\begin{pmatrix}\mu\ \nu\\k\ \ l\end{pmatrix}$$

(3.320)

and

$$\dot{K}\begin{pmatrix}\mu\ \nu\\i\ \ j\end{pmatrix} = \frac{1}{n^{N-1}}\Omega\begin{pmatrix}\mu\ \mu\\i\ \ k\end{pmatrix}K\begin{pmatrix}\mu\ \nu\\k\ \ j\end{pmatrix} +$$
$$\frac{1}{n^{N-1}}\Omega\begin{pmatrix}\nu\ \nu\\j\ \ l\end{pmatrix}K\begin{pmatrix}\mu\ \nu\\i\ \ l\end{pmatrix} +$$
$$\frac{1}{n^{N-1}}Q\begin{pmatrix}\mu\ \mu\ \nu\\i\ \ m\ \ j\end{pmatrix}\lambda\begin{pmatrix}\mu\\m\end{pmatrix} +$$
$$\frac{1}{n^{N-1}}Q\begin{pmatrix}\mu\ \nu\ \nu\\i\ \ j\ \ m\end{pmatrix}\lambda\begin{pmatrix}\nu\\m\end{pmatrix} +$$
$$\frac{1}{2n^{N-2}}Q\begin{pmatrix}\mu\ \mu\ \sigma\\i\ \ k\ \ m\end{pmatrix}K\begin{pmatrix}\mu\ \nu\ \sigma\\k\ \ j\ \ m\end{pmatrix} +$$
$$\frac{1}{2n^{N-2}}Q\begin{pmatrix}\nu\ \nu\ \sigma\\j\ \ k\ \ m\end{pmatrix}K\begin{pmatrix}\mu\ \nu\ \sigma\\i\ \ k\ \ m\end{pmatrix} +$$
$$\frac{1}{2n^{N-2}}D\begin{pmatrix}\mu\ \mu\ \nu\ \nu\\i\ \ k'\ j\ \ l'\end{pmatrix}K\begin{pmatrix}\mu\ \nu\\k'\ l'\end{pmatrix}$$

(3.321)

with $\mu \neq \nu \neq \sigma$. The non-local interactions $Q$ connect to the next higher (factor $1/2n^{N-2}$) and next lower (factor $1/n^{N-1}$) order correlation tensor. In the same way, one finds

$$
\dot{K}\left(\begin{smallmatrix}\mu & \nu & \sigma \\ i & j & m\end{smallmatrix}\right) = \frac{1}{n^{N-1}}\Omega\left(\begin{smallmatrix}\mu & \mu \\ i & k\end{smallmatrix}\right)K\left(\begin{smallmatrix}\mu & \nu & \sigma \\ k & j & m\end{smallmatrix}\right) +
$$

$$
\frac{1}{n^{N-1}}\Omega\left(\begin{smallmatrix}\nu & \nu \\ j & k\end{smallmatrix}\right)K\left(\begin{smallmatrix}\mu & \nu & \sigma \\ i & k & m\end{smallmatrix}\right) +
$$

$$
\frac{1}{n^{N-1}}\Omega\left(\begin{smallmatrix}\sigma & \sigma \\ m & k\end{smallmatrix}\right)K\left(\begin{smallmatrix}\mu & \nu & \sigma \\ i & j & k\end{smallmatrix}\right) +
$$

$$
\frac{1}{n^{N-1}}Q\left(\begin{smallmatrix}\mu & \mu & \sigma \\ i & k & m\end{smallmatrix}\right)K\left(\begin{smallmatrix}\mu & \nu \\ k & j\end{smallmatrix}\right) +
$$

$$
\frac{1}{n^{N-1}}Q\left(\begin{smallmatrix}\nu & \nu & \sigma \\ j & k & m\end{smallmatrix}\right)K\left(\begin{smallmatrix}\mu & \nu \\ i & k\end{smallmatrix}\right) +
$$

$$
\frac{1}{n^{N-1}}Q\left(\begin{smallmatrix}\mu & \nu & \nu \\ i & j & k\end{smallmatrix}\right)K\left(\begin{smallmatrix}\nu & \sigma \\ k & m\end{smallmatrix}\right) +
$$

$$
\frac{1}{n^{N-1}}Q\left(\begin{smallmatrix}\mu & \sigma & \sigma \\ i & m & k\end{smallmatrix}\right)K\left(\begin{smallmatrix}\nu & \sigma \\ j & k\end{smallmatrix}\right) +
$$

$$
\frac{1}{n^{N-1}}Q\left(\begin{smallmatrix}\mu & \mu & \nu \\ i & k & j\end{smallmatrix}\right)K\left(\begin{smallmatrix}\mu & \sigma \\ k & m\end{smallmatrix}\right) + \tag{3.322}
$$

$$
\frac{1}{n^{N-1}}Q\left(\begin{smallmatrix}\nu & \sigma & \sigma \\ j & m & k\end{smallmatrix}\right)K\left(\begin{smallmatrix}\mu & \sigma \\ i & k\end{smallmatrix}\right) +
$$

$$
\frac{1}{2n^{N-2}}Q\left(\begin{smallmatrix}\mu & \mu & \pi \\ i & k & p\end{smallmatrix}\right)K\left(\begin{smallmatrix}\mu & \nu & \sigma & \pi \\ k & j & m & p\end{smallmatrix}\right) +
$$

$$
\frac{1}{2n^{N-2}}Q\left(\begin{smallmatrix}\nu & \nu & \pi \\ j & k & p\end{smallmatrix}\right)K\left(\begin{smallmatrix}\mu & \nu & \sigma & \pi \\ i & k & m & p\end{smallmatrix}\right) +
$$

$$
\frac{1}{2n^{N-2}}Q\left(\begin{smallmatrix}\sigma & \sigma & \pi \\ m & k & p\end{smallmatrix}\right)K\left(\begin{smallmatrix}\mu & \nu & \sigma & \pi \\ i & j & k & p\end{smallmatrix}\right) +
$$

$$
\frac{1}{2^{N-2}}D\left(\begin{smallmatrix}\mu & \mu & \nu & \nu \\ i & k' & j & l'\end{smallmatrix}\right)K\left(\begin{smallmatrix}\mu & \nu & \sigma \\ k' & l' & m\end{smallmatrix}\right) +
$$

$$
\frac{1}{2^{N-2}}D\left(\begin{smallmatrix}\mu & \mu & \sigma & \sigma \\ i & k' & m & m'\end{smallmatrix}\right)K\left(\begin{smallmatrix}\mu & \nu & \sigma \\ k' & j & m'\end{smallmatrix}\right) +
$$

$$
\frac{1}{2^{N-2}}D\left(\begin{smallmatrix}\nu & \nu & \sigma & \sigma \\ j & l' & m & m'\end{smallmatrix}\right)K\left(\begin{smallmatrix}\mu & \nu & \sigma \\ i & l' & m'\end{smallmatrix}\right) .
$$

Thus, the third-order correlation tensor depends on fourth-order and the various second-order correlations. The matrix elements $D\left(\begin{smallmatrix}\mu & \mu & \nu & \nu \\ i & k' & j & l'\end{smallmatrix}\right)$ are identically zero for $n = 2$. The time-evolution of this tensor is determined by (3.322). The corresponding diagrammatic representation is depicted in Fig. 3.20.

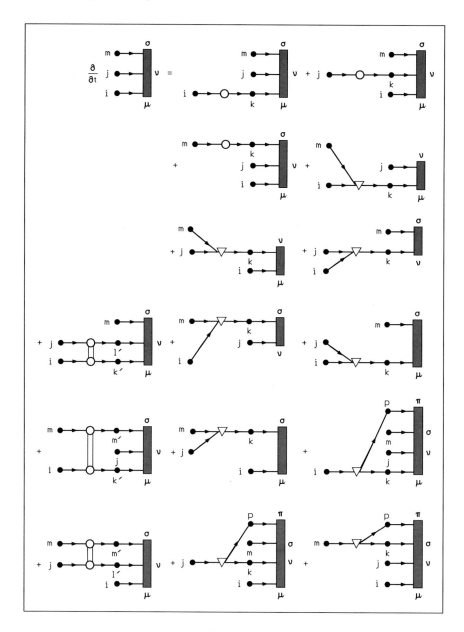

**Fig. 3.20.** Diagrammatic representation of the time-derivative of the third-order correlation tensor

Summarizing, we stress that the dynamics of an (interacting) quantum network is highly non-local. This means that the local driving of one node, for example, affects all other nodes even if they are far apart and do not

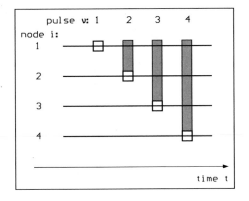

**Fig. 3.21.** Non-locality of quantum networks

interact directly: consider a network with nearest neighbour interaction $C_R$ only. Suppose we start from the ground state (see Fig. 3.21) and subsequently apply a light pulse $g^{(\nu)}T_\nu$ on node $i = \nu$, $i = 1, 2, \ldots$ with its neighbours acting as control. Then, in general (i.e. unless the nodes are in their local eigenstates $\lambda(i) = (0, 0, \pm 1)$), each pulse $i$ changes the state of all previously entangled nodes $j < i$ (indicated in Fig. 3.21 by the shaded strips). This non-locality would disappear with respect to a very special and, in general, very artificial basis set (cf. Sect. 2.2.3.2).

## 3.3 Dynamics of Open Systems

Contents: environment and damping, reduced density operator in open systems, Markovian master equation, relaxation matrix, damping models, spectral density, damped and generalized Bloch equations, damped network equations, quantum beats, semiconductor Bloch equations.

### 3.3.1 Open Systems

So far we have been concerned with *isolated Hamilton models*: interactions with the external world have been restricted to external potential forces (also time-dependent forces like light fields). However, it is obvious that such models will never be a complete description of the actual physical situation, isolation is an idealization. This holds for classical and quantum systems alike. The *influence of the environment* will therefore have to be taken into account at least in some approximate way. Closed quantum systems, in addition, have no "perspective", i.e. there is no reference frame. As groups of various observables are incompatible, the actual elements of reality have to be selected by a *measurement*, i.e. by interaction with a (at least partially) classical environment. The interaction with the environment thus becomes an essential part of modelling (cf. [20, 33]).

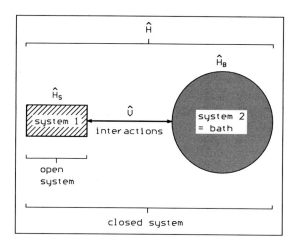

**Fig. 3.22.** Open system 1 and bath as parts of a closed system

There is a continuing discussion of how classical properties might emerge in a world believed to be fundamentally quantum (cf. [74, 109, 143, 155]). Taking such an embedding for granted, we are faced with the complementary question as to how non-classical behaviour can survive in a classical surrounding. The search for macroscopic quantum effects covers only a very special aspect of this field, scenarios in which specific macro-observables still behave quantum mechanically (cf. [9, 123]); here we are concerned with many local observables constituting a network. Furthermore, there is always some kind of ambiguity in defining the system to be treated quantum-mechanically and the classical rest, i.e. in specifying the so-called *Heisenberg cut*. As it will be argued below, this cut should, more properly, be described as a "thick interlayer" between a system and its environment.

In the following discussion we will focus our attention on a composite system consisting of two subsystems, system 1 and system 2. Both subsystems can be considered as parts of a closed system. It is obvious that the behaviour of system 1 will, in general, depend on system 2 (and vice versa): one cannot hope to find a reliable description of system 1 alone. However, if system 2 can be made to act more like a kind of boundary condition (i.e. as a "bath"), a closed description of system 1 may approximately become possible: one necessary condition is that the bath is large so that system 1 couples to a quasi-continuous spectrum. In the following, system 1 will be denoted by S, system 2 (bath) by B (compare Fig. 3.22).

**3.3.1.1 Hamilton Model.** The closed model describing system S and environment B is supposed to be

$$\hat{H} = \hat{H}(S, B) + \hat{V}(S, B) \tag{3.323}$$

with

$$\hat{H}(S, B) = \hat{H}(S) \otimes \hat{1}(B) + \hat{1}(S) \otimes \hat{H}(B) := \hat{H}_{SB} . \tag{3.324}$$

In spectral representation the system S is specified by

$$\hat{H}(S) = \sum_s E_s^{(S)} |s(S)\rangle \langle s(S)| = \sum_s E_s^{(S)} \hat{P}_{ss}(S) := \hat{H}_S , \qquad (3.325)$$

and the bath by

$$\hat{H}(B) = \sum_r E_r^{(B)} |r(B)\rangle \langle r(B)| = \sum_r E_r^{(B)} \hat{P}_{rr}(B) := \hat{H}_B , \qquad (3.326)$$

where the respective Schrödinger equations

$$\hat{H}_S |s(S)\rangle = E_s^{(S)} |s(S)\rangle , \qquad (3.327)$$

$$\hat{H}_B |r(B)\rangle = E_r^{(B)} |r(B)\rangle \qquad (3.328)$$

determine the respective eigenvectors, $|s(S)\rangle$, $|r(B)\rangle$, and energy eigenvalues, $E_s^{(S)}$, $E_r^{(B)}$.

The interaction of the two systems is taken as a bilinear form in operators $\hat{Q}_i$ (defined in system S) and $\hat{F}_i$ (defined in the bath):

$$\hat{V}(S,B) = \sum_i \hat{Q}_i(S) \otimes \hat{F}_i(B) = \sum_i \hat{Q}_i(S)\hat{F}_i(B)$$

$$:= \sum_i \hat{Q}_i \hat{F}_i = \hat{V} \qquad (3.329)$$

with

$$\left[\hat{S}(1) \otimes \hat{F}_i(B), \hat{Q}_i(S) \otimes \hat{1}(B)\right]_- = \left[\hat{F}_i, \hat{Q}_i\right]_- = 0 . \qquad (3.330)$$

We require that

$$\left\langle r \left| \hat{F}_i \right| r \right\rangle = 0 \qquad (3.331)$$

for all $i$.

Any state of the total system can then be represented as a superposition of product states, i.e.

$$|\psi\rangle = \sum_{s,r} c_{sr} |sr\rangle \qquad (3.332)$$

with

$$|sr\rangle = |s(S)\rangle \otimes |r(B)\rangle . \qquad (3.333)$$

**3.3.1.2 The Interaction Picture.** State vectors and operators in the interaction picture (upper index (i)) are defined by the unitary transformation

$$\left|\psi^{(i)}(t)\right\rangle = \hat{U}_{SB} |\psi(t)\rangle , \qquad (3.334)$$

$$\hat{A}^{(i)}(t) = \hat{U}_{SB}\hat{A}\hat{U}_{SB}^\dagger , \qquad (3.335)$$

where $\hat{U}_{SB}$ is given by the non-interacting Hamiltonian

$$\hat{U}_{SB} = e^{i\hat{H}_{SB}t/\hbar} . \tag{3.336}$$

Identifying $\hat{A}$ with the Hamiltonian $\hat{H}$, (3.335) determines the transformation

$$\hat{H}^{(i)}(t) = \hat{U}_{SB}\hat{H}\hat{U}_{SB}^{\dagger} = \hat{H}_{SB} + \hat{U}_{SB}\hat{V}\hat{U}_{SB}^{\dagger}$$
$$= \hat{H}_{SB} + \hat{V}^{(i)}(t) . \tag{3.337}$$

The products $\hat{Q}_i\hat{F}_i$ are transformed according to

$$\hat{Q}_i^{(i)}(t) = e^{i\hat{H}_{S}t/\hbar}\hat{Q}_i e^{-i\hat{H}_{S}t/\hbar} , \tag{3.338}$$

$$\hat{F}_i^{(i)}(t) = e^{i\hat{H}_{B}t/\hbar}\hat{F}_i e^{-i\hat{H}_{B}t/\hbar} \tag{3.339}$$

so that the new interaction operator is given by

$$\hat{V}^{(i)}(t) = \sum_i \hat{Q}_i^{(i)}(t)\hat{F}_i^{(i)}(t) . \tag{3.340}$$

Using (3.338), (3.339), and the power series representation of the operator function $\exp\left(-i\hat{H}_{S}t/\hbar\right)$, the matrix elements

$$\left\langle r \left| \hat{F}_i^{(i)}(t) \right| r' \right\rangle = e^{i\omega_{rr'}t} \left\langle r \left| \hat{F}_i \right| r' \right\rangle , \tag{3.341}$$

$$\left\langle s \left| \hat{Q}_i^{(i)}(t) \right| s' \right\rangle = e^{i\omega_{ss'}t} \left\langle s \left| \hat{Q}_i \right| s' \right\rangle \tag{3.342}$$

follow. Here we have used (3.327), (3.328), while the circular frequency $\omega_{ss'}$ is given by

$$\omega_{ss'} = \frac{E_s^{(S)} - E_{s'}^{(S)}}{\hbar} . \tag{3.343}$$

$\omega_{rr'}$ is defined correspondingly.

### 3.3.1.3 Density Operator and Equation of Motion. The density operator reads in the interaction picture:

$$\hat{\rho}^{(i)}(t) = \hat{U}_{SB}\hat{\rho}(t)\hat{U}_{SB}^{\dagger} \tag{3.344}$$

(compare with (3.335)) with the inverse transformation given by

$$\hat{\rho}(t) = \hat{U}_{SB}^{\dagger}\hat{\rho}^{(i)}(t)\hat{U}_{SB} . \tag{3.345}$$

Making use of the evolution equation

$$i\hbar\frac{\partial\hat{\rho}(t)}{\partial t} = \left[\hat{H}, \hat{\rho}(t)\right]_- \tag{3.346}$$

defining the motion of a density operator $\hat{\rho}(t)$ in a closed system (compare with (3.23)), the evolution equation of the density operator in the interaction picture can be derived by inserting (3.345) and $\hat{H} = \hat{H}_{SB} + \hat{V}$ into (3.346):

$$i\hbar \frac{\partial \hat{\rho}^{(i)}(t)}{\partial t} = \left[\hat{V}^{(i)}(t), \hat{\rho}^{(i)}(t)\right]_- \ . \tag{3.347}$$

This differential equation can be replaced by an integro-differential equation. Integrating the equation of motion (3.347) with respect to time, one obtains

$$\hat{\rho}^{(i)}(t) = \hat{\rho}^{(i)}(0) - \frac{i}{\hbar} \int_0^t \left[\hat{V}^{(i)}(\tau), \hat{\rho}^{(i)}(\tau)\right]_- \mathrm{d}\tau \ . \tag{3.348}$$

Reinserting this formulation into the evolution equation (3.347), (3.347) yields

$$\frac{\partial}{\partial t}\hat{\rho}^{(i)}(t) = -\frac{i}{\hbar}\left[\hat{V}^{(i)}(t), \hat{\rho}^{(i)}(0)\right]_- - \frac{1}{\hbar^2}\int_0^t \mathrm{d}t' \left[\hat{V}^{(i)}(t), \left[\hat{V}^{(i)}(t'), \hat{\rho}^{(i)}(t')\right]_-\right]_- \ . \tag{3.349}$$

This integro-differential equation can be used in an alternative way to determine the motion of the density operator $\hat{\rho}^{(i)}(t)$. Assuming the initial condition

$$\hat{\rho}^{(i)}(0) = \hat{\rho}(0) = \hat{\rho}_S(0) \otimes \hat{\rho}_B(0) \tag{3.350}$$

(no correlation at time $t = 0$), $\hat{\rho}^{(i)}(0)$ occurring in the integro-differential equation (3.349) can be replaced by $\hat{\rho}(0)$.

### 3.3.1.4 Reduced Density Operator in Open Systems (Interaction Picture).
In the case of composite systems, reduced density operators (cf. (2.611)) can be used to describe the behaviour of one subsystem. The reduced density operator $\hat{R}_S^{(i)}(t)$ in the interaction picture reads

$$\hat{R}_S^{(i)}(t) = \mathrm{Tr}_B\left\{\hat{\rho}^{(i)}(t)\right\} = \sum_r \langle r|\,\hat{\rho}^{(i)}(t)\,|r\rangle \ , \quad \hat{R}_S^{(i)}(0) = \hat{\rho}_S(0) \ . \tag{3.351}$$

Differentiating (3.351) with respect to time, and using the above integro-differential equation, one obtains

$$\frac{\partial}{\partial t}\hat{R}_S^{(i)}(t) = -\frac{i}{\hbar}\mathrm{Tr}_B\left[\hat{V}^{(i)}(t), \hat{\rho}_S(0) \otimes \hat{\rho}_B(0)\right]_- - \frac{1}{\hbar^2}\int_0^t \mathrm{d}t'\mathrm{Tr}_B\left\{\left[\hat{V}^{(i)}(t), \left[\hat{V}^{(i)}(t'), \hat{\rho}^{(i)}(t')\right]_-\right]_-\right\} \ . \tag{3.352}$$

The second term of this evolution equation describes *memory effects* in the evolution process. In some cases exact solutions of this integro-differential equation can be derived in a non-numerical way (e.g. see [143]). However, such solutions will not be considered here.

(3.352) determines the exact and still *reversible motion* of the reduced density operator of an open system. However, this equation is not closed, as it still contains $\hat{\rho}^{(i)}(t)$, the density operator of the total system. Approximations will allow us to derive a closed and even local equation for $\hat{R}_S^{(i)}(t)$.

### 3.3.2 Markovian Master Equation

In this section, closed matrix equations resulting from the equation of motion (3.352) are studied. The *Markovian master equation* is derived as an *irreversible equation* which is local in time (cf. [20]).

#### 3.3.2.1 Bath Approximation. Let

$$\hat{\rho}^{(i)}(t) = \hat{R}_S^{(i)}(t) \otimes \hat{R}_B^{(i)}(t) \tag{3.353}$$

(i.e. no correlation at any time). Moreover, consider a stationary bath

$$\hat{R}_B^{(i)}(t) = \hat{\rho}_B(0) \tag{3.354}$$

which is in thermodynamic equilibrium:

$$\hat{\rho}_B(0) = \hat{K}_{\{\hat{H}_B\}} = \frac{1}{Z_{\{\hat{H}_B\}}} \exp\left(-\frac{\hat{H}_B}{k_B T}\right) \tag{3.355}$$

(cf. (2.398)).

Inserting (3.353) and (3.340) into (3.352), the two terms of the r.h.s. of (3.352) yield

$$-\frac{i}{\hbar} \mathrm{Tr}_B \left[\hat{V}^{(i)}(t), \hat{\rho}_S(0) \otimes \hat{\rho}_B(0)\right]_-$$

$$= -\frac{i}{\hbar} \sum_i \left[\hat{Q}_i^{(i)}(t), \hat{\rho}_S(0)\right]_- \mathrm{Tr}_B \left\{\hat{F}_i^{(i)}(t)\hat{\rho}_B(0)\right\} , \tag{3.356}$$

$$-\frac{1}{\hbar^2} \int_0^t dt' \mathrm{Tr}_B \left\{\left[\hat{V}^{(i)}(t), \left[\hat{V}^{(i)}(t'), \hat{\rho}^{(i)}(t')\right]_-\right]_-\right\}$$

$$= -\frac{1}{\hbar^2} \sum_{i,j} \int_0^t dt' \left\{\left[\hat{Q}_i^{(i)}(t), \hat{Q}_j^{(i)}(t')\hat{R}_S^{(i)}(t')\right]_- G_{ij}(t,t') - \right.$$

$$\left. \left[\hat{Q}_i^{(i)}(t), \hat{R}_S^{(i)}(t')\hat{Q}_j^{(i)}(t')\right]_- G_{ji}(t',t)\right\} \tag{3.357}$$

with the *2-time-correlation functions* of the bath:

$$\boxed{\begin{aligned} G_{ij}(t,t') &= \mathrm{Tr}_B \left\{\hat{F}_i^{(i)}(t)\hat{F}_j^{(i)}(t')\hat{\rho}_B(0)\right\} , \\ G_{ji}(t',t) &= \mathrm{Tr}_B \left\{\hat{F}_j^{(i)}(t')\hat{F}_i^{(i)}(t)\hat{\rho}_B(0)\right\} . \end{aligned}} \tag{3.358}$$

We observe that the trace $\mathrm{Tr}_B \left\{\hat{F}_i^{(i)}(t)\hat{\rho}_B(0)\right\}$ vanishes, because $\hat{\rho}_B$ is diagonal in basis $|r\rangle$ and (3.331) applies:

$$\mathrm{Tr}_B \left\{\hat{F}_i^{(i)}(t)\hat{\rho}_B(0)\right\} = \sum_r \left\langle r \left|\hat{F}_i^{(i)}(t)\right| r\right\rangle \langle r|\hat{\rho}_B(0)|r\rangle = 0 . \tag{3.359}$$

Thus, the term (3.356) is identically zero.

Introducing the new time variable $\tau$ with

$$\tau = t - t', \quad d\tau = -dt', \quad \int_0^t dt' = \int_0^t d\tau , \tag{3.360}$$

one finally obtains the evolution equation (3.352) in bath approximation:

$$
\begin{aligned}
\frac{\partial}{\partial t} & \hat{R}_S^{(i)}(t) \\
= & -\frac{1}{\hbar^2} \sum_{i,j} \int_0^t d\tau \left\{ \left[ \hat{Q}_i^{(i)}(t), \hat{Q}_j^{(i)}(t-\tau) \hat{R}_S^{(i)}(t-\tau) \right]_- G_{ij}(\tau) \right. \\
& \left. - \left[ \hat{Q}_i^{(i)}(t), \hat{R}_S^{(i)}(t-\tau) \hat{Q}_j^{(i)}(t-\tau) \right]_- G_{ji}(-\tau) \right\} ,
\end{aligned}
\tag{3.361}
$$

where, exploiting translational invariance of the correlation functions,

$$
\begin{aligned}
G_{ij}(\tau) &= \mathrm{Tr}\left\{ \hat{F}_i^{(i)}(\tau) \hat{F}_j^{(i)}(0) \hat{\rho}_B(0) \right\} \\
&= \mathrm{Tr}\left\{ \hat{F}_i^{(i)}(\tau) \hat{F}_j(0) \hat{\rho}_B(0) \right\} , \\
G_{ji}(-\tau) &= \mathrm{Tr}\left\{ \hat{F}_j^{(i)}(0) \hat{F}_i^{(i)}(\tau) \hat{\rho}_B(0) \right\} \\
&= \mathrm{Tr}\left\{ \hat{F}_j(0) \hat{F}_i^{(i)}(\tau) \hat{\rho}_B(0) \right\} .
\end{aligned}
\tag{3.362}
$$

Here we have used that $\hat{F}_i^{(i)}(0)$, $\hat{F}_j^{(i)}(0)$ are identical with the operators in the Schrödinger picture $\hat{F}_i$, $\hat{F}_j$ (compare (3.339)).

On time scales very much larger than the correlation time $\tau_c$ a further simplification of the equation of motion (3.361) becomes possible.

### 3.3.2.2 Long-Time Behaviour. Let

$$G_{ji}(-\tau), G_{ij}(\tau) \approx 0 \quad \text{for} \quad \tau \gg \tau_c , \tag{3.363}$$

where $\tau_c$ describes the time-range of correlations within the bath and thus is called *correlation time*. Furthermore, it shall be assumed that the reduced density operator $\hat{R}_S^{(i)}(t)$ is only weakly dependent of $\tau$ on the time scale $\tau_c$:

$$\hat{R}_S^{(i)}(t-\tau) \approx \hat{R}_S^{(i)}(t) \quad \text{for} \quad \tau < \tau_c \tag{3.364}$$

(*Markov condition*).

The matrix form of (3.361) reads

$$
\left\langle s' \left| \frac{\partial}{\partial t} \hat{R}_S^{(i)}(t) \right| s \right\rangle = -\frac{1}{\hbar^2} \sum_{i,j} \int_0^t \left\{ M_{s's}^{(1)}(i,j,t,\tau) G_{ij}(\tau) - M_{s's}^{(2)}(i,j,t,\tau) G_{ji}(-\tau) \right\} d\tau
\tag{3.365}
$$

with

$$M^{(1)}_{s's}(i,j,t,\tau) = \left\langle s' \left| \left[ \hat{Q}_i^{(i)}(t), \hat{Q}_j^{(i)}(t-\tau)\hat{R}_S^{(i)}(t-\tau) \right]_- \right| s \right\rangle \tag{3.366}$$

and

$$M^{(2)}_{s's}(i,j,t,\tau) = \left\langle s' \left| \left[ \hat{Q}_i^{(i)}(t), \hat{R}_S^{(i)}(t-\tau)\hat{Q}_j^{(i)}(t-\tau) \right]_- \right| s \right\rangle . \tag{3.367}$$

With (3.364), the matrix elements $M^{(1)}_{ij}(t,\tau)$ can be rewritten as

$$M^{(1)}_{s's}(i,j,t,\tau)$$
$$= \left\langle s' \left| \hat{Q}_i^{(i)}(t)\hat{Q}_j^{(i)}(t-\tau)\hat{R}_S^{(i)}(t) \right| s \right\rangle - \left\langle s' \left| \hat{Q}_j^{(i)}(t-\tau)\hat{R}_S^{(i)}(t)\hat{Q}_i^{(i)}(t) \right| s \right\rangle$$
$$= \sum_{m,n} \left\langle s' \left| \hat{Q}_i^{(i)}(t) \right| m \right\rangle \left\langle m \left| \hat{Q}_j^{(i)}(t-\tau) \right| n \right\rangle \left\langle n \left| \hat{R}_S^{(i)}(t) \right| s \right\rangle -$$
$$\sum_{m,n} \left\langle s' \left| \hat{Q}_j^{(i)}(t-\tau) \right| m \right\rangle \left\langle m \left| \hat{R}_S^{(i)}(t) \right| n \right\rangle \left\langle n \left| \hat{Q}_i^{(i)}(t) \right| s \right\rangle . \tag{3.368}$$

Making use of (3.342), (3.368) can be recast into the form

$$M^{(1)}_{s's}(i,j,t,\tau)$$
$$= \sum_{m,n} m^{(ij)}_{s'mmn} \left\langle n \left| \hat{R}_S^{(i)}(t) \right| s \right\rangle e^{i(\omega_{s'm}+\omega_{mn})t} e^{-i\omega_{mn}\tau} -$$
$$\sum_{m,n} m^{(ij)}_{nss'm} \left\langle m \left| \hat{R}_S^{(i)}(t) \right| n \right\rangle e^{i(\omega_{s'm}+\omega_{ns})t} e^{-i\omega_{s'm}\tau} \tag{3.369}$$

with

$$m^{(ij)}_{klop} = \left\langle k \left| \hat{Q}_i \right| l \right\rangle \left\langle o \left| \hat{Q}_j \right| p \right\rangle . \tag{3.370}$$

Due to (3.343), the relations

$$\omega_{s'm} + \omega_{mn} = \omega_{s'n} , \quad \omega_{s'm} + \omega_{ns} = \omega_{s's} - \omega_{mn} \tag{3.371}$$

hold.

In the same way $M^{(2)}_{s's}(i,j,t,\tau)$ can be calculated:

$$M^{(2)}_{ss'}(i,j,t,\tau)$$
$$= \sum_{m,n} m^{(ij)}_{s'mns} \left\langle m \left| \hat{R}_S^{(i)}(t) \right| n \right\rangle e^{i(\omega_{s's}-\omega_{mn})t} e^{-i\omega_{ns}\tau} -$$
$$\sum_{m,n} m^{(ij)}_{nsmn} \left\langle s' \left| \hat{R}_S^{(i)}(t) \right| m \right\rangle e^{i\omega_{ms}t} e^{-i\omega_{mn}\tau} . \tag{3.372}$$

Inserting the matrix elements $M^{(1)}_{s's}(i,j,t,\tau)$, $M^{(2)}_{s's}(i,j,t,\tau)$ into (3.365), we encounter the terms

$$\boxed{\Gamma^+_{mkln}(t) = \frac{1}{\hbar^2} \sum_{i,j} m^{(ij)}_{mkln} \int_0^t e^{-i\omega_{ln}\tau} G_{ij}(\tau) d\tau ,} \tag{3.373}$$

$$\Gamma^-_{mkln}(t) = \frac{1}{\hbar^2} \sum_{i,j} m^{(ij)}_{lnmk} \int_0^t e^{-i\omega_{mk}\tau} G_{ji}(-\tau) d\tau \ . \tag{3.374}$$

Applying (3.363), we may now replace the upper boundary of the integrals, $t$, for $t \gg \tau_c$ by $\infty$. On this time scale the $\Gamma_{mkln}$ parameters become independent of time. Introducing the abbreviation

$$R^{(i)}_{s's}(t) = \left\langle s' \left| \hat{R}^{(i)}_S(t) \right| s \right\rangle \ , \tag{3.375}$$

the evolution equation (3.365) then yields

$$\dot{R}^{(i)}_{s's}(t) = \sum_{m,n} R_{s'smn}(t) R^{(i)}_{mn}(t) \ , \tag{3.376}$$

where the matrix elements

$$R_{s'smn}(t)$$
$$= \left[ -\sum_k \delta_{sn} \Gamma^+_{s'kkm} + \Gamma^+_{nss'm} + \Gamma^-_{nss'm} - \sum_k \delta_{s'm} \Gamma^-_{nkks} \right] \tag{3.377}$$
$$e^{i(\omega_{s's} - \omega_{mn})t}$$

represent the (still time-dependent) *relaxation matrix* $\tilde{R}$. (The symbols $\delta_{sn}$, $\delta_{s'm}$ denote Kronecker deltas).

The complex parameters $\Gamma^+_{mkln}$, $\Gamma^-_{mkln}$ will be called *complex damping parameters*. They are constrained by

$$\left( \Gamma^-_{nlkm} \right)^* = \Gamma^+_{mkln} \ . \tag{3.378}$$

*Proof.* Due to (3.374), the elements $\Gamma^-_{nlkm}$ can be written in the form

$$\Gamma^-_{nlkm} = \frac{1}{\hbar^2} \sum_{i,j} m^{(ij)}_{kmnl} \int_0^{+\infty} e^{-i\omega_{nl}t''} G_{ji}(-t'') dt'' \tag{3.379}$$

so that the conjugate complex elements $\left( \Gamma^-_{nlkm} \right)^*$ are given by

$$\left( \Gamma^-_{nlkm} \right)^* = \frac{1}{\hbar^2} \sum_{i,j} m^{(ij)}_{mkln} \int_0^{+\infty} e^{-i\omega_{ln}t''} G^*_{ji}(-t'') dt'' \ . \tag{3.380}$$

Due to the fact that the conjugate complex of the expectation value

$$\left\langle \hat{A}\hat{B} \right\rangle = \sum_{i,j,k} A_{ik} B_{kj} \rho_{ji} \tag{3.381}$$

is, for Hermitian matrices $A^*_{ik} = A_{ki}$, given by

$$\left\langle \hat{A}\hat{B} \right\rangle^* = \left\langle \hat{B}\hat{A} \right\rangle \ , \tag{3.382}$$

the correlation function

$$G_{ji}^*(-t'') = \mathrm{Tr_B}\left\{\hat{F}_j^{(i)}(0)\hat{F}_i^{(i)}(t'')\hat{\rho}_B(0)\right\}^*$$
$$= \left\langle \hat{F}_j^{(i)}(0)\hat{F}_i^{(i)}(t'')\right\rangle^* \tag{3.383}$$

is identical with $G_{ij}(t'')$. Inserting

$$\boxed{G_{ji}^*(-t'') = G_{ij}(t'')} \tag{3.384}$$

into (3.380), one obtains

$$\left(\Gamma_{nlkm}^-\right)^* = \frac{1}{\hbar^2}\sum_{i,j} m_{mkln}^{(ij)} \int_0^{+\infty} e^{-i\omega_{ln}t''} G_{ij}(t'')\mathrm{d}t'' \ . \tag{3.385}$$

Comparing this expression with the definition (3.373), equation (3.378) follows.

This means that the parameters $\Gamma_{mkln}^+$ suffice to completely specify the influence of the bath.

**3.3.2.3 Secular Approximation.** The time-dependence of the relaxation matrix $\tilde{\mathsf{R}}$ vanishes if we restrict ourselves to

$$\omega_{s's} - \omega_{mn} = 0 \ . \tag{3.386}$$

Using this condition, only matrix elements $R_{ssss}$, $R_{ssmm}$ $(m \neq s)$, and $R_{s'ss's}$ are unequal zero. These so-called *secular terms* are given by

$s' = s = m = n$ :

$$R_{ssss} = \Gamma_{ssss}^+ + \Gamma_{ssss}^- - \sum_k \left(\Gamma_{skks}^+ + \Gamma_{skks}^-\right) \ , \tag{3.387}$$

$s' = s$ , $m = n$ , $s' \neq m$ :

$$R_{ssmm} = \Gamma_{mssm}^+ + \Gamma_{mssm}^- \ , \tag{3.388}$$

$s' = m$ , $s = n$ , $s' \neq s$ :

$$R_{s'ss's} = \Gamma_{sss's'}^+ + \Gamma_{sss's'}^- - \sum_k \left(\Gamma_{s'kks'}^+ + \Gamma_{skks}^-\right) \ . \tag{3.389}$$

We introduce the *transition probability* (= probability per time unit, in this book also called *transition rate*) from $m$ to $s$ $(m \neq s)$:

$$W_{sm} := \Gamma_{mssm}^+ + \Gamma_{mssm}^- \tag{3.390}$$

which (using (3.378)) can be written as

$$\boxed{W_{sm} = \Gamma_{mssm}^+ + \left(\Gamma_{mssm}^+\right)^* = 2\mathrm{Re}\left\{\Gamma_{mssm}^+\right\} = \mathrm{real} \ ,} \tag{3.391}$$

and the complex so-called *non-adiabatic parameters*

$$\gamma_{s's} = \gamma_{ss'}^* := \sum_k \left(\Gamma_{s'kks'}^+ + \Gamma_{skks}^-\right) \tag{3.392}$$

and the *adiabatic parameters*

$$\tilde{\gamma}_{s's} = \tilde{\gamma}^*_{ss'} := -\left(\Gamma^+_{sss's'} + \Gamma^-_{sss's'}\right) . \tag{3.393}$$

The non-adiabatic parameters can be split into their respective real and imaginary parts:

$$
\begin{aligned}
\mathrm{Re}\,\{\gamma_{s's}\} &= \sum_k \mathrm{Re}\,\{\Gamma^+_{s'kks'}\} + \sum_k \mathrm{Re}\,\{\Gamma^-_{skks}\} \\
&= \sum_k \mathrm{Re}\,\{\Gamma^-_{s'kks'}\} + \sum_k \mathrm{Re}\,\{\Gamma^+_{skks}\}
\end{aligned}
\tag{3.394}
$$

so that

$$\boxed{\mathrm{Re}\,\{\gamma_{s's}\} = \frac{1}{2}\left(\sum_{k,k\neq s'} W_{ks'} + \sum_{k,k\neq s} W_{ks}\right) .} \tag{3.395}$$

Correspondingly we define as adiabatic damping parameters:

$$\boxed{\begin{aligned}\mathrm{Re}\,\{\tilde{\gamma}_{s's}\} &= -\mathrm{Re}\,\{\Gamma^+_{sss's'}\} - \mathrm{Re}\,\{\Gamma^+_{s's'ss}\} = \mathrm{Re}\,\{\tilde{\gamma}_{ss'}\} \\ &:= W^{\mathrm{ad}}_{ss'} .\end{aligned}} \tag{3.396}$$

The imaginary parts are

$$\mathrm{Im}\,\{\gamma_{s's}\} = \mathrm{Im}\left\{\sum_k \Gamma^+_{s'kks'}\right\} + \mathrm{Im}\left\{\sum_k \Gamma^+_{skks}\right\} , \tag{3.397}$$

$$\mathrm{Im}\,\{\tilde{\gamma}_{s's}\} = -\mathrm{Im}\,\{\Gamma^+_{sss's'}\} - \mathrm{Im}\,\{\Gamma^+_{s's'ss}\} . \tag{3.398}$$

Here we have, again, applied (3.378). The imaginary parts will be written as a frequency shift:

$$\boxed{\mathrm{Im}\,\{\gamma_{s's}\} + \mathrm{Im}\,\{\tilde{\gamma}_{s's}\} := \Delta\omega_{s's} .} \tag{3.399}$$

Using these relations, and the abbreviation

$$\boxed{\gamma^{(s's)} = \frac{1}{2}\left(\sum_{k,k\neq s'} W_{ks'} + \sum_{k,k\neq s} W_{ks}\right) + W^{\mathrm{ad}}_{s's} ,} \tag{3.400}$$

the matrix elements (3.387)–(3.389) can be expressed in terms of $W_{ks}$, $W^{\mathrm{ad}}_{s's}$, and $\Delta\omega_{s's}$:

$$
\begin{aligned}
R_{ssss} &= -\sum_{k,k\neq s} W_{ks} , \\
R_{ssmm} &= W_{sm} \ (s \neq m) , \\
R_{s'ss's} &= -\gamma^{(s's)} - \mathrm{i}\Delta\omega_{s's} .
\end{aligned}
\tag{3.401}
$$

This scheme can be condensed into

$$R_{s's\,mn}$$

$$
= \begin{cases}
\delta_{mn}(1 - \delta_{ms})W_{sm} - \delta_{ms}\delta_{ns} \displaystyle\sum_{k,k\neq s} W_{ks} & (s' = s) \\[2mm]
- \left(\gamma^{(s's)} + \mathrm{i}\Delta\omega_{s's}\right)\delta_{ms'}\delta_{ns} & (s' \neq s)
\end{cases}
\tag{3.402}
$$

Thus, the evolution equation (3.376) finally reads

$$\dot{R}^{(i)}_{s's}(t)$$

$$
= \begin{cases}
\displaystyle\sum_{m,m\neq s} W_{sm} R^{(i)}_{mm}(t) - R^{(i)}_{ss}(t) \displaystyle\sum_{m,m\neq s} W_{ms} & (s' = s) \\[2mm]
- \left(\gamma^{(s's)} + \mathrm{i}\Delta\omega_{s's}\right) R^{(i)}_{s's}(t) & (s' \neq s)
\end{cases}
\tag{3.403}
$$

**3.3.2.4 Schrödinger Picture.** Carrying out the transformation

$$
\hat{R}^{(i)}_{\mathrm{S}}(t) = \mathrm{e}^{\mathrm{i}\hat{H}_{\mathrm{S}}t/\hbar}\,\hat{R}_{\mathrm{S}}(t)\mathrm{e}^{-\mathrm{i}\hat{H}_{\mathrm{S}}t/\hbar} \;,
\tag{3.404}
$$

and using the abbreviations

$$
R_{s's}(t) = \left\langle s' \left| \hat{R}_{\mathrm{S}}(t) \right| s \right\rangle
\tag{3.405}
$$

and

$$
D_{s's}(t) = \frac{\mathrm{i}}{\hbar} \left\langle s' \left| \left[\hat{H}_{\mathrm{S}}, \hat{R}_{\mathrm{S}}(t)\right]_{-} \right| s \right\rangle \;,
\tag{3.406}
$$

one obtains the *Markovian master equation*

$$
\dot{R}_{s's}(t) + D_{s's}(t) = \sum_{m,n} R_{s's\,mn} R_{mn}(t) \;.
\tag{3.407}
$$

$\hat{H}_{\mathrm{S}}$ describes the system S under consideration, including – if so required – external potential terms (like coupling to a light field). Due to (3.27), the matrix elements $D_{s's}(t)$ follow from the Hamiltonian model, i.e.

$$
D_{s's}(t) = \frac{\mathrm{i}}{\hbar} \sum_{m,n} D_{s'smn} R_{mn}(t)
\tag{3.408}
$$

with

$$
D_{s's\,mn}(t) = \left(E^{(\mathrm{S})}_{s'} - E^{(\mathrm{S})}_{s}\right)\delta_{ms'}\delta_{ns} + \langle s'| \hat{V}_{\mathrm{ext}} |s\rangle \delta_{ns} -
$$
$$
\langle s'| \hat{V}_{\mathrm{ext}} |s\rangle \delta_{ms'} \;.
\tag{3.409}
$$

Equation (3.407) essentially replaces the evolution equation of a closed system. The additional part in terms of matrix elements $R_{s's\,mn}$ determines *incoherent effects* (*relaxation processes*), while the term containing the matrix elements $D_{s's\,mn}$ describes *coherent effects*. Coherent and incoherent parts are simply additive contributions to the evolution of the (reduced) density matrix.

**3.3.2.5 Constraints on the Relaxation Matrix.** The elements $R_{s'smn}$ of the relaxation matrix $\tilde{R}$ must have the following properties to guarantee that the respective density matrix R remains a density matrix at all times.

*Property 1.* The sum of all diagonal matrix elements $R_{nn}$ defined by (3.408) has to be equal zero, because the diagonal elements $R_{nn}$ define probabilities with $\sum_n R_{nn} = 1$:

$$\sum_n \dot{R}_{nn} = \sum_{m,m'} \left( \sum_n R_{nnmm'} \right) R_{mm'} + \underbrace{\text{coherent part}}_{0} = 0 \,. \qquad (3.410)$$

Assuming $R_{mm'} \neq 0$, it follows that

$$\boxed{\sum_n R_{nnmm'} = 0 \,.} \qquad (3.411)$$

*Property 2.* Due to the fact that the matrix elements $R_{nn'}$ define a Hermitian matrix,

$$R_{nn'}^* = R_{n'n} \,, \qquad (3.412)$$

the complex conjugate form of the equation

$$\dot{R}_{nn'}^* = \sum_{mm'} R_{nn'mm'}^* R_{mm'}^* \qquad (3.413)$$

can be written as

$$\dot{R}_{n'n} = \sum_{mm'} R_{nn'mm'}^* R_{m'm} \,. \qquad (3.414)$$

Comparing with

$$\dot{R}_{n'n} = \sum_{mm'} R_{n'nm'm} R_{m'm} \,, \qquad (3.415)$$

one has to require

$$\boxed{R_{n'nm'm}^* = R_{nn'mm'} \,.} \qquad (3.416)$$

*Property 3.* As $R_{mm}$ describes probabilities, the relation

$$0 \leq R_{mm} \leq 1 \qquad (3.417)$$

holds. Therefore, in the case of $R_{mm} = 1$ at some initial time $t$, the inequalities

$$\begin{aligned} \dot{R}_{nn} &= R_{nnmm} R_{mm} \geq 0 \,, \\ \dot{R}_{mm} &= R_{mmmm} R_{mm} \leq 0 \end{aligned} \qquad (3.418)$$

must hold so that the matrix elements $R_{nnmm}$, $R_{mmmm}$ have to fulfil the inequalities

$$\boxed{R_{nnmm} \geq 0 \,, \quad R_{mmmm} \leq 0 \,, \quad n \neq m \,.} \qquad (3.419)$$

**3.3.2.6 Fermi's Golden Rule.** We write the complex damping parameter $\Gamma^+_{nmmn}$ (cf. (3.373)) in the form

$$\Gamma^+_{nmmn} = \frac{1}{\hbar^2} \sum_{r,r'} \gamma^{nm}_{rr'} \langle r' | \hat{\rho}_B(0) | r' \rangle \int_0^{+\infty} e^{i(E_{r'r} - \hbar\omega_{mn})\tau/\hbar} d\tau \tag{3.420}$$

with

$$\gamma^{nm}_{rr'} = \sum_{i,j} \langle n | \hat{Q}_i | m \rangle \langle m | \hat{Q}_j | n \rangle \langle r' | \hat{F}_i | r \rangle \langle r | \hat{F}_j | r' \rangle . \tag{3.421}$$

Using the identity (cf. (3.329))

$$\sum_i \langle n | \hat{Q}_i | m \rangle \langle r' | \hat{F}_i | r \rangle = \left\langle nr' \left| \sum_i \hat{Q}_i \hat{F}_i \right| mr \right\rangle$$

$$= \langle nr' | \hat{V} | mr \rangle , \tag{3.422}$$

we find that

$$\gamma^{nm}_{rr'} = \langle nr' | \hat{V} | mr \rangle \langle mr | \hat{V} | nr' \rangle$$

$$= \left| \langle mr | \hat{V} | nr' \rangle \right|^2 . \tag{3.423}$$

With the abbreviation

$$\rho_B(0)_{r'r'} = \langle r' | \hat{\rho}_B(0) | r' \rangle , \tag{3.424}$$

and based on the formula

$$\lim_{t \to \infty} \int_0^t e^{i\omega\tau} d\tau = \pi\delta(\omega) + iP\frac{1}{\omega} , \tag{3.425}$$

wherein $P$ denotes the principal value of the integral, the real and imaginary parts of the damping parameters can be separated yielding

$$\boxed{ \begin{aligned} &\mathrm{Re}\left\{\Gamma^+_{nmmn}\right\} \\ &= \frac{\pi}{\hbar} \sum_{rr'} \left| \langle mr | \hat{V} | nr' \rangle \right|^2 \rho_B(0)_{r'r'} \delta\left(E_{r'r} - \hbar\omega_{mn}\right) \end{aligned} } \tag{3.426}$$

and

$$\boxed{ \mathrm{Im}\left\{\Gamma^+_{nmmn}\right\} = \frac{1}{\hbar} P \sum_{rr'} \frac{\left| \langle mr | \hat{V} | nr' \rangle \right|^2 \rho_B(0)_{r'r'}}{E_{r'r} - \hbar\omega_{mn}} , } \tag{3.427}$$

respectively. Based on these results, the transition probabilities

$$W_{mn} = 2\mathrm{Re}\{\Gamma^+_{nmmn}\}$$

$$= 2\mathrm{Re}\left\{\frac{1}{\hbar^2}\sum_{i,j}\langle n|\hat{Q}_i|m\rangle\langle m|\hat{Q}_j|n\rangle\right.$$

$$\left.\int_0^{+\infty} e^{-i\omega_{mn}\tau}G_{ij}(\tau)d\tau\right\}$$

(3.428)

(cf. (3.391), (3.458)) can be recast into

$$W_{mn} = \frac{2\pi}{\hbar}\sum_{rr'}\left|\langle mr|\hat{V}|nr'\rangle\right|^2 \rho_B(0)_{r'r'}$$

$$\delta\left(E_{r'r} - \hbar w_{mn}\right).$$

(3.429)

With the canonical statistical operator (3.355), these matrix elements read

$$W_{mn} = \frac{2\pi}{Z_{\{\hat{H}_B\}}\hbar}\sum_{rr'}\left|\langle mr|\hat{V}|nr'\rangle\right|^2 e^{-E_{r'}^{(B)}/k_B T}$$

$$\delta\left(E_{r'r} - \hbar w_{mn}\right).$$

(3.430)

Equations (3.428)–(3.430) represent three equivalent definitions for transition probabilities $W_{mn}$.

While in (3.428) the influence of the bath is described by a correlation function $G_{ij}(\tau)$, (3.429), (3.430) explicitly refer to the interaction of system S with the bath. Due to these interactions, a coupled transition from the initial state $|nr'\rangle$ into a *final state* $|mr\rangle$ takes place (cf. Fig. 3.23). Due to (3.429), a *reservoir* of states $|r\rangle$, $|r'\rangle$ within the bath has to be considered. (Such a reservoir of states is characterized by a certain spectral density.)

Examples are represented by systems composed of atoms (system S) and a photon bath. Then the l.h.s. of Fig. 3.23 illustrates two states of the atom, and the r.h.s. represents photon states.

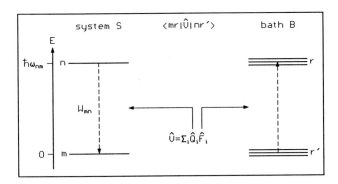

**Fig. 3.23.** Coupled transitions

The result (3.429) represents *Fermi's golden rule*, usually written in the form

$$W_{fi}(t) = \frac{2\pi}{\hbar} \left|V_{fi}\right|^2 \rho_f(E_i) \,, \tag{3.431}$$

where the index $i$ denotes the initial state, $f$ the final state of the total open system. The matrix elements are

$$V_{fi} = \left\langle f \left| \hat{V} \right| i \right\rangle \,, \tag{3.432}$$

and $\rho_f(E_i)$ is the density of final states.

One easily convinces oneself that the inverse transition probability is, according to (3.430) (exchange $m$ and $n$, $r$ and $r'$),

$$
\begin{aligned}
W_{nm} &= \frac{2\pi}{Z_{\{\hat{H}_{\mathrm{B}}\}}\hbar} \sum_{rr'} \left|\left\langle nr' \left| \hat{V} \right| mr \right\rangle\right|^2 e^{-E_r^{(\mathrm{B})}/k_{\mathrm{B}}T} \\
&\quad \delta\left(E_{rr'} - \hbar w_{nm}\right) \\
&= \frac{2\pi}{Z_{\{\hat{H}_{\mathrm{B}}\}}\hbar} \sum_{rr'} \left|\left\langle mr \left| \hat{V} \right| nr' \right\rangle\right|^2 e^{-E_r^{(\mathrm{B})}/k_{\mathrm{B}}T} \\
&\quad \delta\left(E_{r'r} - \hbar w_{mn}\right)
\end{aligned}
\tag{3.433}
$$

so that

$$\frac{W_{nm}}{W_{mn}} = e^{-\hbar\omega_{mn}/k_{\mathrm{B}}T} \,, \tag{3.434}$$

which should be contrasted with $W_{nm} = W_{mn}$ valid for closed systems.

Similarly, the real and imaginary parts of the damping parameters $\tilde{\gamma}_{mn}$ yield (cf. (3.396), (3.398))

$$
\begin{aligned}
&\mathrm{Re}\left\{\tilde{\gamma}_{mn}\right\} \\
&= -\mathrm{Re}\left\{\Gamma^+_{nnmm}\right\} - \mathrm{Re}\left\{\Gamma^+_{mmnn}\right\} \\
&= -\frac{2\pi}{\hbar} \sum_{rr'} \left\langle nr' \left| \hat{V} \right| nr \right\rangle \left\langle mr \left| \hat{V} \right| mr' \right\rangle \rho_{\mathrm{B}}(0)_{r'r'} \delta\left(E_{r'r}\right) \\
&= W_{mn}^{\mathrm{ad}}
\end{aligned}
\tag{3.435}
$$

and

$$
\begin{aligned}
&\mathrm{Im}\left\{\tilde{\gamma}_{mn}\right\} \\
&= -\frac{2}{\hbar}P\sum_{rr'} \frac{\left\langle nr' \left| \hat{V} \right| nr \right\rangle \left\langle mr \left| \hat{V} \right| mr' \right\rangle \rho_{\mathrm{B}}(0)_{r'r'}}{E_{r'r}} \,,
\end{aligned}
\tag{3.436}
$$

respectively. The damping parameters $\mathrm{Im}\left\{\gamma_{mn}\right\}$ read

$$\boxed{\begin{aligned}
&\mathrm{Im}\,\{\gamma_{mn}\} \\
&= \mathrm{Im}\left\{\sum_k \Gamma^+_{mkkm}\right\} + \mathrm{Im}\left\{\sum_k \Gamma^+_{nkkn}\right\} \\
&= \frac{1}{\hbar}P\sum_{rr'}\sum_k \frac{\left\langle mr'\left|\hat{V}\right|kr\right\rangle\left\langle kr\left|\hat{V}\right|mr'\right\rangle \rho_B(0)_{r'r}}{E_{r'r}-\hbar\omega_{km}} + \\
&\quad \frac{1}{\hbar}P\sum_{rr'}\sum_k \frac{\left\langle nr'\left|\hat{V}\right|kr\right\rangle\left\langle kr\left|\hat{V}\right|nr'\right\rangle \rho_B(0)_{r'r}}{E_{r'r}-\hbar\omega_{kn}}.
\end{aligned}}$$
(3.437)

These terms give rise to *frequency shifts* (compare (3.399)).

### 3.3.3 Quantum Dynamical Semigroup: The Lindblad Form

The dynamics of a closed quantum mechanical system can be represented by

$$\hat{\rho}(t) := \hat{G}(t)\hat{\rho}(0) \ \ (t \geq 0) \,.$$
(3.438)

$\hat{G}(t)$ is a "super operator" acting on the operator $\hat{\rho}(0)$, where this super operator is specified by (cf. [115])

$$\hat{G}(t)\hat{\rho}(0) := \hat{U}(t)\hat{\rho}(0)\hat{U}^\dagger(t) \,.$$
(3.439)

$\hat{U}(t)$ defines a one-parameter unitary group (with infinitesimal generator $\hat{H}$). The master equation derived in the preceding sections can be considered as a special case for the irreversible dynamics generated by a super operator $\hat{G}(t)$:

$$\hat{\rho}(t) = \hat{G}(t)\hat{\rho}(0) \ \ (t \geq 0) \,,$$
(3.440)

where $\hat{G}(t)$ defines a one-parameter *semigroup*, i.e.

$$\hat{G}(t_1+t_2) = \hat{G}(t_1)\hat{G}(t_2) \ \ (t_1,t_2 \geq 0) \,.$$
(3.441)

If $\hat{G}$ is required to be linear, positive, trace-preserving, and strongly continuous, we may write

$$\hat{G}(t) = \exp\left\{\hat{\mathfrak{L}}t\right\} \,,$$
(3.442)

where $\hat{\mathfrak{L}}$ is the linear independent infinitesimal generator of the semigroup $\hat{G}$, and (3.440) can be written as

$$\frac{\mathrm{d}}{\mathrm{d}t}\hat{\rho}(t) = \hat{\mathfrak{L}}\hat{\rho}(t) \,.$$
(3.443)

According to a theorem of Gorini, Kossakowski, and Sudarshan (cf. [59]), the super operator $\hat{\mathfrak{L}}$ must have the following structure ($s = n^2 - 1$):

$$\boxed{\hat{\mathfrak{L}}\hat{\rho} = -\frac{i}{\hbar}\left[\hat{H},\hat{\rho}\right]_- + \frac{1}{2}\sum_{i,j=1}^s A_{ij}\left(\left[\hat{F}_i,\hat{\rho}\hat{F}_j^+\right]_- + \left[\hat{F}_i\hat{\rho},\hat{F}_j^+\right]_-\right) \,,}$$
(3.444)

where

$$\text{Tr}\left\{\hat{F}_i\right\} = 0 , \tag{3.445}$$

$$\text{Tr}\left\{\hat{F}_i\hat{F}_j^+\right\} = \delta_{ij} . \tag{3.446}$$

A is positive semidefinite, i.e.

$$A_{ii} \geq 0 \quad (i = 1, 2, \ldots, s) , \tag{3.447}$$

$$|A_{ik}|^2 \leq A_{ii}A_{kk} . \tag{3.448}$$

$\hat{\mathcal{L}}$ is also known as the *Lindblad operator* (cf. [86]).

For later reference we split the incoherent term in (3.444) into two parts:

$$\begin{aligned}
\hat{\mathcal{L}}_{\text{inc}}\hat{\rho} &= \frac{1}{2}\sum_{i,j=1}^{s} A_{ij}\left(2\hat{F}_i\rho\hat{F}_j^+ - \rho\hat{F}_j^+\hat{F}_i - \hat{F}_j^+\hat{F}_i\rho\right) \\
&:= \hat{\mathcal{L}}_{\text{inc}}^{(1)}\hat{\rho} + \hat{\mathcal{L}}_{\text{inc}}^{(2)}\hat{\rho} ,
\end{aligned} \tag{3.449}$$

where

$$\boxed{\begin{aligned}
\hat{\mathcal{L}}_{\text{inc}}^{(1)}\hat{\rho} &= \sum_{i,j=1}^{s} A_{ij}\hat{F}_i\rho\hat{F}_j^+ , \\
\hat{\mathcal{L}}_{\text{inc}}^{(2)}\hat{\rho} &= -\frac{1}{2}\sum_{i,j=1}^{s} A_{ij}\left(\rho\hat{F}_j^+\hat{F}_i + \hat{F}_j^+\hat{F}_i\rho\right) .
\end{aligned}} \tag{3.450}$$

The raising and lowering operators (see Sect. 2.2.4)

$$\begin{aligned}
\hat{\sigma}_{ij}^+ &= \frac{1}{2}\left(\hat{u}_{ij} + i\hat{v}_{ij}\right) = \hat{P}_{ji} = \hat{\sigma}_{ji}^- , \\
\hat{\sigma}_{ij}^- &= \frac{1}{2}\left(\hat{u}_{ij} - i\hat{v}_{ij}\right) = \hat{P}_{ij} = \hat{\sigma}_{ji}^+
\end{aligned} \tag{3.451}$$

with $i < j$ and

$$\begin{aligned}
\hat{\sigma}_{ij}^+\hat{\sigma}_{ij}^- &= \hat{P}_{jj} , \\
\hat{\sigma}_{ij}^-\hat{\sigma}_{ij}^+ &= \hat{P}_{ii}
\end{aligned} \tag{3.452}$$

have the properties required for $\hat{F}_j^+$, $\hat{F}_j$. In the simplest case, $A_{ij}$ will be diagonal.

*Example 3.3.1.* For a 2-level system with a single decay channel $|2\rangle \to |1\rangle$ we would have according to (3.449)

$$\hat{\mathcal{L}}_{\text{inc}}\hat{\rho} = \frac{1}{2}W_{12}\left(2\hat{\sigma}_{12}^-\rho\hat{\sigma}_{12}^+ - \rho\hat{\sigma}_{12}^+\hat{\sigma}_{12}^- - \hat{\sigma}_{12}^+\hat{\sigma}_{12}^-\rho\right) . \tag{3.453}$$

Observing that

$$\hat{\sigma}_{12}^{+}\hat{\sigma}_{12}^{-} = \hat{P}_{22} ,$$
$$\hat{\sigma}_{12}^{-}\hat{\sigma}_{12}^{+} = \hat{P}_{11} ,$$

(3.454)

we find

$$\hat{\mathcal{L}}_{\text{inc}}^{(1)}\hat{\rho} = W_{12}\hat{P}_{12}\hat{\rho}\hat{P}_{21} = W_{12}\hat{P}_{11}\rho_{22} ,$$

(3.455)

$$\hat{\mathcal{L}}_{\text{inc}}^{(2)}\hat{\rho} = -\frac{1}{2}W_{12}\left(\hat{\rho}\hat{P}_{22} - \hat{P}_{22}\hat{\rho}\right) ,$$

(3.456)

and the incoherent part of the master equation reads in matrix notation:

$$(\dot{\rho}_{11})_{\text{inc}} = W_{12}\rho_{22} ,$$

$$(\dot{\rho}_{22})_{\text{inc}} = -W_{12}\rho_{22} ,$$

$$(\dot{\rho}_{12})_{\text{inc}} = -\frac{1}{2}W_{12}\rho_{12} ,$$

$$(\dot{\rho}_{21})_{\text{inc}} = -\frac{1}{2}W_{12}\rho_{21} .$$

(3.457)

The first equation results from $\hat{\mathcal{L}}_{\text{inc}}^{(1)}\hat{\rho}$, the following three equations are a consequence of $\hat{\mathcal{L}}_{\text{inc}}^{(2)}\hat{\rho}$.

### 3.3.4 Damping Channels

A *Markov bath* (with bath correlation time $\tau_c \to 0$) may be characterized as *broad-band damping*, i.e. a *non-selective coupling* in frequency space. For applications other than the most primitive treatment of just a single (optical) transition this model is too restrictive. Extensions to be discussed now include

a) *Resonant damping*: a damping channel will often be selective in frequency space. Such a frequency dependence implies memory effects. These can be modelled by an enlarged environment consisting of a broad-band bath coupled to a 2-level system via Förster-type coupling (cf. [135]).

b) *Selective multi-channels*: distinguishable channels (distinguishable in frequency space) cannot be of broad-band type: high frequency resolution requires narrow bands and, consequently, large environment correlation times. Pertinent scenarios may include specially designed measurement environments with various frequency filters.

c) *Overlapping channels* (indistinguishability): with respect to different transitions within one node these give rise to *non-selective damping terms*. With respect to the same type of transition and different nodes that indistinguishability leads to non-local damping (cf. Sect. 4.4.7). The most general case (different transitions in different nodes) will not be discussed here.

Selection can also be introduced in terms of an enlarged state space (involving, e.g., polarization, momentum, position). We will return to information aspects of the damping channels in Chap. 4; it should be clear that selectivity is essential for the "logic" allowing us to interpret measurement protocols.

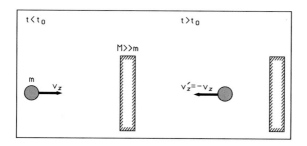

**Fig. 3.24.** Mechanical reflection

**3.3.4.1 Environments of Classical Systems.** In Fig. 3.24 we show a ball (free particle of mass $m$ and velocity $v_z$) impinging on a wall (mass $M \gg m$). In the usual picture of reflection, the ball will hit (time $t_0$) and then return with velocity $-v_z$. The wall acts as a *static boundary condition*. It is obvious that from looking at the mechanical state of the wall we cannot infer that anything might have happened at all.

Similarly, if we put a body of mass $m$ and temperature $T_0$ into a large reservoir of water (mass $M \gg m$, temperature $T_B < T_0$) (see Fig. 3.25), the reservoir will, as a boundary condition, let the mass $m$ cool down to the equilibrium state $T_0 = T_B$. Again, there is (in this model) no way to infer from a thermodynamic measurement of the reservoir that such a non-equilibrium state has been present at all: no information is available.

We now turn to what may be called a *reacting environment*. In Fig. 3.26 the wall has been supplied with an additional subsystem $m' \approx m$, a buffer consisting of a spring and plate at rest. After impact (at time $t_0$), the ball will also reflect ($|v'_z| < v_z$), but leaving the buffer in oscillatory motion (no damping at present). A *memory* has been built up in the mechanical environment. From a complete *measurement protocol* of the buffer dynamics we can infer when and what has happend; classically this protocol could be generated without back action (damping). If we were going to repeat such an experiment at some later time, the reflection would depend on the time elapsed since the first encounter. So, restricting our attention to the ball only (i.e. we refrain from enlarging the system to include the $m'$ subsystem), the reacting environment renders the ball dynamics non-local in time. As there will, in general, be quite a number of reacting coordinates in the environment, it may be a good idea to do so. However, the price we have to pay is *dynamical equations with memory*.

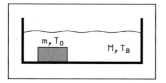

**Fig. 3.25.** Reservoir of water as boundary condition

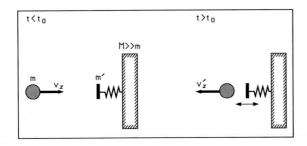

**Fig. 3.26.** Mechanical reflection with memory

Memory effects would disappear if the buffer was overdamped. Then the reflection is partly inelastic; but, as the buffer would come to rest "immediately", the next impact would find the same well-controlled initial state. As far as the subsystem $m'$ is concerned, its response as a damped harmonic oscillator has a broad frequency distribution.

These considerations will now be transcribed to open quantum mechanical systems.

**3.3.4.2 Spectral Density.** In the case of an open system S, the bath causes incoherent transitions within S. On the level of a master equation, these damping effects are described by various kinds of damping parameters which define transition probabilities and renormalized energies. These damping parameters are now analyzed in more detail.

Let us write the complex damping parameters $\Gamma^+_{mkln}$ (cf. (3.373)) in the form

$$\Gamma^+_{mkln} = \sum_{i,j} \Gamma^+_{mkln}(i,j) \,, \tag{3.458}$$

$$
\begin{aligned}
\Gamma^+_{mkln}(i,j) &= \frac{1}{\hbar^2} m^{(ij)}_{mkln} \int_0^{+\infty} e^{-i\omega_{ln}\tau} G_{ij}(\tau)\mathrm{d}\tau \\
&= \frac{1}{\hbar^2} \left\langle m \left| \hat{Q}_i \right| k \right\rangle \left\langle l \left| \hat{Q}_j \right| n \right\rangle \int_0^{+\infty} e^{-i\omega_{ln}\tau} G_{ij}(\tau)\mathrm{d}\tau
\end{aligned}
\tag{3.459}
$$

with (according to (3.362))

$$
\begin{aligned}
G_{ij}(\tau) &\\
&= \mathrm{Tr}\left\{ \hat{F}^{(\mathrm{i})}_i(\tau)\hat{F}_j(0)\hat{\rho}_\mathrm{B}(0) \right\} \\
&= \sum_{rr'} \left\langle r' \left| \hat{F}^{(\mathrm{i})}_i(\tau) \right| r \right\rangle \left\langle r \left| \hat{F}_j(0) \right| r' \right\rangle \langle r' |\hat{\rho}_\mathrm{B}(0)| r' \rangle \\
&= \sum_{rr'} \left\langle r' \left| \hat{F}_i(0) \right| r \right\rangle \left\langle r \left| \hat{F}_j(0) \right| r' \right\rangle \langle r' |\hat{\rho}_\mathrm{B}(0)| r' \rangle\, e^{iE_{r'r}\tau/\hbar} \\
&= \sum_{rr'} \left\langle r' \left| \hat{F}_i \right| r \right\rangle \left\langle r \left| \hat{F}_j \right| r' \right\rangle \langle r' |\hat{\rho}_\mathrm{B}(0)| r' \rangle\, e^{iE_{r'r}\tau/\hbar} \,,
\end{aligned}
\tag{3.460}
$$

$$\boxed{E_{r'r} = E_{r'}^{(B)} - E_r^{(B)} \; .}$$

(3.461)

While the matrix elements $\langle m | \hat{Q}_i | k \rangle$, occurring in these complex damping parameters, describe transitions within the system S, the 2-time-correlation functions $G_{ij}(\tau)$ characterize the bath.

Introducing the *spectral density*

$$\boxed{G_{ij}(\omega) = \sum_{rr'} \langle r' | \hat{F}_i | r \rangle \langle r | \hat{F}_j | r' \rangle \langle r' | \hat{\rho}_B(0) | r' \rangle \delta(\omega - E_{r'r}/\hbar) \, ,}$$

(3.462)

the 2-time-correlation function $G_{ij}(\tau)$ is recovered as the Fourier integral

$$G_{ij}(\tau) = \int_{-\infty}^{+\infty} G_{ij}(\omega) e^{-i\omega\tau} d\omega \; .$$

(3.463)

In particular cases one will find an explicit solution for (3.462). However, very often details of $G_{ij}(\omega)$ are irrelevant and one may resort to analytical models: *Ohmic damping*, e.g., is described by $G(\omega) \sim \omega$.

In the following, the interaction with the bath will be modelled by

$$\boxed{G_{ij}(\omega) \sim \frac{K}{(\omega - \omega_{ij})^2 + K^2} \; .}$$

(3.464)

This *Lorentz distribution* is illustrated in a Figure which will be considered later (cf. Fig. 3.28). The correlation functions $G_{ij}(\tau)$ are then given by ($\tau \geq 0$)

$$\boxed{G_{ij}(\tau) \sim e^{i\omega_{ij}\tau} e^{-\tau K} \; .}$$

(3.465)

The decay of the correlation is thus determined by the *correlation time*

$$\tau_c = 1/K \; .$$

(3.466)

$\hbar\omega_{ij}$ represents the *central frequency* of the respective spectrum. $K$ denotes the *half-width* (*bandwidth*) of the distribution. If the central frequency of the spectral density and the transition energy of a coupled 2-level system, $\hbar\omega_0$, are identical, we speak of *resonant coupling*. For $K \to \infty$ the bath coupling becomes frequency independent and the Markov limit "exact".

Due to relation (3.466), a small width $K$ in the energy distribution causes a large correlation time $\tau_c$, i.e. in such a case the correlation survives a relatively long time. High *frequency resolution* implies loss of locality in time (*coarse-graining*). This is in accord with the *energy time uncertainty relation*

$$\Delta E \Delta t \geq \hbar = 6.6262 \cdot 10^{-16} \, \text{eVs} \; .$$

(3.467)

In the optical regime we can avoid dealing with memory effects if we enlarge the system under consideration. We may approximate a peaked bath coupling as of (3.464) by a (2-level) system coupled via a Förster-type interaction in $\hat{H}(1,2)$ to a second 2-level system representing an *interlayer*

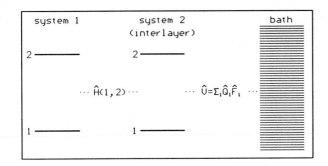

**Fig. 3.27.** System coupling via interlayer

(the buffer in the classical example), which is then coupled to a broad-band reservoir. In Fig. 3.27 this scenario is sketched.

In this way, system 1 is only indirectly damped and, depending on the pertinent parameters, the dynamics of system 2 is either frozen out (*overdamped case*) or creates a memory. Our network model thus allows us also to treat extended environments: whether we choose to associate a particular node with the system or with the environment becomes ambiguous. For low frequency transitions coupled to a reservoir confined to positive frequencies, this model extension does not work, though: the system cannot avoid feeling the edge of the bath spectrum.

**3.3.4.3 2-Level Model.** The coupling is specified by

$$\hat{V} = \sum_i \hat{Q}_i \hat{F}_i \ . \tag{3.468}$$

For non-adiabatic transitions the operators in system S, $\hat{Q}_i$, are projection operators of the corresponding transition $|n\rangle \rightarrow |m\rangle$. In a 2-level node we just have

$$\hat{V} = \hat{Q}_1 \hat{F}_1 + \hat{Q}_2 \hat{F}_2 \tag{3.469}$$

with

$$\hat{Q}_1 = \hat{P}_{12} \ , \quad \hat{Q}_2 = \hat{P}_{21} \ . \tag{3.470}$$

The corresponding operators in the bath (quasi-continuum) may then be written as

$$\hat{F}_1 = \sum_{m'} g^*_{m'} \hat{b}^+_{m'} \ , \tag{3.471}$$

$$\hat{F}_2 = \sum_m g_m \hat{b}_m \ , \tag{3.472}$$

where $\hat{b}^+_m$ is a creation operator of a photon mode $m$ with energy $\hbar\Omega_m$, say. $\hat{V}$ is Hermitian even though $\hat{F}_j$ is not.

We thus obtain

$$G_{21}(\tau) = \sum_{r,r'} \sum_{m,m'} g_m g_{m'}^* \left\langle r' \left| \hat{b}_m \right| r \right\rangle \left\langle r \left| \hat{b}_{m'}^+ \right| r' \right\rangle e^{iE_{r'r}\tau/\hbar}$$
$$\rho_B(0)_{r'r'} \tag{3.473}$$

with the Fourier transform

$$G_{21}(\omega) = \sum_{r,r'} \sum_{m,m'} g_m g_{m'}^* \left\langle r' \left| \hat{b}_m \right| r \right\rangle \left\langle r \left| \hat{b}_{m'}^+ \right| r' \right\rangle \rho_B(0)_{r'r'}$$
$$\delta(\omega + E_{r'r}/\hbar) . \tag{3.474}$$

For a thermal state at $T_B = 0$ (vacuum) with

$$\hat{\rho}_B(0) = |0\rangle \langle 0| \tag{3.475}$$

and

$$\hat{\rho}_B(0)_{r'r'} = \begin{cases} 1 \\ 0 \end{cases} \text{ for } \begin{matrix} r' = 0 \\ \text{otherwise} \end{matrix} , \tag{3.476}$$

one thus finds with $r = m'$, $m = m'$, $E_{0m} = -\hbar\Omega_m$:

$$G_{21}(\omega) = \sum_m |g_m|^2 \delta(\omega - \Omega_m) \tag{3.477}$$

while (because $\hat{b}_m |0\rangle = 0$)

$$G_{12}(\omega) = G_{11}(\omega) = G_{22}(\omega) = 0 . \tag{3.478}$$

This implies that there is only one non-zero damping parameter (cf. (3.459)):

$$\Gamma_{2112}^+ = \frac{1}{\hbar^2} \left\langle 2 \left| \hat{P}_{21} \right| 1 \right\rangle \left\langle 1 \left| \hat{P}_{12} \right| 2 \right\rangle \int_0^\infty e^{i\omega_{21}\tau} G_{21}(\tau) d\tau$$
$$= \frac{1}{\hbar^2} \int_0^\infty e^{i\omega_{21}\tau} G_{21}(\tau) d\tau . \tag{3.479}$$

In Fig. 3.28 we illustrate the case of a 2-level node coupled to its bath via a Lorentzian $G_{21}(\omega)$. Memory effects are then negligible only for $\Delta t \gg \tau_c = 1/K$. For a bounded bath spectrum, $\omega \geq 0$, we should have $K \lesssim \omega_{21}$ so that with $\tau_c \gtrsim 1/\omega_{21}$ the bath correlation time can conveniently be small for transitions in the optical region.

**3.3.4.4 3-Level Model: Cross Terms.** We restrict ourselves to two damping channels and write

$$\hat{V} = \hat{Q}_1 \hat{F}_1 + \hat{Q}_2 \hat{F}_2 + \hat{Q}_3 \hat{F}_3 + \hat{Q}_4 \hat{F}_4 \tag{3.480}$$

with

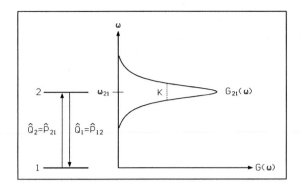

Fig. 3.28. A 2-level system coupled to its environment by a Lorentzian

$$\hat{Q}_1 = \hat{P}_{12} \; , \quad \hat{F}_1 = \sum_{m'} g_{m'}^{(1)*} \hat{b}_{m'}^+ \; ,$$

$$\hat{Q}_2 = \hat{P}_{21} \; , \quad \hat{F}_2 = \sum_{m} g_{m}^{(1)} \hat{b}_{m} \; ,$$

$$\hat{Q}_3 = \hat{P}_{13} \; , \quad \hat{F}_3 = \sum_{m'} g_{m'}^{(2)*} \hat{b}_{m'}^+ \; , \tag{3.481}$$

$$\hat{Q}_4 = \hat{P}_{31} \; , \quad \hat{F}_4 = \sum_{m} g_{m}^{(2)} \hat{b}_{m} \; .$$

In addition to the bath correlation functions $G_{21}(\tau)$ and $G_{43}(\tau)$ of the type discussed already in the preceding section, one encounters now also *cross terms*. For $T_{\mathrm{B}} = 0$ they can be written as

$$G_{41}(\omega) = \sum_{m} g_{m}^{(2)} g_{m}^{(1)*} \delta(\omega - \Omega_m) \; , \tag{3.482}$$

$$G_{23}(\omega) = \sum_{m} g_{m}^{(1)} g_{m}^{(2)*} \delta(\omega - \Omega_m) \; . \tag{3.483}$$

One notes that

$$|G_{41}(\omega)|^2 \, , |G_{23}(\omega)|^2 \le G_{43}, G_{21} \; . \tag{3.484}$$

If the two frequency channels do not overlap (see Fig. 3.29), these cross terms are zero. The resulting damping parameters are $\Gamma_{2112}^+$ and $\Gamma_{3113}^+$. If they do overlap, however, (which means that the environment cannot reliably distinguish between the two transitions) additional damping parameters have to be considered:

$$\Gamma_{3112}^+ = \frac{1}{\hbar^2} \left\langle 3 \left| \hat{P}_{31} \right| 1 \right\rangle \left\langle 1 \left| \hat{P}_{12} \right| 2 \right\rangle \int_0^\infty e^{i\omega_{21}\tau} G_{41}(\tau) d\tau \; , \tag{3.485}$$

$$\Gamma_{2113}^+ = \frac{1}{\hbar^2} \left\langle 2 \left| \hat{P}_{21} \right| 1 \right\rangle \left\langle 1 \left| \hat{P}_{13} \right| 3 \right\rangle \int_0^\infty e^{i\omega_{31}\tau} G_{23}(\tau) d\tau \; . \tag{3.486}$$

In Fig. 3.30 we sketch such a situation with $g_m^{(1)} = g_m^{(2)}$, i.e.

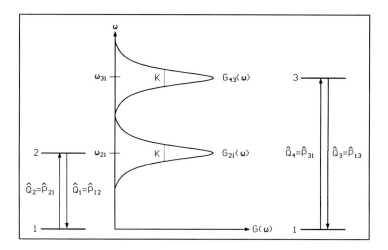

**Fig. 3.29.** A 3-level system with (in frequency space) local coupling to heat baths. The transitions are thus distinguishable

$$G_{43}(\omega) = G_{21}(\omega) = G_{41}(\omega) = G_{23}(\omega) \ . \tag{3.487}$$

The corresponding damping parameters are therefore all identical, the cross terms cannot be neglected. These additional parameters enter (as so-called *non-secular terms*) into the equations of motion giving rise to *beat phenomena* (non-exponential decay).

In both cases the frequency dependence of the bath coupling can be modelled by interactions with additional 2-level nodes coupled to a broad-band

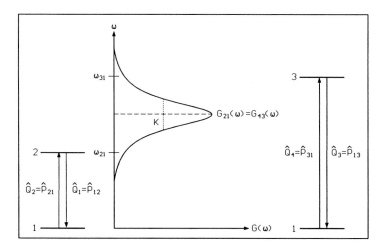

**Fig. 3.30.** A 3-level system with (in frequency space) non-local damping. The two transitions are indistinguishable

bath (cf. Sect. 3.3.7.5): in this enlarged system non-secular terms do not appear explicitly.

### 3.3.4.5 Glauber Model. The Boltzmann distribution implies that

$$\frac{W_{s,s'}}{W_{s',s}} = \exp\left(-\frac{E_{s',s}}{k_B T}\right) , \tag{3.488}$$

$$E_{s',s} = E_{s'} - E_s \tag{3.489}$$

(cf. 3.434). For a 2-level system with $s = \pm 1$, the *Glauber model* (cf. [58]), consistent with the above ratio, postulates the transition rates

$$\boxed{W_{s,-s} = \frac{1}{2\tau_0}\left[1 - \tanh\left(\frac{E_{1,-1}}{2k_B T}s\right)\right] ,} \tag{3.490}$$

wherein $\tau_0$ determines the time scale of the process. Replacing $s$ by $-s$ one obtains $W_{-s,s}$. (3.490) defines a non-selective thermal damping channel.

### 3.3.4.6 Incoherent Optical Driving. The coupling to a coherent optical driving field has already been studied. In the following we take the respective general master equation as a starting point and show under which conditions an effective rate equation results.

Consider the master equation for a $n$-level system coupled to an electro-magnetic field $\epsilon(t)$:

$$\frac{d}{dt}R_{ij}(t) = i\omega_{ij}R_{ij}(t) + \frac{i}{\hbar}\epsilon(t)\sum_{k=1}^{n}[d_{ik}R_{kj}(t) - d_{kj}R_{ik}(t)] -$$
$$(1 - \delta_{ij})\gamma_{ij}R_{ij}(t) \tag{3.491}$$

($d_{ij}$ = matrix elements of the dipole operator in the direction of the applied electrical field, $\gamma_{ij}$ = damping coefficients controlling the respective natural line-widths). The coherence of the pump light $\epsilon$ is specified by the correlation function

$$g(\tau) = \langle\epsilon(t)\epsilon(t+\tau)\rangle = \frac{1}{\Delta t}\int_{t-\Delta t/2}^{t+\Delta t/2}\epsilon(t')\epsilon(t'+\tau)dt' \tag{3.492}$$

($\Delta t \gg$ typical oscillation times of $\epsilon$), where – introducing the Fourier trans-form $U(\omega)$ with bandwidth $\Delta\omega \sim 1/\tau_c$ – this correlation function can be written as a Fourier integral:

$$g(\tau) = \frac{1}{\epsilon_0}\int_0^\infty U(\omega)\cos(\omega\tau)d\omega \tag{3.493}$$

($\epsilon_0$ = dielectric constant). Assuming

$$|R_{ij}(t)| \ll R_{ii}(t), R_{jj}(t) , \tag{3.494}$$

the evolution equation above reduces for $i \neq j$ to

$$\frac{d}{dt}R_{ij}(t) = (i\omega_{ij} - \gamma_{ij})\,R_{ij}(t) + \frac{i}{\hbar}\epsilon(t)\left[d_{ij}R_{jj}(t) - d_{ij}R_{ii}(t)\right] . \qquad (3.495)$$

Integrating with respect to time $t$, one obtains the solution

$$R_{ij}(t) = \frac{i}{\hbar}d_{ij}\int_{-\infty}^{t} e^{(i\omega_{ij} - \gamma_{ij})(t - t')}\epsilon(t')\left[R_{jj}(t') - R_{ii}(t')\right]dt' . \qquad (3.496)$$

Inserting this solution into

$$\frac{d}{dt}R_{ii}(t) = \frac{i}{\hbar}\epsilon(t)\sum_{j=1}^{n}\left[d_{ij}R_{ji}(t) - d_{ji}R_{ij}(t)\right] \qquad (3.497)$$

yields

$$\frac{d}{dt}R_{ii}(t) = \frac{2}{\hbar}\sum_{j=1}^{n}|d_{ij}|^2\int_{0}^{\infty}\epsilon(t)\epsilon(t - \tau)\cos(\omega_{ij}\tau)\,e^{-\gamma_{ij}\tau}$$

$$[R_{jj}(t - \tau) - R_{ii}(t - \tau)]\,d\tau \qquad (3.498)$$

with

$$\tau = t - t' . \qquad (3.499)$$

Replacing $\epsilon(t)\epsilon(t - \tau)$ by $\langle\epsilon(t)\epsilon(t - \tau)\rangle = g(\tau)$ and considering Markov processes so that $[R_{jj}(t - \tau) - R_{ii}(t - \tau)]$ can be replaced by $[R_{jj}(t) - R_{ii}(t)]$, one finds

$$\frac{d}{dt}R_{ii}(t) = \frac{2}{\hbar}\sum_{j=1}^{n}\left\{|d_{ij}|^2\left[R_{jj}(t) - R_{ii}(t)\right]\right.$$

$$\left.\int_{0}^{\infty}g(\tau)\cos(\omega_{ij}\tau)\,e^{-\gamma_{ij}\tau}d\tau\right\} . \qquad (3.500)$$

Replacing $g(\tau)$ by the Fourier integral (3.493), the $\tau$ integration can be carried out:

$$\frac{d}{dt}R_{ii}(t)$$

$$= \frac{1}{\epsilon_0\hbar^2}\sum_{j=1}^{n}\left\{|d_{ij}|^2\left[R_{jj}(t) - R_{ii}(t)\right]\right.$$

$$\left.\int_{0}^{\infty}U(\omega)\left[\frac{\gamma_{ij}}{(\omega_{ij} - \omega)^2 + \gamma_{ij}^2} + \frac{\gamma_{ij}}{(\omega_{ij} + \omega)^2 + \gamma_{ij}^2}\right]d\omega\right\} . \qquad (3.501)$$

In this integral, only the peak $\omega = |\omega_{ij}|$ will contribute significantly. If $\Delta\omega$ of $U(\omega)$ is large compared to $\gamma_{ij}$, we can take

$$U(\omega) \approx U\left(|\omega_{ij}|\right) \qquad (3.502)$$

in front of the integral:

$$\frac{d}{dt}R_{ii}(t) = \frac{\pi}{\epsilon_0 \hbar^2} \sum_{j=1}^{n} |d_{ij}|^2 U(|\omega_{ij}|) [R_{jj}(t) - R_{ii}(t)] \; . \tag{3.503}$$

Finally, introducing the abbreviation

$$\boxed{I_{ij} = B_{ij} U(|\omega_{ij}|) = \frac{\pi}{\epsilon_0 \hbar^2} |d_{ij}|^2 U(|\omega_{ij}|) = I_{ji} \; ,} \tag{3.504}$$

one obtains the master equation

$$\boxed{\frac{d}{dt}R_{ii}(t) = \sum_{j=1}^{n} [I_{ij} R_{jj}(t) - I_{ji} R_{ii}(t)]} \tag{3.505}$$

where the parameters $I_{ij}$ represent the transition rates in the optical model. $B_{ij}$ denotes the so-called *Einstein coefficients*. Due to the Markov condition used, this master equation is valid only on a time scale $\tau \gg \tau_c$. Under this condition the electromagnetic field acts like a heat bath (*incoherent optical driving*), which can be controlled by $U(|\omega_{ij}|)$ instead of by the temperature. Its coupling can therefore be made selective in frequency (and/or polarization) space. This scenario reminds us that only the coupling to a (partially) coherent light field will transfer coherence into the material system.

### 3.3.5 Damped Bloch Equations in $SU(2)$

In this section the Markovian master equation is transformed into the $SU(n)$ representation. For $n = 2$ this leads to the *damped Bloch equations*. Stationary and non-stationary solutions are studied. They apply to ensembles of non-interacting nodes.

**3.3.5.1 Generalized Bloch Equations.** The equations of motion for the coherence vector

$$\begin{aligned}
\boldsymbol{\lambda} &= \{\lambda_1, \lambda_2, \ldots, \lambda_k, \ldots \lambda_s\} \\
&= \{u_{12}, u_{13}, u_{23}, \ldots, v_{12}, v_{13}, v_{23}, \ldots, w_1, w_2, \ldots, w_{n-1}\}
\end{aligned} \tag{3.506}$$

$(k = 1, 2, \ldots, s = n^2 - 1)$ can be obtained from

$$\dot{\boldsymbol{\lambda}} = \mathrm{Tr}\left\{ \dot{\hat{R}}\hat{\lambda}_i \right\} = \sum_{n,m} \dot{R}_{nm} \left\langle m \left| \hat{\lambda}_i \right| n \right\rangle \; . \tag{3.507}$$

The matrix elements $\left\langle m \left| \hat{\lambda}_i \right| n \right\rangle$ are easily calculated based on (2.80). Then, using (3.407), and after expressing the density matrix elements on the r.h.s. by components of the coherence vector, the *generalized Bloch equations* result in the form

$$\boxed{\dot{\lambda}_i - \sum_{k=1}^{s} \Omega_{ik} \lambda_k = \sum_{k=1}^{s} \xi_{ik} \lambda_k + \eta_i \; .} \tag{3.508}$$

Alternatively one can start from the Lindblad operator. Here, $\xi_{ik}$ represents elements of the (Bloch) *damping matrix* $(\xi_{ik})$, and $\eta_i$ denotes elements of the (Bloch) *damping vector* $\boldsymbol{\eta}$. These equations describe the *dissipative dynamics* of the coherence vector. $\Omega_{ik}$ denotes the elements of the rotation matrix discussed previously. Examples for this equation will be discussed now. In particular, we derive $(\xi_{ik})$ and $\eta_i$ from the respective relaxation matrix.

**3.3.5.2 Damping Matrix.** We specialize (3.407) and (3.402) for $SU(2)$ to obtain

$$
\begin{aligned}
&\dot{R}_{12}(t) + D_{12}(t) \\
&\quad = -\frac{1}{2}\left(W_{12} + W_{21}\right) R_{12}(t) - \operatorname{Re}\left\{\tilde{\gamma}_{12}\right\} R_{12}(t) \\
&\qquad -i\Delta\omega_{12} R_{12}(t) \, , \\
&\dot{R}_{21}(t) + D_{21}(t) \\
&\quad = -\frac{1}{2}\left(W_{12} + W_{21}\right) R_{21}(t) - \operatorname{Re}\left\{\tilde{\gamma}_{21}\right\} R_{21}(t) \\
&\qquad -i\Delta\omega_{21} R_{21}(t) \, , \\
&\dot{R}_{11}(t) + D_{11}(t) \\
&\quad = W_{12} R_{22}(t) - W_{21} R_{11}(t) \, , \\
&\dot{R}_{22}(t) + D_{22}(t) \\
&\quad = W_{21} R_{11}(t) - W_{12} R_{22}(t)
\end{aligned}
\tag{3.509}
$$

with

$$
D_{s's}(t) = \frac{i}{\hbar}\left\langle s' \left| \left[\hat{H}_S, \hat{R}_S(t)\right]_- \right| s \right\rangle \, .
\tag{3.510}
$$

$D_{s's}$ can be inferred from (3.408)–(3.409). The damping model is indicated in Fig. 3.31. We are now going to derive the corresponding damping matrix $(\xi_{ij})$ and the damping vector $\boldsymbol{\eta}$ from (3.509) by setting

$$
D_{ij} = 0 \, .
\tag{3.511}
$$

Observing

$$
\operatorname{Re}\left\{\tilde{\gamma}_{12}\right\} = \operatorname{Re}\left\{\tilde{\gamma}_{21}\right\}
\tag{3.512}
$$

(cf. (3.390)–(3.399)), we can rewrite the first two equations of (3.509):

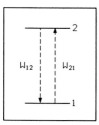

**Fig. 3.31.** An example of incoherent transitions in a 2-level system

$$\dot{R}_{12}(t) = -\frac{1}{2}\left(W_{12} + W_{21}\right)R_{12}(t) - \text{Re}\left\{\tilde{\gamma}_{12}\right\}R_{12}(t)$$
$$\qquad - i\Delta\omega_{12}R_{12}(t)\,,$$
$$\dot{R}_{21}(t) = -\frac{1}{2}\left(W_{12} + W_{21}\right)R_{21}(t) - \text{Re}\left\{\tilde{\gamma}_{12}\right\}R_{21}(t)$$
$$\qquad + i\Delta\omega_{12}R_{21}(t)\,. \tag{3.513}$$

In the present case we have

$$\dot{u}_{12} = \dot{R}_{12}(t) + \dot{R}_{21}(t)\,,$$
$$\dot{v}_{12} = i\left[\dot{R}_{12}(t) - \dot{R}_{21}(t)\right]\,, \tag{3.514}$$
$$\dot{w}_1 = -\dot{R}_{11}(t) + \dot{R}_{22}(t)$$

so that, based on the model (3.509) and

$$\mathsf{R} = \frac{1}{2}\begin{pmatrix} 1 - w_1 & u_{12} + iv_{12} \\ u_{12} - iv_{12} & 1 + w_1 \end{pmatrix}\,, \tag{3.515}$$

we find

$$\dot{u}_{12} = -\frac{1}{2}\left(W_{12} + W_{21}\right)\left[R_{12}(t) + R_{21}(t)\right]$$
$$\qquad -\text{Re}\left\{\tilde{\gamma}_{12}\right\}\left[R_{12}(t) + R_{21}(t)\right] - i\Delta\omega_{12}\left[R_{12}(t) - R_{21}(t)\right]$$
$$\qquad = -\frac{1}{2}\left(W_{12} + W_{21}\right)u_{12} - \text{Re}\left\{\tilde{\gamma}_{12}\right\}u_{12} - \Delta\omega_{12}v_{12}\,, \tag{3.516}$$

$$\dot{v}_{12} = -\frac{1}{2}\left(W_{12} + W_{21}\right)\left\{i\left[R_{12}(t) - R_{21}(t)\right]\right\}$$
$$\qquad -\text{Re}\left\{\tilde{\gamma}_{12}\right\}\left\{i\left[R_{12}(t) - R_{21}(t)\right]\right\} + \Delta\omega_{12}\left[R_{12}(t) + R_{21}(t)\right]$$
$$\qquad = -\frac{1}{2}\left(W_{12} + W_{21}\right)v_{12} - \text{Re}\left\{\tilde{\gamma}_{12}\right\}v_{12} + \Delta\omega_{12}u_{12}\,, \tag{3.517}$$

$$\dot{w}_1 = -\left(W_{12} + W_{21}\right)\left[-R_{11}(t) + R_{22}(t)\right] - W_{12} + W_{21}$$
$$\qquad = -\left(W_{12} + W_{21}\right)w_1 - W_{12} + W_{21}\,. \tag{3.518}$$

Inserting the abbreviations

$$\xi = W_{12} + W_{21}\,,$$
$$\gamma = \frac{1}{2}\left(W_{12} + W_{21} + 2\text{Re}\left\{\tilde{\gamma}_{12}\right\}\right)\,, \tag{3.519}$$
$$\eta_z = W_{21} - W_{12} = \eta_3\,,$$

this system of equations reduces to

$$\dot{u}_{12} + \Delta\omega_{12}v_{12} = -\gamma u_{12}\,,$$
$$\dot{v}_{12} - \Delta\omega_{12}u_{12} = -\gamma v_{12}\,, \tag{3.520}$$
$$\dot{w}_1 = -\xi w_1 + \eta_z$$

so that the damping matrix is given by

$$(\xi_{ik}) = \begin{pmatrix} -\gamma & 0 & 0 \\ 0 & -\gamma & 0 \\ 0 & 0 & -\xi \end{pmatrix} \tag{3.521}$$

(with $\gamma$, $\xi$ according to (3.519)), and the damping vector by

$$\boldsymbol{\eta} = (0, 0, \eta_z) = (0, 0, W_{21} - W_{12}) \ . \tag{3.522}$$

For an optically driven system the rotation matrix has been shown to be

$$(\Omega_{ij}) = \begin{pmatrix} 0 & -\delta & 0 \\ +\delta & 0 & -g \\ 0 & +g & 0 \end{pmatrix} \tag{3.523}$$

(cf. (3.97)) so that, defining the *shifted detuning* as

$$\tilde{\delta} = \omega_{21} + \Delta\omega_{12} - \omega \tag{3.524}$$

(compare (3.95)) the damped Bloch equations read

$$\begin{aligned} \dot{P}_x + \tilde{\delta}P_y &= -\gamma P_x \ , \\ \dot{P}_y - \tilde{\delta}P_x + gP_z &= -\gamma P_y \ , \\ \dot{P}_z - gP_y &= -\xi P_z + \eta_z \end{aligned} \tag{3.525}$$

(compare (3.98)). $\gamma$ is the reciprocal *transverse relaxation time* $T_2$,

$$\gamma = \frac{1}{T_2} \ , \tag{3.526}$$

$\xi$ the reciprocal *longitudinal relaxation time* $T_1$,

$$\xi = \frac{1}{T_1} \ . \tag{3.527}$$

For zero temperature, $W_{21} \ll W_{12}$, one can write

$$\frac{2}{T_2} = \frac{1}{T_1} + \frac{1}{\tau^*} \ , \tag{3.528}$$

$$\frac{1}{\tau^*} = \operatorname{Re}\{\tilde{\gamma}_{12}\} = W_{12}^{\mathrm{ad}} \ . \tag{3.529}$$

$1/\tau^*$ is the adiabatic contribution (cf. (3.396)), which can be dominating: in this case $1/T_2 \gg 1/T_1$.

### 3.3.5.3 Comparison: Density Matrix Versus Coherence Vector. A
simple time-dependent numerical solution of the master equations (3.509) and of the Bloch equations (3.525) is shown in Fig. 3.32. It has been assumed that at time $t = 0$ only the excited state (upper level, cf. Fig. 3.31)

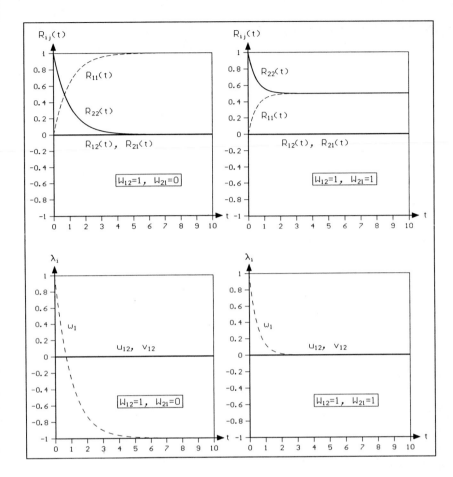

**Fig. 3.32.** 2-level system: time-evolution of density matrix and corresponding coherence vector. Initial State of the density matrix: $R_{11}(t) = 0$, $R_{22}(t) = 1$, $R_{12}(t) = R_{21}(t) = 0$. Corresponding initial state of the coherence vector: $u_{12} = v_{12} = 0$, $w_1 = 1$

of the members of the ensemble considered is populated. In the density matrix picture this is described by $R_{22}(t = 0) = 1$ and $R_{11}(t = 0) = 0$ ("inverse population"), and in the coherence vector picture by $w_1 = 1$ ("inverse macroscopic polarization"). The non-diagonal matrix elements $R_{12}(t)$, $R_{21}(t)$ and the coherence vector components $u_{12}$, $v_{12}$ are zero (and remain zero). Depending on the incoherent transition rates $W_{12}$, $W_{21}$, a decay of the initial population emerges: if only decay rates $W_{12}$ have to be taken into account, for $t \to \infty$ only the ground state (lower level, cf. Fig. 3.31) of the ensemble members is populated. In the density matrix picture this is described by $R_{22}(t \to \infty) = 0$ and $R_{11}(t \to \infty) = 1$ ("equilibrium population"), and in the coherence vector picture by $w_1 = -1$ ("macroscopic polarization"). If both transition rates are equal, for $t \to \infty$ the matrix elements reach the values $R_{11}(t \to \infty) = R_{22}(t \to \infty) = 0.5$ ("equilibrium population for high temperatures"), and $w_1$ reaches the value 0 ("no macroscopic polarization").

### 3.3.5.4 Density Matrix Form of the Damped Bloch Equations.
For later reference we transform the Bloch equations (3.525) for a zero-temperature bath ($\gamma = W_{12}/2$, $\xi = W_{12}$, $\eta_3 = W_{12}$) into density matrix form (cf. [2]):

$$
\begin{aligned}
\dot{R}_{11} &= -\frac{i}{2}g\left(R_{12} - R_{21}\right) + W_{12}R_{22}\;, \\
\dot{R}_{12} &= -\frac{1}{2}W_{12}R_{12} + i\frac{g}{2}\left(R_{22} - R_{11}\right) - i\tilde{\delta}R_{12}\;, \\
\dot{R}_{21} &= -\frac{1}{2}W_{12}R_{21} - i\frac{g}{2}\left(R_{22} - R_{11}\right) + i\tilde{\delta}R_{21}\;, \\
\dot{R}_{22} &= \frac{i}{2}g\left(R_{12} - R_{21}\right) - W_{12}R_{22}\;.
\end{aligned}
\tag{3.530}
$$

### 3.3.5.5 Stationary Solutions.
Stationary solutions of the damped Bloch equations (3.525) are defined by

$$
\dot{P}_x = \dot{P}_y = \dot{P}_z = 0\;.
\tag{3.531}
$$

Under this condition, the Bloch equations reduce to

$$
\begin{aligned}
\delta P_y^{(s)} &= -\frac{1}{T_2}P_x^{(s)}\;, \\
-\delta P_x^{(s)} + gP_z^{(s)} &= -\frac{1}{T_2}P_y^{(s)}\;, \\
-gP_y^{(s)} &= -\frac{1}{T_1}\left(P_z^{(s)} - P_0\right)
\end{aligned}
\tag{3.532}
$$

with

$$
\eta_z = \frac{P_0}{T_1}\;,
\tag{3.533}
$$

$$
P_0 = \frac{\eta_z}{\xi} = \frac{W_{21} - W_{12}}{W_{21} + W_{12}}\;.
\tag{3.534}
$$

This is a set of linear equations for the stationary solutions $P_x^{(s)}$, $P_y^{(s)}$, $P_z^{(s)}$. One readily gets

$$
P_x^{(s)} = \delta g L(\delta)\;, \quad P_y^{(s)} = -\frac{g}{T_2}L(\delta)\;, \quad P_z^{(s)} = \left(\delta^2 + \frac{1}{T_2^2}\right)L(\delta)\;,
\tag{3.535}
$$

$$
L(\delta) = \frac{P_0}{\delta^2 + g^2\dfrac{T_1}{T_2} + \dfrac{1}{T_2^2}}\;.
\tag{3.536}
$$

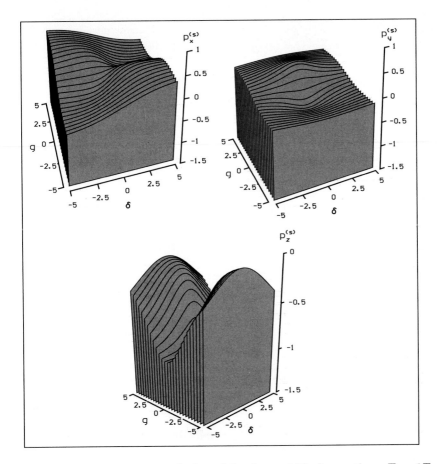

**Fig. 3.33.** The stationary solution of the damped Bloch equations. $T_1 = 2T_2 = 1$, $P_0 = -1$

Figure 3.33 illustrates the solution. In the overdamped case

$$\frac{1}{T_2} \gg \delta, g , \tag{3.537}$$

this scheme reduces to

$$P_x^{(s)} \approx 0 , \quad P_y^{(s)} \approx 0 , \quad P_z^{(s)} \approx \frac{1}{T_2^2} L(\delta) \approx P_0 . \tag{3.538}$$

For zero temperature the relation $P_0 \simeq -1$ holds, because in such a case $W_{21} \ll W_{12}$ (cf. (3.519)).

**3.3.5.6 Pulse Excitation in $SU(2)$.** Consider an ensemble of non-interacting $SU(2)$ nodes at time $t = 0$ in *thermal equilibrium* characterized by the polarization vector

$$P_x(0) = P_y(0) = 0 , \quad P_z(0) = P_0 < 0 . \tag{3.539}$$

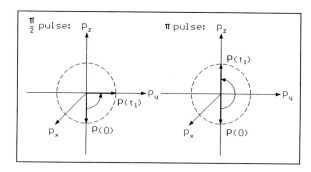

**Fig. 3.34.** $\pi/2$ pulse and $\pi$ pulse excitations in the case of thermal excitation states $P_x(0) = P_y(0) = 0$, $P_z(0) = P_0$

*Optical excitation pulses* are then used to change this initial state. On time scales with $\Delta t$ small compared to any relaxation times, such an excitation process is coherent. It plays the same role as the static devices for implementing unitary transformations as discussed in Sect. 2.3.6.3. Two types of excitation pulses are commonly used: the so-called $\pi/2$ *pulse* is defined by

$$g\Delta t = \pi/2 \ , \tag{3.540}$$

where $\Delta t$ is the pulse duration, and $\delta = 0$ (zero detuning). Its effect is, from the undamped Bloch equations:

$$\boxed{\begin{aligned} P_x(t_1) &= P_x(0) = 0 \ , \\ P_y(t_1) &= -P_z(0) \ , \\ P_z(t_1) &= P_y(0) = 0 \ . \end{aligned}} \tag{3.541}$$

Here, $P_i(t_1)$ $(i = x, y, z)$ represent the components of the polarization vector after the pulse, $t_1 = \Delta t$. The so-called $\pi$ *pulse* is defined by

$$g\Delta t = \pi \ . \tag{3.542}$$

It mirrors the initial vector at the $P_x$ axis:

$$\boxed{\begin{aligned} P_x(t_1) &= P_x(0) = 0 \ , \\ P_y(t_1) &= -P_y(0) = 0 \ , \\ P_z(t_1) &= -P_z(0) \ . \end{aligned}} \tag{3.543}$$

Using the thermal initial state determined by (3.539), one obtains the examples sketched in Fig. 3.34.

**3.3.5.7 Free Induction Decay.** After $\pi/2$ pulse excitation, the prepared state decays for $t > t_1$ (*free induction decay*). This decay process is described by the damped Bloch equations (cf. (3.525)) for $g = 0$: they read (we are still working in a rotating reference frame, $\delta = (E_2 - E_1)/\hbar - \omega)$

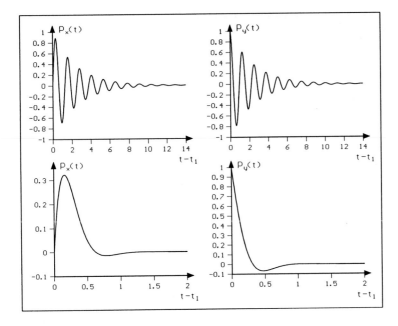

**Fig. 3.35.** Damped oscillations of the polarization vector in the $x, y$ plane after $\pi/2$ pulse. Above: oscillations $P_x(t)$, $P_y(t)$ for $\delta = -5$, $1/T_2 = 0.4$. Below: strongly damped motion for $\delta = -5$, $1/T_2 = 5$. In this figure, $P_x(t_1) = 0$, $P_y(t_1) = 1$ has been used

$$
\begin{aligned}
\dot{P}_x + \delta P_y &= -\frac{1}{T_2} P_x \,, \\
\dot{P}_y - \delta P_x &= -\frac{1}{T_2} P_y \,, \\
\dot{P}_z &= -\frac{1}{T_1} (P_z - P_0) \,.
\end{aligned}
$$

(3.544)

The last equation is readily solved:

$$
P_z(t) = P_z(0) + [P_z(t_1) - P_z(0)] \, \mathrm{e}^{-\frac{t-t_1}{T_1}} \,.
$$

(3.545)

This is the decay of the population inversion.

Differentiating the first equation of (3.544) with respect to time, inserting for $\dot{P}_y$ the second equation, differentiating the second equation, and inserting for $\dot{P}_x$ the first, the system of equations of first order, (3.544), is replaced by a system of second order,

$$
\begin{aligned}
\ddot{P}_x + \frac{1}{T_2} \dot{P}_x + \delta^2 P_x &= \frac{\delta}{T_2} P_y \,, \\
\ddot{P}_y + \frac{1}{T_2} \dot{P}_y + \delta^2 P_y &= -\frac{\delta}{T_2} P_x \,,
\end{aligned}
$$

(3.546)

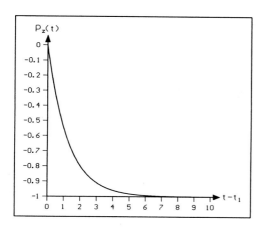

**Fig. 3.36.** The decay of the population inversion, $1/T_1 = 0.8$

which describes two mutually coupled *harmonic oscillators*. It is easily solved observing the initial state for $t = t_1$:

$$
\boxed{
\begin{aligned}
P_x(t) &= \{P_x(t_1) \cos\left[\delta(t - t_1)\right] - P_y(t_1) \sin\left[\delta(t - t_1)\right]\} \, e^{-\frac{t-t_1}{T_2}}, \\
P_y(t) &= \{P_x(t_1) \sin\left[\delta(t - t_1)\right] + P_y(t_1) \cos\left[\delta(t - t_1)\right]\} \, e^{-\frac{t-t_1}{T_2}}
\end{aligned}
}
\tag{3.547}
$$

$(t \geq t_1)$. Equation (3.547) describes a precession of the polarization vector in the $x, y$ plane. The exponential function implies a decay of the amplitudes of this oscillation.

In Fig. 3.35 the polarization components $P_x(t)$, $P_y(t)$ for $1/T_2 = 0.4, 5$ and $\delta = -5$ are illustrated. In Fig. 3.36 the decay of the population inversion is shown for the initial condition $P_z(t_1) = 0$ $(P_z(0) = -1)$.

### 3.3.5.8 Inhomogeneous Ensemble: Spin Echo.

Any Hamiltonian is defined by a specific set of model parameters. Inhomogeneous ensembles (cf. [121]) are characterized by a distribution of such parameters rather than unique values. For example, due to spatially varying magnetic fields, in spin systems the single spins do not have the same resonance frequency. 2-level defects in a solid state matrix may have different resonance frequencies resulting from different local environments. A typical distribution function is

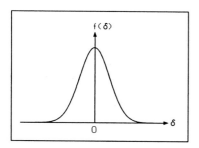

**Fig. 3.37.** Inhomogeneous ensemble. Distribution $f$ of resonance frequencies $\omega_{21}$ and thus of the detuning $\delta = \omega_{21} - \omega$. $f$ is measured as the so-called inhomogeneous line width. The centre frequency is $\omega_0 = \omega$

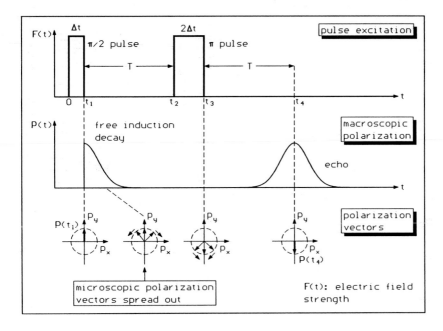

**Fig. 3.38.** Spin or photon echo in the rotating frame

sketched in Fig. 3.37. The rotating frame will be chosen with $\omega = \omega_0$, the *centre frequency*. *Photon* or *spin echos* demonstrate that the apparent decay of coherence due to an inhomogeneously distributed detuning parameter $\delta$ can be reversed ("undone"). The experimental scenario is sketched in Fig. 3.38. In the upper part of this figure, the sequence of excitation pulses (a $\pi/2$ pulse, and after time $T$, an additional $\pi$ pulse) is shown. Below, the behaviour of the polarization vector is illustrated. The first excitation pulse flips the thermal state into the state $P_x(t_1) = P_z(t_1) = 0$, $P_y(t_1) = P_0$. After this flip-over a free induction decay takes place. (Strictly speaking, this decay process starts during the pulse, however, this shall be neglected.) This decay process represents the decay of the polarization vector $\boldsymbol{P}$ (which is a macroscopic quantity describing the behaviour of a whole ensemble), due to different precession velocities of the microscopic polarization vectors of the individual ensemble members. In the lower part of Fig. 3.38 the spreading of the polarization vectors observed in the rotating reference frame is sketched. The $\pi$ pulse at time $T$ now mirrors the various polarization vectors at the $x$ axis. The individual polarization vectors then continue to rotate and thus converge again after time $T$, when the ensemble members are in phase to produce a maximal radiation signal. This "answer" of the photon or spin system is called *photon* or *spin echo* (cf. [121]).

The experiment can be described by means of the Bloch equations. At the beginning of the experiment the ensemble is in the thermal state defined by (3.539). After the $\pi/2$ pulse, the state is

$$P_x(t_1) = P_x(\Delta t) = P_x(0) \quad = 0 \;,$$
$$P_y(t_1) = P_y(\Delta t) = -P_z(0) = -P_0 > 0 \;, \tag{3.548}$$
$$P_z(t_1) = P_z(\Delta t) = P_y(0) \quad = 0 \;.$$

Then free induction decay begins and can be described by solutions of the damped Bloch equations (3.547)–(3.545). Inserting the initial conditions (3.548), this decay process is

$$P_x(t) = P_0 \sin[\delta(t - \Delta t)] e^{-\frac{t-\Delta t}{T_2}} \;,$$
$$P_y(t) = -P_0 \cos[\delta(t - \Delta t)] e^{-\frac{t-\Delta t}{T_2}} \;, \tag{3.549}$$
$$P_z(t) = P_0 \left(1 - e^{-\frac{t-\Delta t}{T_1}}\right)$$

($\Delta t \le t < \Delta t + T$). After a $\pi$ pulse at time $t = t_2 = t_1 + T = \Delta t + T$, the polarization vector turns into

$$P_x(3\Delta t + T) = P_0 \sin[\delta(3\Delta t + T)] e^{-\frac{3\Delta t+T}{T_2}} \;,$$
$$P_y(3\Delta t + T) = P_0 \cos[\delta(3\Delta t + T)] e^{-\frac{3\Delta t+T}{T_2}} \;, \tag{3.550}$$
$$P_z(3\Delta t + T) = -P_0 \left(1 - e^{-\frac{3\Delta t+T}{T_1}}\right) \;.$$

This system of equations defines the initial conditions of the now starting second induction decay, again described by the system of solutions (3.545)–(3.547):

$$P_x(t) = \big\{ P_x(3\Delta t + T) \cos[\delta(t - 3\Delta t - T)] +$$
$$P_y(3\Delta t + T) \sin[\delta(t - 3\Delta t - T)] \big\} e^{-\frac{t-3\Delta t-T}{T_2}} \;,$$
$$P_y(t) = \big\{ P_x(3\Delta t + T) \sin[\delta(t - 3\Delta t - T)] - \tag{3.551}$$
$$P_y(3\Delta t + T) \cos[\delta(t - 3\Delta t - T)] \big\} e^{-\frac{t-3\Delta t-T}{T_2}} \;,$$
$$P_z(t) = P_0 - [P_0 + P_z(3\Delta t + T)] e^{-\frac{t-3\Delta t-T}{T_1}}$$

($3\Delta t + T \le t \le 3\Delta t + 2T$). Inserting the initial conditions (3.550) into those solutions, and using the trigonometric relations

$$\sin \alpha \cos \beta + \cos \alpha \sin \beta = \sin(\alpha + \beta) \;,$$
$$\sin \alpha \sin \beta + \cos \alpha \cos \beta = \cos(\alpha - \beta) \;, \tag{3.552}$$

one obtains

$$P_x(t) = P_0 \sin(\delta t) e^{-\frac{t}{T_2}} \;,$$
$$P_y(t) = P_0 \cos[\delta(t - 6\Delta t - 2T)] e^{-\frac{t}{T_2}} \;, \tag{3.553}$$
$$P_z(t) = P_0 - 2P_0 e^{-\frac{t-3\Delta t-T}{T_1}} + P_0 e^{-\frac{t}{T_1}}$$

($3\Delta t + T \le t \le 3\Delta t + 2T$). Assuming small pulse times,

$$\Delta t \ll T \;, \tag{3.554}$$

$\Delta t$ can be neglected. In this approximation the time $t = 2T$ marks the maximal echo amplitude, for which (3.553) reduces to

$$P_x(2T) = P_0 \sin(\delta 2T) e^{-\frac{2T}{T_2}} ,$$
$$P_y(2T) = P_0 e^{-\frac{2T}{T_2}} , \qquad (3.555)$$
$$P_z(2T) = P_0 - 2P_0 e^{-\frac{T}{T_1}} + P_0 e^{-\frac{2T}{T_1}} .$$

With the distribution of Fig. 3.37 the ensemble average of $P_x(2T)$ is zero. Considering delay times $T$ very much smaller than the relaxation times $T_1$, $T_2$, one finally obtains

$$\boxed{\begin{aligned} P_x(2T) &= P_x(t_4) = P_x(\Delta t) &= 0 , \\ P_y(2T) &= P_y(t_4) = -P_y(\Delta t) = P_0 , \\ P_z(2T) &= P_z(t_4) = P_z(\Delta t) &= 0 . \end{aligned}} \qquad (3.556)$$

The photon or spin echo is reduced by damping.

### 3.3.5.9 Oscillation of Dipole Moment.

The *dipole operator* can be represented by

$$\widehat{\boldsymbol{d}} = \hat{P}_{11}\boldsymbol{d}_{11} + \hat{P}_{22}\boldsymbol{d}_{22} + \hat{P}_{12}\boldsymbol{d}_{12} + \hat{P}_{21}\boldsymbol{d}_{21} \qquad (3.557)$$

or, in terms of the generating operators of $SU(2)$, by

$$\boxed{\begin{aligned} \widehat{\boldsymbol{d}} = \tfrac{1}{2}\big[ (\boldsymbol{d}_{11} + \boldsymbol{d}_{22})\,\hat{1} + (\boldsymbol{d}_{22} - \boldsymbol{d}_{11})\,\hat{w}_1 \big] + \\ \operatorname{Re}\{\boldsymbol{d}_{12}\}\,\hat{u}_{12} + \operatorname{Im}\{\boldsymbol{d}_{12}\}\,\hat{v}_{12} . \end{aligned}} \qquad (3.558)$$

For $\boldsymbol{d}_{11} = \boldsymbol{d}_{22} = 0$ this representation reduces to

$$\widehat{\boldsymbol{d}} = \operatorname{Re}\{\boldsymbol{d}_{12}\}\,\hat{u}_{12} + \operatorname{Im}\{\boldsymbol{d}_{12}\}\,\hat{v}_{12} \qquad (3.559)$$

so that the expectation value is given by

$$\boxed{\left\langle \widehat{\boldsymbol{d}} \right\rangle = \operatorname{Re}\{\boldsymbol{d}_{12}\}\,u_{12} + \operatorname{Im}\{\boldsymbol{d}_{12}\}\,v_{12} .} \qquad (3.560)$$

This representation is valid in the laboratory frame. Transforming those expectation values into a reference frame which rotates with the frequency $\omega = \omega_0$ by using the transformation rules already introduced (cf. (3.123)), we get

$$\left\langle \widehat{\boldsymbol{d}} \right\rangle = \operatorname{Re}\{\boldsymbol{d}_{12}\}\,[P_x(t)\cos(\omega_0 t) + P_y(t)\sin(\omega_0 t)] \qquad (3.561)$$

if the imaginary part is identically zero: $\operatorname{Im}\{\boldsymbol{d}_{12}\} = 0$.

In the expression (3.561), the components $P_x(t)$, $P_y(t)$ are components of the polarization vector in the rotating reference frame, which we have just calculated (cf. (3.549), (3.553)). As the $P_x$ polarization vanishes in the ensemble average, we find, neglecting $\Delta t$:

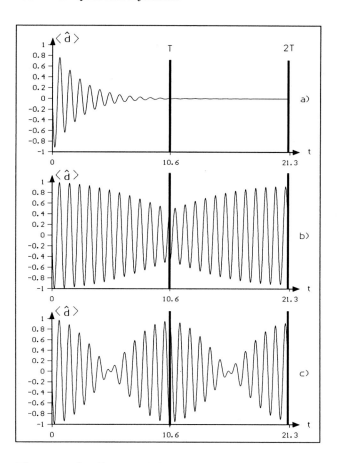

**Fig. 3.39.** Oscillations of the dipole expectation value fot homogeneous subensembles ($T = 10.6$, $\omega_0 = 7$, $P_0 = 1$). a) $1/T_2 = 0.4$, $\delta = 0.1$. b): $1/T_2 = 0.004$, $\delta = 0.1$. c) $1/T_2 = 0.004$, $\delta = 0.3$

$$\langle \hat{d} \rangle = \begin{cases} -\mathrm{Re}\,\{d_{12}\}\, P_0 e^{-\frac{t}{T_2}} \sin(\omega_0 t)\cos(\delta t) & : \ 0 \le t < T \\ +\mathrm{Re}\,\{d_{12}\}\, P_0 e^{-\frac{t}{T_2}} \sin(\omega_0 t)\cos[\delta(t - 2T)] & : \ T \le t \le 2T \end{cases} \tag{3.562}$$

Examples for $\mathrm{Re}\,\{d_{12}\} = 1$ are sketched in Fig. 3.39. The upper part shows an *irreversible evolution*, i.e. the dipole moment vanishes before the $\pi$ pulse at time $T$. Below, almost *reversible motions* are illustrated, i.e. after time $2T$ *constructive interference* occurs between the subensembles with different $\delta$.

### 3.3.6 Damped Bloch Equations in $SU(3)$

In the last section, the Markovian master equation has been specialized in 2-level nodes. In this section, damped Bloch equations of 3-level nodes will be studied; various driving–damping scenarios will be taken into account.

**3.3.6.1 Damping Matrix.** We consider 3 separate (non-overlapping) damping channels as sketched in Fig. 3.40:

channel 1 :  parameters $W_{12}, W_{21}, \mathrm{Re}\{\tilde{\gamma}_{12}\}, \Delta\omega_{12}$ ,

channel 2 :  parameters $W_{13}, W_{31}, \mathrm{Re}\{\tilde{\gamma}_{13}\}, \Delta\omega_{13}$ ,          (3.563)

channel 3 :  parameters $W_{23}, W_{32}, \mathrm{Re}\{\tilde{\gamma}_{23}\}, \Delta\omega_{23}$ .

We now derive the damping matrix as defined in (3.508). Observing

$$\mathrm{Re}\{\tilde{\gamma}_{op}\} = \mathrm{Re}\{\tilde{\gamma}_{po}\} \quad (o, p = 1, 2, 3) , \qquad (3.564)$$

one obtains the systems of equations (cf. (3.407), (3.402))

$$\boxed{\begin{aligned}
\dot{R}_{12}(t) + D_{12}(t) &= -\gamma^{(12)} R_{12}(t) - i\Delta\omega_{12} R_{12}(t) , \\
\dot{R}_{21}(t) + D_{21}(t) &= -\gamma^{(12)} R_{21}(t) - i\Delta\omega_{21} R_{21}(t) , \\
\dot{R}_{13}(t) + D_{13}(t) &= -\gamma^{(13)} R_{13}(t) - i\Delta\omega_{13} R_{13}(t) , \\
\dot{R}_{31}(t) + D_{31}(t) &= -\gamma^{(13)} R_{31}(t) - i\Delta\omega_{31} R_{31}(t) , \\
\dot{R}_{23}(t) + D_{23}(t) &= -\gamma^{(23)} R_{23}(t) - i\Delta\omega_{23} R_{23}(t) , \\
\dot{R}_{32}(t) + D_{32}(t) &= -\gamma^{(23)} R_{32}(t) - i\Delta\omega_{32} R_{32}(t) ,
\end{aligned}}$$
(3.565)

$$\boxed{\begin{aligned}
\dot{R}_{11}(t) + D_{11}(t) &= - (W_{21} + W_{31}) R_{11}(t) + \\
&\quad W_{12} R_{22}(t) + W_{13} R_{33}(t) , \\
\dot{R}_{22}(t) + D_{22}(t) &= W_{21} R_{11}(t) - (W_{12} + W_{31}) R_{22}(t) + \\
&\quad W_{23} R_{33}(t) , \\
\dot{R}_{33}(t) + D_{33}(t) &= W_{31} R_{11}(t) + W_{32} R_{22}(t) - \\
&\quad (W_{13} + W_{23}) R_{33}(t) ,
\end{aligned}}$$
(3.566)

where, due to (3.400), the parameters $\gamma^{(s's)}$ are given by

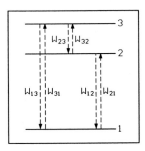

Fig. 3.40. Incoherent transition processes in a 3-level system

$$\gamma^{(12)} = \frac{1}{2}\left(W_{12} + W_{21} + W_{31} + W_{32}\right) + \text{Re}\left\{\tilde{\gamma}_{12}\right\},$$

$$\gamma^{(13)} = \frac{1}{2}\left(W_{13} + W_{21} + W_{23} + W_{31}\right) + \text{Re}\left\{\tilde{\gamma}_{13}\right\}, \tag{3.567}$$

$$\gamma^{(23)} = \frac{1}{2}\left(W_{12} + W_{13} + W_{23} + W_{32}\right) + \text{Re}\left\{\tilde{\gamma}_{23}\right\}.$$

The coherent terms $D_{s's}(t)$, again, are defined by (3.510). For the following considerations they will be put to zero.

In $SU(3)$, the connection between the reduced density matrix R and the expectation values $u_{12}$, $u_{13}$, $u_{23}$, $v_{12}$, $v_{13}$, $v_{23}$, $w_1$, $w_2$ of the generating operators $\hat{\lambda}_1 = \hat{u}_{12}$, $\hat{\lambda}_2 = \hat{u}_{13}$, $\hat{\lambda}_3 = \hat{u}_{23}$, $\hat{\lambda}_4 = \hat{v}_{12}$, $\hat{\lambda}_5 = \hat{v}_{13}$, $\hat{\lambda}_6 = \hat{v}_{23}$, $\hat{\lambda}_7 = \hat{w}_1$, $\hat{\lambda}_8 = \hat{w}_2$ is given by

$$\text{R} = \frac{1}{2}\begin{pmatrix} \frac{2}{3} - w_1 - \frac{1}{\sqrt{3}}w_2 & u_{12} + iv_{12} & u_{13} + iv_{13} \\ u_{12} - iv_{12} & \frac{2}{3} + w_1 - \frac{1}{\sqrt{3}}w_2 & u_{23} + iv_{23} \\ u_{13} - iv_{13} & u_{23} - iv_{23} & \frac{2}{3} + \frac{2}{\sqrt{3}}w_2 \end{pmatrix} \tag{3.568}$$

(cf. (2.284). Inserting (3.565)-(3.566) into

$$\dot{u}_{12} = \dot{R}_{12}(t) + \dot{R}_{21}(t),$$

$$\dot{u}_{13} = \dot{R}_{13}(t) + \dot{R}_{31}(t),$$

$$\dot{u}_{23} = \dot{R}_{23}(t) + \dot{R}_{32}(t),$$

$$\dot{v}_{12} = i\left[\dot{R}_{12}(t) - \dot{R}_{21}(t)\right],$$

$$\dot{v}_{13} = i\left[\dot{R}_{13}(t) - \dot{R}_{31}(t)\right], \tag{3.569}$$

$$\dot{v}_{23} = i\left[\dot{R}_{23}(t) - \dot{R}_{32}(t)\right],$$

$$\dot{w}_1 = -\dot{R}_{11}(t) + \dot{R}_{22}(t),$$

$$\dot{w}_2 = -\frac{1}{\sqrt{3}}\left[\dot{R}_{11}(t) + \dot{R}_{22}(t) - 2\dot{R}_{33}(t)\right],$$

one obtains a system of equations in $u, v$ subspace,

$$\boxed{\begin{aligned} \dot{u}_{jk} &= -\gamma^{(jk)}u_{jk} - \Delta\omega_{jk}v_{jk} \quad (jk = 12, 13, 23), \\ \dot{v}_{jk} &= -\gamma^{(jk)}v_{jk} + \Delta\omega_{jk}u_{jk} \quad (jk = 12, 13, 23) \end{aligned}} \tag{3.570}$$

implying

$$\begin{aligned} \xi_{11} &= \xi_{44} = -\gamma^{(12)}, \\ \xi_{22} &= \xi_{55} = -\gamma^{(13)}, \\ \xi_{33} &= \xi_{66} = -\gamma^{(23)} \end{aligned} \tag{3.571}$$

($\Delta\omega_{jk}$ is incorporated into a shifted transition frequency), and a system of equations in $w$ subspace,

$$\boxed{\begin{aligned} \dot{w}_1 &= \xi_{77}w_1 + \xi_{78}w_2 + \eta_7, \\ \dot{w}_2 &= \xi_{87}w_1 + \xi_{88}w_2 + \eta_8 \end{aligned}} \tag{3.572}$$

with

$$\xi_{77} = -\left(W_{12} + \frac{1}{2}W_{31} + \frac{1}{2}W_{32} + W_{21}\right) ,$$

$$\xi_{78} = \frac{1}{\sqrt{3}}\left(W_{12} - W_{13} + W_{23} + \frac{1}{2}W_{32} - \frac{1}{2}W_{31} - W_{21}\right) ,$$

$$\xi_{87} = \frac{\sqrt{3}}{2}\left(-W_{31} + W_{32}\right) ,$$

$$\xi_{88} = -W_{13} - W_{23} - \frac{1}{2}\left(W_{31} + W_{32}\right) ,$$

$$\eta_7 = -\frac{1}{3}\left(2W_{12} + W_{13} - W_{23} + W_{32} - W_{31} - 2W_{21}\right) ,$$

$$\eta_8 = -\frac{1}{\sqrt{3}}\left(W_{13} + W_{23} - W_{32} - W_{31}\right) .$$

(3.573)

Therefore, the damping matrix $(\xi_{ij})$ and the damping vector $\boldsymbol{\eta}$ read in $SU(3)$ (cf. [115]):

$$(\xi_{ij}) = \begin{pmatrix} -\gamma^{(12)} & 0 & 0 & 0 & 0 & 0 & | & 0 & 0 \\ 0 & -\gamma^{(13)} & 0 & 0 & 0 & 0 & | & 0 & 0 \\ 0 & 0 & -\gamma^{(23)} & 0 & 0 & 0 & | & 0 & 0 \\ 0 & 0 & 0 & -\gamma^{(12)} & \cdot\,0 & 0 & | & 0 & 0 \\ 0 & 0 & 0 & 0 & -\gamma^{(13)} & 0 & | & 0 & 0 \\ 0 & 0 & 0 & 0 & 0 & -\gamma^{(23)} & | & 0 & 0 \\ - & - & - & - & - & - & & - & - \\ 0 & 0 & 0 & 0 & 0 & 0 & | & \xi_{77} & \xi_{78} \\ 0 & 0 & 0 & 0 & 0 & 0 & | & \xi_{87} & \xi_{88} \end{pmatrix}$$

(3.574)

and

$$\boldsymbol{\eta} = (0,0,0,0,0,0,\eta_7,\eta_8) .$$

(3.575)

Without non-secular terms, the damping matrix is block diagonal with respect to the $u, v$ and the $w$ subspace.

For zero temperature we have with $W_{21} \ll W_{12}$, $W_{31} \ll W_{13}$, $W_{32} \ll W_{23}$:

$$\eta_7 = -\frac{1}{3}\left(2W_{12} + W_{13} - W_{23}\right) ,$$

$$\eta_8 = -\frac{1}{\sqrt{3}}\left(W_{13} + W_{23}\right)$$

(3.576)

implying as the *attractor state* $(\dot{w}_1 = \dot{w}_2 = 0)$ the ground state

$$w_1 = -1 , \quad w_2 = -\frac{1}{\sqrt{3}} .$$

(3.577)

**3.3.6.2 Influence of Non-secular Terms.** Let us consider for a 3-level system *non-secular terms* with respect to the states $|2\rangle$ and $|3\rangle$, i.e. we include time-dependent relaxation matrix elements (see (3.377)) proportional to $\exp(\pm i\omega_{32}t)$. These additional terms supplement the equations of motion (in the interaction picture, cf. (3.376)):

$$\dot{R}_{23}^{(i)}(t) = R_{22}^{(i)}(t)e^{i\omega_{23}t}\left(-\sum_{k=1}^{3}\Gamma_{2kk3}^- + \Gamma_{2322}^+ + \Gamma_{2322}^-\right) +$$
$$R_{33}^{(i)}(t)e^{i\omega_{23}t}\left(-\sum_{k=1}^{3}\Gamma_{2kk3}^- + \Gamma_{3323}^+ + \Gamma_{3323}^-\right) . \tag{3.578}$$

In the following we restrict ourselves to

$$\dot{R}_{23}^{(i)}(t) = -e^{i\omega_{23}t}\left[\Gamma_{2113}^- R_{22}^{(i)}(t) + \Gamma_{2113}^- R_{33}^{(i)}(t)\right] . \tag{3.579}$$

Correspondingly we obtain

$$\dot{R}_{32}^{(i)}(t) = -e^{i\omega_{32}t}\left[\Gamma_{3112}^- R_{22}^{(i)}(t) + \Gamma_{3112}^- R_{33}^{(i)}(t)\right] . \tag{3.580}$$

The parameters $\Gamma_{3112}^\pm$, $\Gamma_{2113}^\pm$ have been defined in (3.485)–(3.486).

Transforming back to the Schrödinger picture,

$$R_{23} = R_{23}^{(i)}e^{-i\omega_{23}t} , \quad R_{32} = R_{32}^{(i)}e^{-i\omega_{32}t} ,$$
$$R_{22} = R_{22}^{(i)} , \quad R_{33} = R_{33}^{(i)} , \tag{3.581}$$

we thus find

$$\boxed{\begin{aligned}\dot{u}_{23} &= \dot{R}_{23} + \dot{R}_{32} \\ &\quad -2\text{Re}\left\{\Gamma_{2113}^-\right\}R_{22} - 2\text{Re}\left\{\Gamma_{3112}^-\right\}R_{33} + \omega_{32}v_{23} , \\ \dot{v}_{23} &= -\omega_{32}u_{23}\end{aligned}} \tag{3.582}$$

and

$$\boxed{\begin{aligned}\dot{R}_{11} &= R_{23}\left(\Gamma_{3112}^+ + \Gamma_{3112}^-\right) + R_{32}\left(\Gamma_{2113}^+ + \Gamma_{2113}^-\right) , \\ \dot{R}_{22} &= -R_{23}\Gamma_{3112}^- - R_{32}\Gamma_{2113}^+ , \\ \dot{R}_{33} &= -R_{23}\Gamma_{3112}^+ - R_{32}\Gamma_{2113}^- .\end{aligned}} \tag{3.583}$$

Assuming, for simplicity,

$$\kappa = 2\text{Re}\left\{\Gamma_{2113}^\pm\right\} = 2\text{Re}\left\{\Gamma_{3112}^\pm\right\} , \tag{3.584}$$

$$\text{Im}\left\{\Gamma_{2113}^\pm\right\} = \text{Im}\left\{\Gamma_{3112}^\pm\right\} = 0 , \tag{3.585}$$

these equations of motion reduce to

$$\dot{u}_{23} = -2\kappa\left(R_{22} + R_{33}\right) + \omega_{32}v_{23} ,$$
$$\dot{v}_{23} = -\omega_{32}u_{23} \tag{3.586}$$

and

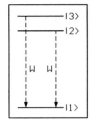

**Fig. 3.41.** 3-level node with two damping channels

$$\dot{R}_{11} = 2\kappa\,(R_{23} + R_{32})\ ,$$

$$\dot{R}_{22} = -\kappa\,(R_{23} + R_{32})\ ,$$

$$\dot{R}_{33} = -\kappa\,(R_{23} + R_{32})\ . \tag{3.587}$$

*Example 3.3.2 (Quantum Beats).* Let us consider the effect of these non-secular terms on the relaxation of the excited doublet state sketched in Fig. 3.41. We assume zero temperature, $W_{31} = W_{21} = W_{32} = 0$, and $W_{13} = W_{12} = W$, $W_{23} = 0$.

Transforming to the coherence vector in $SU(3)$, it is convenient to reverse the order of states, $|3\rangle$, $|2\rangle$, $|1\rangle$, and thus to define

$$w_1 = R_{22} - R_{33}\ ,$$

$$w_2 = \frac{1}{\sqrt{3}}\,(2R_{11} - R_{22} - R_{33})\ , \tag{3.588}$$

$$u_{ji} = u_{ji}\ ,\quad v_{ji} = -v_{ij}\ \ (i < j)\ .$$

Then $w_1$ describes the internal inversion of the doublet, which is a constant of motion for $W_{23} = W_{32} = 0$. As there is no coherence involving the ground state initially,

$$u_{31} = v_{31} = u_{21} = v_{21} = 0\ , \tag{3.589}$$

this remains so at all times and we are left with the three equations

$$\dot{u}_{32} = -\omega_{32}v_{32} - W u_{32} + \frac{\kappa}{\sqrt{3}}w_2 - \frac{2\kappa}{3}\ ,$$

$$\dot{v}_{32} = \omega_{32}u_{32} - W v_{32}\ , \tag{3.590}$$

$$\dot{w}_2 = -W w_2 + \frac{2}{\sqrt{3}}W + \kappa\sqrt{3}u_{32}\ .$$

We see that the damping matrix (3.574) gets two additional off-diagonal entries; also the damping vector is modified. Contrary to free induction decay of a 2-level system,

$$\dot{u}_{12} = -\omega_{21}v_{12} - \frac{1}{T_2}u_{12} ,$$

$$\dot{v}_{12} = \omega_{21}u_{12} - \frac{1}{T_2}v_{12} , \qquad (3.591)$$

$$\dot{w}_1 = -\frac{1}{T_1}w_1 + \frac{1}{T_1}$$

(cf. (3.544)), the present equations dissipatively couple $w_2$ with the coherence term $u_{32}$: for $\omega_{32} \neq 0$ this implies a non-exponential decay (cf. [65]). Note that transient coherence is generated even if $u_{32} = 0$ initially. Quantum beats will be studied from a different point of view in Sect. 3.3.7.5.

**3.3.6.3 Equations of Motion: $\nu$ Scenario.** According to (3.147), the Hamiltonian for a $\nu$ scenario is specified by the $SU(3)$ vector

$$(\mathcal{H}_j) = \left(\hbar g_{21}, \hbar g_{31}, 0, 0, 0, 0, \hbar\delta_{21}, \frac{\hbar}{\sqrt{3}}(\delta^* + \delta_{31})\right) , \qquad (3.592)$$

$$\delta^* = \delta_{31} - \delta_{21} . \qquad (3.593)$$

The factors $g_{21}$, $g_{31}$ define the coupling to the light field, and $\delta_{21}$, $\delta_{31}$ are detuning parameters. With the components of this $SU(3)$ vector, and with the structure constants $f_{ijk}$ for $SU(3)$ (see (2.78)), we find for

$$\Omega_{ik} = \frac{1}{\hbar}\sum_{j=1}^{8} f_{ijk}\mathcal{H}_j \qquad (3.594)$$

the result

$$(\Omega_{ik}) = \left(\begin{array}{ccc|ccc|cc}
0 & 0 & 0 & -\delta_{21} & 0 & -\frac{g_{31}}{2} & 0 & 0 \\
0 & 0 & 0 & 0 & -\delta_{31} & \frac{g_{21}}{2} & 0 & 0 \\
0 & 0 & 0 & \frac{g_{31}}{2} & \frac{g_{21}}{2} & -\delta^* & 0 & 0 \\
\hline
\delta_{21} & 0 & -\frac{g_{31}}{2} & 0 & 0 & 0 & -g_{21} & 0 \\
0 & \delta_{31} & -\frac{g_{21}}{2} & 0 & 0 & 0 & -\frac{g_{31}}{2} & -\frac{\sqrt{3}g_{31}}{2} \\
\frac{g_{31}}{2} & -\frac{g_{21}}{2} & \delta^* & 0 & 0 & 0 & 0 & 0 \\
\hline
0 & 0 & 0 & g_{21} & \frac{g_{31}}{2} & 0 & 0 & 0 \\
0 & 0 & 0 & 0 & \frac{\sqrt{3}g_{31}}{2} & 0 & 0 & 0
\end{array}\right) . \qquad (3.595)$$

Inserting this rotation matrix, the damping matrix (3.574), and the damping vector (3.575) into the generalized Bloch equations (3.508), one obtains the equations of motion

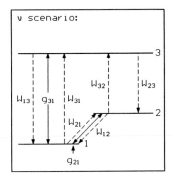

**Fig. 3.42.** $\nu$ scenario for vanishing detuning parameters

$$\dot{\lambda}_1 = -\delta_{21}\lambda_4 - \frac{1}{2}g_{31}\lambda_6 - \gamma^{(12)}\lambda_1 \,,$$

$$\dot{\lambda}_2 = -\delta_{31}\lambda_5 + \frac{1}{2}g_{21}\lambda_6 - \gamma^{(13)}\lambda_2 \,,$$

$$\dot{\lambda}_3 = \frac{1}{2}g_{31}\lambda_4 + \frac{1}{2}g_{21}\lambda_5 + (\delta_{21} - \delta_{31})\lambda_6 - \gamma^{(23)}\lambda_3 \,,$$

$$\dot{\lambda}_4 = \delta_{21}\lambda_1 - \frac{1}{2}g_{31}\lambda_3 - g_{21}\lambda_7 - \gamma^{(12)}\lambda_4 \,,$$

$$\dot{\lambda}_5 = \delta_{31}\lambda_2 - \frac{1}{2}g_{21}\lambda_3 - \frac{1}{2}g_{31}\lambda_7 - \frac{1}{2}\sqrt{3}g_{31}\lambda_8 - \gamma^{(13)}\lambda_5 \,,$$

$$\dot{\lambda}_6 = \frac{1}{2}g_{31}\lambda_1 - \frac{1}{2}g_{21}\lambda_2 - (\delta_{21} - \delta_{31})\lambda_3 - \gamma^{(23)}\lambda_6 \,,$$

$$\dot{\lambda}_7 = g_{21}\lambda_4 + \frac{1}{2}g_{31}\lambda_5 + \xi_{77}\lambda_7 + \xi_{78}\lambda_8 + \eta_7 \,,$$

$$\dot{\lambda}_8 = \frac{1}{2}\sqrt{3}g_{31}\lambda_5 + \xi_{87}\lambda_7 + \xi_{88}\lambda_8 + \eta_8 \,.$$

$$(3.596)$$

This system of equations describes the dissipative dynamics of an $SU(3)$ node in the $\nu$ scenario. The motion of the components $\lambda_i$ of the coherence vector $\boldsymbol{\lambda}$ is determined by transition probabilities $W_{12}$, $W_{21}$, $W_{13}$, $W_{31}$, $W_{23}$, $W_{32}$ implicitly contained in $\gamma^{(ij)}$, $\xi_{kl}$, $\eta_k$ (cf. Sect. 3.3.6.1), and by optical driving forces determined by the coupling constants $g_{21}$, $g_{31}$.

If we specialize to the resonance condition, i.e.

$$\delta_{21} = \delta_{31} = 0 \tag{3.597}$$

(cf. Fig. 3.42), and if we restrict ourselves to only one zero-temperature damping channel $W_{23}$ (cf. Fig. 3.43), we end up with the Bloch equations

$$\dot{\lambda}_1 = -\frac{1}{2}g_{31}\lambda_6 \ ,$$

$$\dot{\lambda}_2 = \frac{1}{2}g_{21}\lambda_6 - \frac{1}{2}W_{23}\lambda_2 \ ,$$

$$\dot{\lambda}_3 = \frac{1}{2}g_{31}\lambda_4 + \frac{1}{2}g_{21}\lambda_5 - \frac{1}{2}W_{23}\lambda_3 \ ,$$

$$\dot{\lambda}_4 = -\frac{1}{2}g_{31}\lambda_3 - g_{21}\lambda_7 \ ,$$

$$\dot{\lambda}_5 = -\frac{1}{2}g_{21}\lambda_3 - \frac{1}{2}g_{31}\lambda_7 - \frac{1}{2}\sqrt{3}g_{31}\lambda_8 - \frac{1}{2}W_{23}\lambda_5 \ ,$$

$$\dot{\lambda}_6 = \frac{1}{2}g_{31}\lambda_1 - \frac{1}{2}g_{21}\lambda_2 - \frac{1}{2}W_{23}\lambda_6 \ ,$$

$$\dot{\lambda}_7 = g_{21}\lambda_4 + \frac{1}{2}g_{31}\lambda_5 + \frac{1}{\sqrt{3}}W_{23}\lambda_8 + \frac{1}{3}W_{23} \ ,$$

$$\dot{\lambda}_8 = \frac{1}{2}\sqrt{3}g_{31}\lambda_5 - W_{23}\lambda_8 - \frac{1}{\sqrt{3}}W_{23} \ .$$

$$(3.598)$$

This system of equations has the stationary solution

$$\lambda_1 = 0 \ ,$$

$$\lambda_2 = 0 \ ,$$

$$\lambda_3 = -\frac{g_{31}g_{21}\left(g_{31}^2 - g_{21}^2\right)}{\frac{1}{2}g_{31}^4 + g_{21}^4 + g_{21}^2 W_{23}^2} \ ,$$

$$\lambda_4 = -\frac{g_{31}^2 g_{21} W_{23}}{\frac{1}{2}g_{31}^4 + g_{21}^4 + g_{21}^2 W_{23}^2} \ ,$$

$$\lambda_5 = \frac{g_{31}g_{21}^2 W_{23}}{\frac{1}{2}g_{31}^4 + g_{21}^4 + g_{21}^2 W_{23}^2} \ ,$$

$$\lambda_6 = 0 \ ,$$

$$\lambda_7 = \frac{g_{31}^2\left(g_{31}^2 - g_{21}^2\right)}{g_{31}^4 + 2g_{21}^4 + 2g_{21}^2 W_{23}^2} \ ,$$

$$\lambda_8 = -\frac{g_{31}^4 + 2g_{21}^4 - 3g_{31}^2 g_{21}^2 + 2g_{21}^2 W_{23}^2}{\sqrt{3}g_{31}^4 + 2\sqrt{3}g_{21}^4 + 2\sqrt{3}g_{21}^2 W_{23}^2} \ .$$

$$(3.599)$$

so that, assuming the parameters

$$W_{23} = 1 \ , \quad g_{31} = g_{21} = 1.2 \ , \tag{3.600}$$

one obtains the stationary solution

$$\lambda_1 = 0 \ , \quad \lambda_2 = 0 \ , \quad \lambda_3 = 0 \ , \quad \lambda_4 \approx -0.38 \ , \quad \lambda_5 \approx 0.38 \ , \quad \lambda_6 = 0 \ ,$$
$$\lambda_7 = 0 \ , \quad \lambda_8 \approx -0.183 \ . \tag{3.601}$$

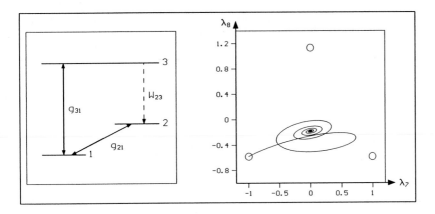

**Fig. 3.43.** Approach to the attractor state. L.h.s: energy spectrum of the transition scenario. R.h.s.: numerical solution for the parameters $W_{23} = 1$, $g_{12} = g_{13} = 1.2$. Simulation by R. Wawer (cf. [149])

Fig. 3.43 shows the approach to this attractor state in the $\lambda_7, \lambda_8$ plane out of the ground state. The circles indicate the three eigenstates, in the $\lambda_7, \lambda_8$ plane described by (cf. Fig. 2.11)

$$\{(\lambda_7, \lambda_8)\} = \left\{\left(-1, -1/\sqrt{3}\right), \left(1, -1/\sqrt{3}\right), \left(0, 2/\sqrt{3}\right)\right\} . \qquad (3.602)$$

The ground state is the pure state $\left(-1, -1/\sqrt{3}\right)$ (cf. (3.577)).

**3.3.6.4 Equations of Motion: $\Lambda$ Scenario.** In the $\Lambda$ scenario (cf. Fig. 3.44) the Hamiltonian is specified by the $SU(3)$ vector

$$(\mathcal{H}_j) = \left(0, \hbar g_{31}, \hbar g_{32}, 0, 0, 0, \hbar \delta^{(-)}, \frac{\hbar}{\sqrt{3}} \delta^{(+)}\right) , \qquad (3.603)$$

$$\delta^{(-)} = \delta_{31} - \delta_{32} , \quad \delta^{(+)} = \delta_{31} + \delta_{32} \qquad (3.604)$$

(see (3.154)) so that the respective rotation matrix reads

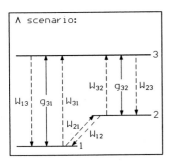

**Fig. 3.44.** $\Lambda$ scenario for vanishing detuning parameters

$$(\Omega_{ik}) = \begin{pmatrix} 0 & 0 & 0 & | & -\delta^{(-)} & -\frac{g_{32}}{2} & -\frac{g_{31}}{2} & | & 0 & 0 \\ 0 & 0 & 0 & | & -\frac{g_{32}}{2} & -\delta_{31} & 0 & | & 0 & 0 \\ 0 & 0 & 0 & | & \frac{g_{31}}{2} & 0 & -\delta_{32} & | & 0 & 0 \\ - & - & - & - & - & - & - & - & - & - \\ \delta^{(-)} & \frac{g_{32}}{2} & -\frac{g_{31}}{2} & | & 0 & 0 & 0 & | & 0 & 0 \\ \frac{g_{32}}{2} & \delta_{31} & 0 & | & 0 & 0 & 0 & | & \frac{g_{31}}{2} & -\frac{\sqrt{3}g_{31}}{2} \\ \frac{g_{31}}{2} & 0 & \delta_{32} & | & 0 & 0 & 0 & | & \frac{g_{32}}{2} & -\frac{\sqrt{3}}{2}g_{32} \\ - & - & - & - & - & - & - & - & - & - \\ 0 & 0 & 0 & | & 0 & -\frac{g_{31}}{2} & -\frac{g_{32}}{2} & | & 0 & 0 \\ 0 & 0 & 0 & | & 0 & \frac{\sqrt{3}g_{31}}{2} & \frac{\sqrt{3}g_{32}}{2} & | & 0 & 0 \end{pmatrix} \cdot \qquad (3.605)$$

Inserting this rotation matrix as well as the damping matrix (3.574) and the damping vector (3.575) into the generalized Bloch equations (3.508), one obtains the equations of motion

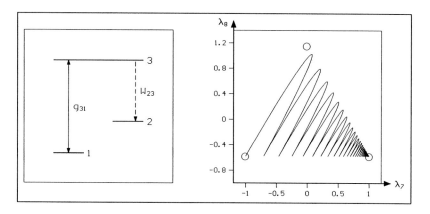

**Fig. 3.45.** Attractor state in $SU(3)$: $W_{23} = 1$, $g_{31} = 20$. Simulation by R. Wawer (cf. [149])

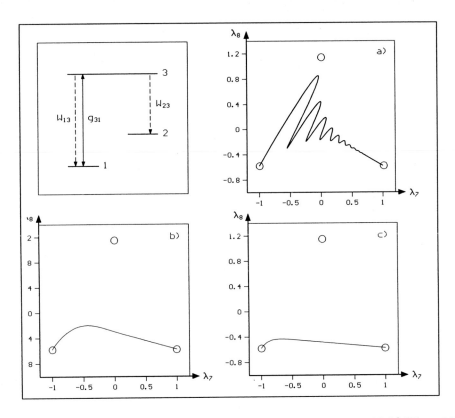

**Fig. 3.46.** Attractor states in $SU(3)$. a) $W_{13} = 2$, $W_{23} = 1$, $g_{31} = 20$. b) $W_{13} = 20$, $W_{23} = 10$, $g_{31} = 20$. c) $W_{13} = 40$, $W_{23} = 20$, $g_{31} = 20$. Simulation by R. Wawer (cf. [149])

$$\dot{\lambda}_1 = (\delta_{32} - \delta_{31})\,\lambda_4 - \frac{1}{2}g_{32}\lambda_5 - \frac{1}{2}g_{31}\lambda_6 - \gamma^{(12)}\lambda_1\ ,$$

$$\dot{\lambda}_2 = -\frac{1}{2}g_{32}\lambda_4 - \delta_{31}\lambda_5 - \gamma^{(13)}\lambda_2\ ,$$

$$\dot{\lambda}_3 = \frac{1}{2}g_{31}\lambda_4 - \delta_{32}\lambda_6 - \gamma^{(23)}\lambda_3\ ,$$

$$\dot{\lambda}_4 = (\delta_{31} - \delta_{32})\,\lambda_1 + \frac{1}{2}g_{32}\lambda_2 - \frac{1}{2}g_{31}\lambda_3 - \gamma^{(12)}\lambda_4\ ,$$

$$\dot{\lambda}_5 = \frac{1}{2}g_{32}\lambda_1 + \delta_{31}\lambda_2 + \frac{1}{2}g_{31}\lambda_7 - \frac{1}{2}\sqrt{3}g_{31}\lambda_8 - \gamma^{(13)}\lambda_5\ , \qquad (3.606)$$

$$\dot{\lambda}_6 = \frac{1}{2}g_{31}\lambda_1 + \delta_{32}\lambda_3 + \frac{1}{2}g_{32}\lambda_7 - \frac{1}{2}\sqrt{3}g_{32}\lambda_8 - \gamma^{(23)}\lambda_6\ ,$$

$$\dot{\lambda}_7 = -\frac{1}{2}g_{31}\lambda_5 - \frac{1}{2}g_{32}\lambda_6 + \xi_{77}\lambda_7 + \xi_{78}\lambda_8 + \eta_7\ ,$$

$$\dot{\lambda}_8 = \frac{1}{2}\sqrt{3}g_{31}\lambda_5 + \frac{1}{2}\sqrt{3}g_{32}\lambda_6 + \xi_{87}\lambda_7 + \xi_{88}\lambda_8 + \eta_8\ .$$

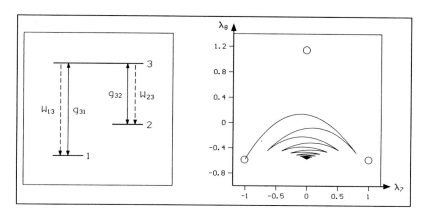

**Fig. 3.47.** Attractor state in $SU(3)$: $W_{13} = W_{23} = 1$, $g_{31} = g_{32} = 10$. Simulation by R. Wawer (cf. [149])

These equations include all damping channels represented by $W_{12}, W_{21}, W_{13}$, $W_{31}, W_{23}, W_{32}$:

$$\gamma^{(12)} = \frac{1}{2}\left(W_{12} + W_{21} + W_{31} + W_{32}\right) + \mathrm{Re}\left\{\tilde{\gamma}_{12}\right\} ,$$

$$\gamma^{(13)} = \frac{1}{2}\left(W_{13} + W_{21} + W_{23} + W_{31}\right) + \mathrm{Re}\left\{\tilde{\gamma}_{13}\right\} , \qquad (3.607)$$

$$\gamma^{(23)} = \frac{1}{2}\left(W_{12} + W_{13} + W_{23} + W_{32}\right) + \mathrm{Re}\left\{\tilde{\gamma}_{23}\right\} ,$$

$$\xi_{77} = -\left(W_{12} + \frac{1}{2}W_{31} + \frac{1}{2}W_{32} + W_{21}\right) ,$$

$$\xi_{78} = \frac{1}{\sqrt{3}}\left(W_{12} - W_{13} + W_{23} + \frac{1}{2}W_{32} - \frac{1}{2}W_{31} - W_{21}\right) , \qquad (3.608)$$

$$\xi_{87} = \frac{\sqrt{3}}{2}\left(-W_{31} + W_{32}\right) ,$$

$$\xi_{88} = -W_{13} - W_{23} - \frac{1}{2}\left(W_{31} + W_{32}\right) , \qquad (3.609)$$

$$\eta_7 = -\frac{1}{3}\left(2W_{12} + W_{13} - W_{23} + W_{32} - W_{31} - 2W_{21}\right) ,$$

$$\eta_8 = -\frac{1}{\sqrt{3}}\left(W_{13} + W_{23} - W_{32} - W_{31}\right) . \qquad (3.610)$$

In Figs. 3.45–3.47 special $\Lambda$ scenarios and corresponding numerical solutions in the $\lambda_7, \lambda_8$ plane are shown. As an initial state the ground state 1 (in the $\lambda_7, \lambda_8$ plane represented by $\left(-1, -1/\sqrt{3}\right)$) was used: if only one driving force is applied (see Figs. 3.45, 3.46), the system evolves from the ground state 1 into the metastable state 2 (in the $\lambda_7, \lambda_8$ plane represented by $(1, -1/\sqrt{3})$), i.e. the system represents an optically controlled *quantum switch*. If only one damping channel is applied (see Fig. 3.45), the motion is fully developed. Figure 3.46 illustrates the motion if two damping channels are applied: for weak damping the motion evolves in the whole $\lambda_7, \lambda_8$ plane (see Fig. 3.46, a)). For strong damping the motion of the coherence vector is essentially confined to $\lambda_7$ (see Fig. 3.46, b), c)). If a second driving force is added (see Fig. 3.47), the attractor is symmetric with respect to state 1 and 2: $\lambda_7(t \to \infty) = 0$.

### 3.3.7 Open Networks

We finally study the influence of damping on small networks. In the simplest case, the damping channels are locally coupled to the individual nodes of the network (cf. Fig. 3.48). Then the environment tends to validate the local picture in the sense that non-local properties are damped, and the respective damping models valid in $SU(2)$ or $SU(3)$ can directly be applied. Situations in which damping becomes non-local will be taken up in Sect. 4.4.7.

**3.3.7.1 Local Damping.** We want to derive network equations for 2-node networks of the type studied in Sect. 3.2.5.3, but now including damping.

We first note that damping is additive (e.g., see (3.508)) and characterized by a local damping matrix $(\xi_{ij}(\nu))$ and a local damping vector $\boldsymbol{\eta}(\nu)$, where $\nu$ is the node index. The influence of the damping matrix is easily included

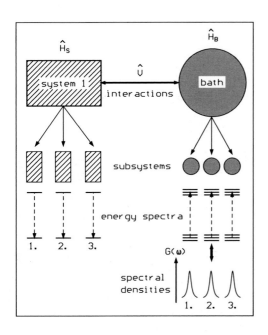

**Fig. 3.48.** Local damping: system 1 and its surrounding (bath) are both composed of subsystems. Here, system 1 is a network of 2-level nodes. A transition $n \to m$ within a node $\nu$ is coupled with a set of transitions $r' \to r$ within a channel $\nu$ of the bath. The transitions are thus taken to be distinguishable in $\nu$ space even though they might overlap in frequency space

in the equations for the coherence vectors, (3.283), and the equation for the correlation tensor, (3.284), if we replace $\frac{1}{n_2}\Omega\begin{pmatrix} 1 & 1 \\ i & j \end{pmatrix}$ by

$$\frac{1}{n_2}\Omega'\begin{pmatrix} 1 & 1 \\ i & j \end{pmatrix} = \frac{1}{n_2}\Omega\begin{pmatrix} 1 & 1 \\ i & j \end{pmatrix} + \xi\begin{pmatrix} 1 & 1 \\ i & j \end{pmatrix} \tag{3.611}$$

and $\frac{1}{n_1}\Omega\begin{pmatrix} 2 & 2 \\ i & j \end{pmatrix}$ correspondingly. (Only $\Omega$ is scaled by the dimension of the other node as it contains a trace over the total Hamiltonian. $\xi$ is defined locally from the start.) The damping vector has no analogue in the coherent evolution; the damping vector components connect the time-derivative of a tensor component of order $m$ with those of order $m - 1$:

$$m = 1: \quad \dot{\lambda}\begin{pmatrix} 1 \\ i \end{pmatrix} = \eta\begin{pmatrix} 1 \\ i \end{pmatrix} ,$$

$$m = 2: \quad \dot{K}\begin{pmatrix} 1 & 2 \\ i & j \end{pmatrix} = \eta\begin{pmatrix} 1 \\ i \end{pmatrix}\lambda\begin{pmatrix} 2 \\ j \end{pmatrix} + \eta\begin{pmatrix} 2 \\ j \end{pmatrix}\lambda\begin{pmatrix} 1 \\ i \end{pmatrix} , \tag{3.612}$$

etc.

### 3.3.7.2 Damped Network Equations in $SU(n_1) \otimes SU(n_2)$. Supplementing the network equations (3.283)–(3.284) on the basis of these relations, one obtains

$$\dot{\lambda}\begin{pmatrix} 1 \\ i \end{pmatrix} = \frac{1}{n_2}\Omega\begin{pmatrix} 1 & 1 \\ i & k \end{pmatrix}\lambda\begin{pmatrix} 1 \\ k \end{pmatrix} + \frac{1}{2}Q\begin{pmatrix} 1 & 1 & 2 \\ i & k & l \end{pmatrix}K\begin{pmatrix} 1 & 2 \\ k & l \end{pmatrix} +$$
$$\xi\begin{pmatrix} 1 & 1 \\ i & k \end{pmatrix}\lambda\begin{pmatrix} 1 \\ k \end{pmatrix} + \eta\begin{pmatrix} 1 \\ i \end{pmatrix} ,$$

$$\dot{\lambda}\begin{pmatrix} 2 \\ j \end{pmatrix} = \frac{1}{n_1}\Omega\begin{pmatrix} 2 & 2 \\ j & l \end{pmatrix}\lambda\begin{pmatrix} 2 \\ l \end{pmatrix} + \frac{1}{2}Q\begin{pmatrix} 1 & 2 & 2 \\ l & j & k \end{pmatrix}K\begin{pmatrix} 1 & 2 \\ l & k \end{pmatrix} +$$
$$\xi\begin{pmatrix} 2 & 2 \\ j & l \end{pmatrix}\lambda\begin{pmatrix} 2 \\ l \end{pmatrix} + \eta\begin{pmatrix} 2 \\ j \end{pmatrix} , \tag{3.613}$$

$$\dot{K}\begin{pmatrix} 1 & 2 \\ i & j \end{pmatrix} = \frac{1}{n_2}\Omega\begin{pmatrix} 1 & 1 \\ i & k \end{pmatrix}K\begin{pmatrix} 1 & 2 \\ k & j \end{pmatrix} + \frac{1}{n_1}\Omega\begin{pmatrix} 2 & 2 \\ j & l \end{pmatrix}K\begin{pmatrix} 1 & 2 \\ i & l \end{pmatrix} +$$
$$\frac{1}{n_2}Q\begin{pmatrix} 1 & 1 & 2 \\ i & m & j \end{pmatrix}\lambda\begin{pmatrix} 1 \\ m \end{pmatrix} + \frac{1}{n_1}Q\begin{pmatrix} 1 & 2 & 2 \\ i & j & m \end{pmatrix}\lambda\begin{pmatrix} 2 \\ m \end{pmatrix} +$$
$$\frac{1}{2}D\begin{pmatrix} 1 & 1 & 2 & 2 \\ i & k' & j & l' \end{pmatrix}K\begin{pmatrix} 1 & 2 \\ k' & l' \end{pmatrix} + \tag{3.614}$$
$$\xi\begin{pmatrix} 1 & 1 \\ i & k \end{pmatrix}K\begin{pmatrix} 1 & 2 \\ k & j \end{pmatrix} + \xi\begin{pmatrix} 2 & 2 \\ j & l \end{pmatrix}K\begin{pmatrix} 1 & 2 \\ i & l \end{pmatrix} +$$
$$\eta\begin{pmatrix} 1 \\ i \end{pmatrix}\lambda\begin{pmatrix} 2 \\ j \end{pmatrix} + \eta\begin{pmatrix} 2 \\ j \end{pmatrix}\lambda\begin{pmatrix} 1 \\ i \end{pmatrix} .$$

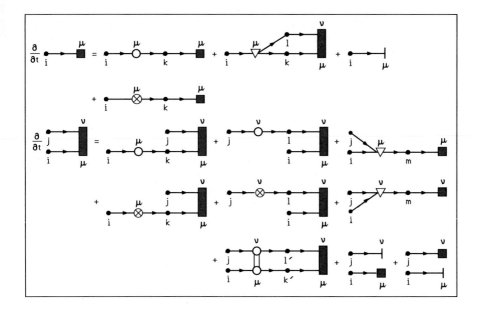

**Fig. 3.49.** Diagrammatic representation of damped network equations

These coupled equations describe the time-evolution of 2 interacting $SU(n)$ nodes. They replace the evolution equations (3.283)–(3.284) in the presence of damping. We note that the 2-node correlation tensor is additively damped by the local damping channels: the damping of the matrix elements of a $N$-th order correlation tensor thus scales linearly with $N$. Higher order correlations thus become increasingly fragile, as they are damped by all local channels to which their respective nodes are connected (compare Sect. 3.2.5.7 for $\Omega$ terms supplemented by $\xi$). It may be understood as justification of the neglect of many-node correlations in macroscopic systems, where the number of nodes would be extremely large. In Fig. 3.49 the diagrammatic representation of the network equations (3.613) and (3.614) is depicted. The corresponding elements for the representation of the damping terms are shown in Fig. 3.50.

**3.3.7.3 Dynamics in $SU(2) \otimes SU(2)$.** The coupled Bloch equations and equations for the coherence tensors will now be applied to a special 2-level-2-node system. The corresponding subsystems may be driven by external

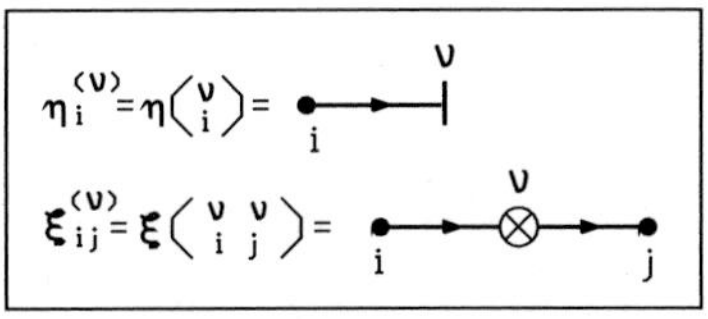

$$\eta_i^{(\nu)} = \eta\binom{\nu}{i} =$$

$$\xi_{ij}^{(\nu)} = \xi\binom{\nu\ \nu}{i\ j} =$$

**Fig. 3.50.** Elements for a diagrammatic representation of damping terms

electromagnetic fields. The communication between the two nodes takes place via Coulomb-type interaction; damping effects are considered explicitly.

*State Space.* In $SU(n_1) \otimes SU(n_2)$ (where node 1 represents a system with $n_1$ energy levels, and node 2 a system with $n_2$ levels) the density operator reads ($s_1 = n_1^2 - 1$, $s_2 = n_2^2 - 1$)

$$\hat{\rho} = \frac{1}{n_1 n_2} \hat{1} + \frac{1}{2n_2} \sum_{j=1}^{s_1} \lambda_j(1) \left[ \hat{\lambda}_j(1) \otimes \hat{1}(2) \right]$$

$$+ \frac{1}{2n_1} \sum_{k=1}^{s_2} \lambda_k(2) \left[ \hat{1}(1) \otimes \hat{\lambda}_k(2) \right]$$

$$+ \frac{1}{4} \sum_{j,k=1}^{s_1,s_2} K_{jk}(1,2) \left[ \hat{\lambda}_j(1) \otimes \hat{\lambda}_k(2) \right] \qquad (3.615)$$

(compare (2.599)). $K_{jk}(1,2)$ denotes the elements of the correlation tensor K. The operators $\hat{\lambda}_j(v)$ $(v = 1,2)$ are the generating operators of the $SU(n)$ algebra, the respective expectation values are in subsystem $v$, and

$$\lambda_j(1) = \left\langle \hat{\lambda}_j(1) \right\rangle = \mathrm{Tr}_1 \left\{ \hat{\rho} \left[ \hat{\lambda}_j(1) \otimes \hat{1}(2) \right] \right\} = \mathrm{Tr}_1 \left\{ \hat{R}(1) \hat{\lambda}_j(1) \right\} , \quad (3.616)$$

$$\lambda_k(2) = \left\langle \hat{\lambda}_k(2) \right\rangle = \mathrm{Tr}_2 \left\{ \hat{\rho} \left[ \hat{1}(1) \otimes \hat{\lambda}_k(2) \right] \right\} = \mathrm{Tr}_1 \left\{ \hat{R}(2) \hat{\lambda}_k(2) \right\} , \quad (3.617)$$

$$K_{jk}(1,2) = \left\langle \hat{\lambda}_j(1) \otimes \hat{\lambda}_k(2) \right\rangle \qquad (3.618)$$

(compare (2.272), (2.600), (2.601), $\hat{R}(v)$ denotes reduced density operators). The correlation tensor proper, M(1,2), then is defined by

$$M_{jk}(1,2) = K_{jk}(1,2) - \lambda_j(1)\lambda_k(2) . \qquad (3.619)$$

This tensor is zero (no entanglement) if 2-node product states describe the state of the network.

*Hamilton Model.* Assuming a quantum network consisting of 2-level nodes ($n_1 = n_2 = n = 2$) and their interactions ("edges"), the Hamilton operator can be written as

$$\hat{H} = \frac{1}{4} \mathrm{Tr}\left\{ \hat{H} \right\} \hat{1} + \frac{1}{4} \sum_{j=1}^{s_1} \mathcal{H}_j(1) \left[ \hat{\lambda}_j(1) \otimes \hat{1}(2) \right]$$

$$+ \frac{1}{4} \sum_{k=1}^{s_2} \mathcal{H}_k(2) \left[ \hat{1}(1) \otimes \hat{\lambda}_k(2) \right]$$

$$+ \frac{1}{4} \sum_{j=1}^{s_1} \sum_{k=1}^{s_2} \mathcal{H}_{jk}(1,2) \left[ \hat{\lambda}_j(1) \otimes \hat{\lambda}_k(2) \right] . \qquad (3.620)$$

For an optically driven pair of 2-level systems with field coupling $g^{(v)}$ and detuning

$$\delta^{(v)} = \frac{E_2^{(v)} - E_1^{(v)}}{\hbar} - \omega \tag{3.621}$$

we have

$$\mathrm{Tr}\left\{\hat{H}\right\} = E_1^{(1)} + E_2^{(1)} + E_1^{(2)} + E_2^{(2)}$$
$$= \mathrm{Tr}_1\left\{\hat{H}(1)\right\} + \mathrm{Tr}_2\left\{\hat{H}(2)\right\} \tag{3.622}$$

and

$$\mathcal{H}_j(1) = \mathrm{Tr}\left\{\hat{H}\hat{\lambda}_j(1)\right\} = \mathrm{Tr}_1\left\{\hat{H}(1)\hat{\lambda}_j(1)\right\}\mathrm{Tr}_2\left\{\hat{1}(2)\right\}$$
$$= \mathrm{Tr}_1\left\{\hat{H}(1)\hat{\lambda}_j(1)\right\}n\;,$$
$$\mathcal{H}_k(2) = \mathrm{Tr}\left\{\hat{H}\hat{\lambda}_k(2)\right\} = \mathrm{Tr}_2\left\{\hat{H}(2)\hat{\lambda}_k(2)\right\}\mathrm{Tr}_1\left\{\hat{1}(1)\right\}$$
$$= \mathrm{Tr}_2\left\{\hat{H}(2)\hat{\lambda}_k(2)\right\}n \tag{3.623}$$

so that the $SU(2)$ vectors read

$$(\mathcal{H}_j(v)) = \left(\hbar g^{(v)}, 0, \hbar\delta^{(v)}\right)n\;. \tag{3.624}$$

Typical (2-node) interactions can be represented by the $SU(2){\otimes}SU(2)$ matrix

$$(\mathcal{H}_{jk}(1,2)) = 2\hbar \begin{pmatrix} C_F & 0 & 0 \\ 0 & C_F & 0 \\ 0 & 0 & -C_R \end{pmatrix}\;. \tag{3.625}$$

The 2-level-2-node quantum network determined by the Hamiltonian (3.620) will be taken as a basis during the following considerations. In Fig. 3.51 we give a schematic representation of the network considered.

*Network Equations without Damping.* In the case of a 2-level-2-node system defined by the $SU(2) \otimes SU(2)$ Hamiltonian (3.620), the Liouville equation leads to ($n = 2$):

$$\dot{\lambda}_i(1) = \frac{1}{n}\sum_{k=1}^{3}\Omega_{ik}^{(1)}\lambda_k(1) + \frac{1}{2}\sum_{k=1}^{3}\sum_{l=1}^{3}Q_{ikl}^{(1)}K_{kl}(1,2)$$
$$= \dot{\lambda}_i^c(1)\;,$$
$$\dot{\lambda}_i(2) = \frac{1}{n}\sum_{k=1}^{3}\Omega_{ik}^{(2)}\lambda_k(2) + \frac{1}{2}\sum_{k=1}^{3}\sum_{l=1}^{3}Q_{lik}^{(2)}K_{lk}(1,2)$$
$$= \dot{\lambda}_i^c(2) \tag{3.626}$$

and

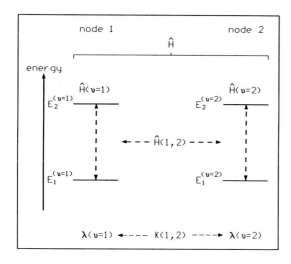

**Fig. 3.51.** A schematic representation of the considered 2-level-2-node network

$$\dot{K}_{ij}(1,2) = \frac{1}{2}\sum_{l=1}^{3} Q_{ilj}^{(1)}\lambda_l(1) + \frac{1}{2}\sum_{l=1}^{3} Q_{ijl}^{(2)}\lambda_l(2) +$$

$$\frac{1}{n}\sum_{k=1}^{3} \Omega_{ik}^{(1)} K_{kj}(1,2) + \frac{1}{n}\sum_{l=1}^{3} \Omega_{jl}^{(2)} K_{il}(1,2)$$

$$= \dot{K}_{ij}^{c}(1,2) \tag{3.627}$$

with the matrix elements

$$\Omega_{ik}^{(1)} = \frac{1}{\hbar}\sum_{j=1}^{3} f_{ijk}(1)\mathcal{H}_j(1)\,, \quad \Omega_{ik}^{(2)} = \frac{1}{\hbar}\sum_{j=1}^{3} f_{ijk}(2)\mathcal{H}_j(2) \tag{3.628}$$

and

$$Q_{ijk}^{(1)} = \frac{1}{\hbar}\sum_{l=1}^{3} f_{ilj}(1)\mathcal{H}_{lk}(1,2)\,, \quad Q_{ijk}^{(2)} = \frac{1}{\hbar}\sum_{l=1}^{3} f_{jlk}(2)\mathcal{H}_{il}(1,2)\,. \tag{3.629}$$

These matrix elements contain specific parameters like $C_F$, $C_R$, $\delta^{(v)}$, $g^{(v)}$ in an implicit way. In particular, the matrix elements $\Omega_{ij}^{(v)}$ are represented by

$$\left(\Omega_{ij}^{(v)}\right) = n\begin{pmatrix} 0 & -\delta^{(v)} & 0 \\ +\delta^{(v)} & 0 & -g^{(v)} \\ 0 & +g^{(v)} & 0 \end{pmatrix}\,. \tag{3.630}$$

The derivation procedure of this coupled set of equations was presented in Sect. 3.2.5.3. An explicit formulation of the evolution equations (3.626)–(3.627) was presented in Sect. 3.2.5.4 (see (3.295)–(3.296)).

*Network Equations with Damping.* Local damping is specified by the damping matrix

$$\left( \xi_{ij}^{(v)} \right) = \begin{pmatrix} -\gamma^{(v)} & 0 & 0 \\ 0 & -\gamma^{(v)} & 0 \\ 0 & 0 & -\xi^{(v)} \end{pmatrix} \tag{3.631}$$

(compare (3.521)), and the damping vector

$$\eta^{(v)} = \left( 0, 0, \eta_3^{(v)} \right) = \left( 0, 0, W_{21}^{(v)} - W_{12}^{(v)} \right) \tag{3.632}$$

(compare (3.522)). Additionally, the definitions

$$\xi^{(v)} = W_{12}^{(v)} + W_{21}^{(v)} , \quad \gamma^{(v)} = \frac{1}{2} \left( W_{12}^{(v)} + W_{21}^{(v)} + 2\mathrm{Re} \left\{ \tilde{\gamma}_{12}^{(v)} \right\} \right) \tag{3.633}$$

have to be taken into account.

The equations of motion in the presence of local damping follow from the undamped equations (3.626), (3.627) by replacing

$$\frac{1}{n_2} \Omega_{ik}^{(1)} \rightarrow \frac{1}{n_2} \Omega_{ik}^{(1)'} = \frac{1}{n_2} \Omega_{ik}^{(1)} + \xi_{ik}^{(1)} ,$$

$$\frac{1}{n_1} \Omega_{ik}^{(2)} \rightarrow \frac{1}{n_1} \Omega_{ik}^{(2)'} = \frac{1}{n_1} \Omega_{ik}^{(2)} + \xi_{ik}^{(2)} \tag{3.634}$$

and adding the damping vector contribution. We thus obtain

$$\boxed{\begin{aligned} \dot{\lambda}_1(v) - \dot{\lambda}_1^c(v) &= -\gamma^{(v)} \lambda_1(v) , \\ \dot{\lambda}_2(v) - \dot{\lambda}_2^c(v) &= -\gamma^{(v)} \lambda_2(v) , \\ \dot{\lambda}_3(v) - \dot{\lambda}_3^c(v) &= -\xi^{(v)} \lambda_3(v) + \eta_3^{(v)} \end{aligned}} \tag{3.635}$$

and

$$\boxed{\begin{aligned} \dot{K}_{kl}(1,2) - \dot{K}_{kl}^c(1,2) &= - \left( \gamma^{(1)} + \gamma^{(2)} \right) K_{kl}(1,2) , \\ \dot{K}_{k3}(1,2) - \dot{K}_{k3}^c(1,2) &= - \left( \gamma^{(1)} + \xi^{(2)} \right) K_{k3}(1,2) \\ &\quad - \eta_3^{(2)} \lambda_k(1) , \\ \dot{K}_{3l}(1,2) - \dot{K}_{3l}^c(1,2) &= - \left( \xi^{(1)} + \gamma^{(2)} \right) K_{3l}(1,2) \\ &\quad - \eta_3^{(1)} \lambda_l(2) , \\ \dot{K}_{33}(1,2) - \dot{K}_{33}^c(1,2) &= - \left( \xi^{(1)} + \xi^{(2)} \right) K_{33}(1,2) \\ &\quad + \eta_3^{(1)} \lambda_3(2) + \eta_3^{(2)} \lambda_3(1) \end{aligned}} \tag{3.636}$$

with

$$k, l = 1, 2 . \tag{3.637}$$

Equations (3.635)–(3.636) describe the dynamics of the 2-level-2-node system including coherent driving forces and damping. Such an $SU(2) \otimes SU(2)$ representation describes the dynamics of expectation values represented by the coherence vectors and the correlation tensor.

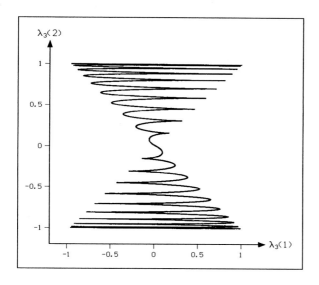

**Fig. 3.52.** Portrait of the ensemble dynamics of two coupled 2-level nodes in the $\lambda_3(1), \lambda_3(2)$ plane. The ground state is the initial state. Parameters: $g^{(1)} = 2$, $C_F = 0.05$, other parameters zero (unitary motion). Simulation by M. Keller (cf. [76])

*Ensemble Dynamics in the* $\lambda_3(1), \lambda_3(2)$ *Plane.* In $SU(2) \otimes SU(2)$, the dynamics of the 2-node network is described by (3.635)–(3.636). Assuming a special damping scenario defined by

$$W_{12}^{(v)} \neq 0 \, , \quad W_{21}^{(v)} = \operatorname{Re}\left\{ \tilde{\gamma}_{12}^{(v)} \right\} = 0 \, , \tag{3.638}$$

(3.635)–(3.636) reduce to

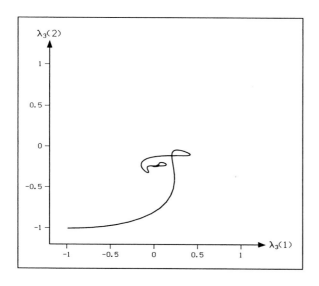

**Fig. 3.53.** As before but with the parameters $g^{(1)} = 2$, $C_F = 1$, $W_{12}^{(2)} = 1$, other parameters zero. Simulation by M. Keller (cf. [76])

$$\dot{\lambda}_1(v) - \dot{\lambda}_1^c(v) = -\frac{1}{2}W_{12}^{(v)}\lambda_1(v)\,,$$
$$\dot{\lambda}_2(v) - \dot{\lambda}_2^c(v) = -\frac{1}{2}W_{12}^{(v)}\lambda_2(v)\,,$$
$$\dot{\lambda}_3(v) - \dot{\lambda}_3^c(v) = -W_{12}^{(v)}\lambda_3(v) - W_{12}^{(v)}$$

(3.639)

and $(k, l = 1, 2)$

$$\dot{K}_{kl}(1,2) - \dot{K}_{kl}^c(1,2) = -\frac{1}{2}\left(W_{12}^{(1)} + W_{12}^{(2)}\right)K_{kl}(1,2)\,,$$
$$\dot{K}_{k3}(1,2) - \dot{K}_{k3}^c(1,2) = -\left(\frac{1}{2}W_{12}^{(1)} + W_{12}^{(2)}\right)K_{k3}(1,2)$$
$$-W_{12}^{(2)}\lambda_k(1)\,,$$
$$\dot{K}_{3l}(1,2) - \dot{K}_{3l}^c(1,2) = -\left(W_{12}^{(1)} + \frac{1}{2}W_{12}^{(2)}\right)K_{3l}(1,2)$$
$$-W_{12}^{(1)}\lambda_l(2)\,,$$
$$\dot{K}_{33}(1,2) - \dot{K}_{33}^c(1,2) = -\left(W_{12}^{(1)} + W_{12}^{(2)}\right)K_{33}(1,2)$$
$$-W_{12}^{(1)}\lambda_3(2) - W_{12}^{(2)}\lambda_3(1)\,.$$

(3.640)

A numerical example projected on the $\lambda_3(1), \lambda_3(2)$ plane is shown in Fig. 3.53. $g^{(1)}$ represents the Rabi frequency which indicates that coherent driving forces are applied, $C_F$ represents the Coulomb interaction between both nodes, and $W_{12}^{(2)}$ represents the intensity of damping. ($g^{(1)}$ and $C_F$ are implicitly included in the terms $\dot{\lambda}_i^c(v)$, $\dot{K}_{ij}^c(1,2)$ which are defined by (3.626) and (3.627).) Contrary to the unitary evolution (motion without damping) in Fig. 3.52, the ensemble dynamics approaches an attractor state.

**3.3.7.4 Network Equations in $SU(3) \otimes SU(2)$.** The network equations of a damped $SU(n_1) \otimes SU(n_2)$ system were derived in Sect. 3.3.7.2 (cf. (3.613)-(3.614)). Using the notation $\lambda_i(v)$, $K_{ij}(1,2)$ for the components of the coherence vector and the correlation tensor, and specializing to

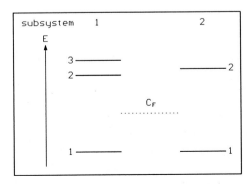

**Fig. 3.54.** $SU(3)$ and $SU(2)$ node coupled via $C_F$

$$s_1 = n_1^2 - 1 \,, \quad s_2 = n_2^2 - 1 \,, \quad n_1 = 3 \,, \quad n_2 = 2 \tag{3.641}$$

(compare Fig. 3.54), one obtains

$$
\begin{aligned}
\dot{\lambda}_i(1) &= \frac{1}{2} \sum_{k=1}^{8} \Omega_{ik}^{(1)} \lambda_k(1) + \frac{1}{2} \sum_{k=1}^{8} \sum_{l=1}^{3} Q_{ikl}^{(1)} K_{kl}(1,2) + \\
&\quad \sum_{k=1}^{8} \xi_{ik}^{(1)} \lambda_k(1) + \eta_i^{(1)} \,, \\
\dot{\lambda}_j(2) &= \frac{1}{3} \sum_{l=1}^{3} \Omega_{jl}^{(2)} \lambda_l(2) + \frac{1}{2} \sum_{k=1}^{3} \sum_{l=1}^{8} Q_{ljk}^{(2)} K_{lk}(1,2) + \\
&\quad \sum_{l=1}^{3} \xi_{jl}^{(2)} \lambda_l(2) + \eta_j^{(2)} \,,
\end{aligned}
\tag{3.642}
$$

which describe the time-evolution of the coherence vector of subsystem 1,

$$
\begin{aligned}
\boldsymbol{\lambda}(1) &= \{u_{12}(1), u_{13}(1), u_{23}(1), v_{12}(1), v_{13}(1), v_{23}(1), w_1(1), w_2(1)\} \\
&= \{\lambda_1(1), \lambda_2(1), \lambda_3(1), \lambda_4(1), \lambda_5(1), \lambda_6(1), \lambda_7(1), \lambda_8(1)\}
\end{aligned}
\tag{3.643}
$$

and of the coherence vector of subsystem 2,

$$
\begin{aligned}
\boldsymbol{\lambda}(2) &= \{u_{12}(2), v_{12}(2), w_1(2)\} \\
&= \{\lambda_1(2), \lambda_2(2), \lambda_3(2)\} \,,
\end{aligned}
\tag{3.644}
$$

and the equations

$$
\begin{aligned}
\dot{K}_{ij}(1,2) &= \frac{1}{2} \sum_{m=1}^{8} Q_{imj}^{(1)} \lambda_m(1) + \frac{1}{3} \sum_{m=1}^{3} Q_{ijm}^{(2)} \lambda_m(2) + \\
&\quad \frac{1}{2} \sum_{k=1}^{8} \Omega_{ik}^{(1)} K_{kj}(1,2) + \frac{1}{3} \sum_{l=1}^{3} \Omega_{jl}^{(2)} K_{il}(1,2) + \\
&\quad \frac{1}{2} \sum_{k'=1}^{8} \sum_{l'=1}^{3} D_{ik'jl'}^{(12)} K_{k'l'}(1,2) + \\
&\quad \sum_{k=1}^{8} \xi_{ik}^{(1)} K_{kj}(1,2) + \sum_{l=1}^{3} \xi_{jl}^{(2)} K_{il}(1,2) + \\
&\quad \eta_i^{(1)} \lambda_j(2) + \eta_j^{(2)} \lambda_i(1) \,,
\end{aligned}
\tag{3.645}
$$

which describe the evolution of the $3 \times 8$ correlation tensor K.

Specifying the matrix elements $\Omega_{ik}^{(v)}$, $Q_{ijk}^{(v)}$, $D_{ik'jl'}^{(12)}$, the damping matrices $\left(\xi_{ij}^{(v)}\right)$, and the damping vectors $\eta^{(v)}$, one obtains network equations for the considered $SU(3) \otimes SU(2)$ system: due to (3.275)–(3.276) and (3.285)–(3.286), the matrix elements $\Omega_{ik}^{(v)}$, $Q_{ijk}^{(v)}$ are given by

$$\Omega_{ik}^{(1)} = \frac{1}{\hbar} \sum_{j=1}^{8} f_{ijk}(1) \mathcal{H}_j(1) , \quad \Omega_{ik}^{(2)} = \frac{1}{\hbar} \sum_{j=1}^{3} f_{ijk}(2) \mathcal{H}_j(2) ,$$

$$Q_{ijk}^{(1)} = \frac{1}{\hbar} \sum_{l=1}^{8} f_{ilj}(1) \mathcal{H}_{lk}(1,2) , \quad Q_{ijk}^{(2)} = \frac{1}{\hbar} \sum_{l=1}^{3} f_{jlk}(2) \mathcal{H}_{il}(1,2) ,$$

(3.646)

and the matrix elements $D_{ik'jl'}^{(12)}$ are defined by

$$D_{ik'jl'}^{(12)} = \frac{1}{\hbar} \sum_{k,l=1}^{s_1,s_2} [d_{kk'i}(1) f_{ll'j}(2) + d_{ll'j}(2) f_{kk'i}(1)] \mathcal{H}_{kl}(1,2) , \qquad (3.647)$$

where the related $SU(3)$ vector $(\mathcal{H}_i(1))$ and the $SU(2)$ vector $(\mathcal{H}_j(2))$ are given by

$$(\mathcal{H}_i(1))$$
$$= 2 \left( 0,0,0,0,0,0, E_2^{(1)} - E_1^{(1)}, \frac{1}{\sqrt{3}} \left( E_1^{(1)} + E_2^{(1)} - 2E_3^{(1)} \right) \right) , \quad (3.648)$$

$$(\mathcal{H}_j(2)) = 3 \left( 0,0, E_2^{(2)} - E_1^{(2)} \right) \qquad (3.649)$$

(see Sect. 2.2.5), and where the $SU(3) \otimes SU(2)$ matrix is given by (cf. Sect. 2.4.3)

$$(\mathcal{H}_{ij}) = 2\hbar C_F \begin{pmatrix} 1 & 0 & 0 \\ 1 & 0 & 0 \\ 0 & 0 & 0 \\ 0 & 1 & 0 \\ 0 & 1 & 0 \\ 0 & 0 & 0 \\ 0 & 0 & 0 \\ 0 & 0 & 0 \end{pmatrix} . \qquad (3.650)$$

(Here we have assumed that there is no coupling between transition $|3\rangle \to |2\rangle$ in subsystem 1 and the 2-level system.) Due to the fact that the structure constants $f_{ijk}(1)$ of the $SU(3)$ system, $f_{ijk}(2)$ of the $SU(2)$ system, and the elements $d_{ijk}(1)$, $d_{ijk}(2)$ were defined in Sect. 2.2.2.3 and Sect. 2.2.2.5, the matrix elements $\Omega_{ik}^{(1)}$, $\Omega_{ik}^{(2)}$, $Q_{ijk}^{(1)}$, $Q_{ijk}^{(2)}$, $D_{ik'jl'}^{(12)}$ can be calculated.

For simplicity we will assume only one damping channel. While subsystem 1 is undamped,

$$\left( \xi_{ik}^{(1)} \right) = 0 , \quad \eta^{(1)} = 0 , \qquad (3.651)$$

the damping of subsystem 2 is described by the transition probability $W$ so that

$$\left( \xi_{ik}^{(2)} \right) = \begin{pmatrix} -W/2 & 0 & 0 \\ 0 & -W/2 & 0 \\ 0 & 0 & -W \end{pmatrix} , \quad \eta^{(2)} = (0,0,-W) \qquad (3.652)$$

holds (cf. (3.521), (3.522)).

Inserting these parameters into the network equations above, one obtains a system of equations determining the time-evolution of the coherence vectors and of the correlation tensor. Using the abbreviations

$$\omega_{ij}^{(\nu)} = \frac{E_i^{(\nu)} - E_j^{(\nu)}}{\hbar} \, , \tag{3.653}$$

the time-evolution of the coherence vector of subsystem 1 can be described by

$$
\begin{aligned}
\dot{\lambda}_1(1) &= -\omega_{21}^{(1)}\lambda_4(1) + \frac{1}{2}C_F\left(K_{32} - K_{61} + 2K_{72}\right) \, , \\
\dot{\lambda}_2(1) &= -\omega_{31}^{(1)}\lambda_5(1) + \frac{1}{2}C_F\left(K_{32} + K_{61} + K_{72} + \sqrt{3}K_{82}\right) \, , \\
\dot{\lambda}_3(1) &= -\left(\omega_{31}^{(1)} - \omega_{21}^{(1)}\right)\lambda_6(1) + \\
&\qquad \frac{1}{2}C_F\left(-K_{12} - K_{22} + K_{41} + K_{51}\right) \, , \\
\dot{\lambda}_4(1) &= \omega_{21}^{(1)}\lambda_1(1) + \frac{1}{2}C_F\left(-K_{31} - K_{62} + 2K_{71}\right) \, , \\
\dot{\lambda}_5(1) &= \omega_{31}^{(1)}\lambda_2(1) - \\
&\qquad \frac{1}{2}C_F\left(K_{31} - K_{62} + K_{71} + \sqrt{3}K_{81}\right) \, , \\
\dot{\lambda}_6(1) &= \left(\omega_{31}^{(1)} - \omega_{21}^{(1)}\right)\lambda_3(1) + \\
&\qquad \frac{1}{2}C_F\left(K_{11} - K_{21} + K_{42} - K_{52}\right) \, , \\
\dot{\lambda}_7(1) &= \frac{1}{2}C_F\left(-2K_{12} - K_{22} + 2K_{41} + K_{51}\right) \, , \\
\dot{\lambda}_8(1) &= \frac{\sqrt{3}}{2}C_F\left(-K_{22} + K_{51}\right) \, ,
\end{aligned}
\tag{3.654}
$$

the time-evolution of the coherence vector of subsystem 2 is given by

$$
\begin{aligned}
\dot{\lambda}_1(2) &= -\omega_{21}^{(2)}\lambda_2(2) - \frac{1}{2}W\lambda_1(2) + C_F\left(K_{43} + K_{53}\right) \, , \\
\dot{\lambda}_2(2) &= \omega_{21}^{(2)}\lambda_1(2) - \frac{1}{2}W\lambda_2(2) + C_F\left(-K_{13} + K_{23}\right) \, , \\
\dot{\lambda}_3(2) &= -W - W\lambda_3(2) + C_F\left(K_{12} + K_{22} - K_{41} - K_{51}\right) \, ,
\end{aligned}
\tag{3.655}
$$

and the evolution equations of the correlation tensor read

$$\dot{K}_{11} = -\omega_{21}^{(2)} K_{12} - \omega_{21}^{(1)} K_{41} - \frac{1}{2} W K_{11} +$$

$$\frac{1}{2} C_F \left[ -\lambda_6(1) + K_{63} \right] ,$$

$$\dot{K}_{12} = \omega_{21}^{(2)} K_{11} - \omega_{21}^{(1)} K_{42} - \frac{1}{2} W K_{12} +$$

$$\frac{1}{6} C_F \left[ 3\lambda_3(1) + 6\lambda_7(1) - 4\lambda_3(2) + \right.$$

$$\left. 3 K_{33} + 2\sqrt{3} K_{83} \right] ,$$

$$\dot{K}_{13} = -\omega_{21}^{(1)} K_{43} - W\lambda_1(1) - W K_{13} +$$

$$\frac{1}{6} C_F \left[ 4\lambda_2(2) + 3K_{33} - 3K_{61} - 2\sqrt{3} K_{82} \right] ,$$

$$\dot{K}_{21} = -\omega_{21}^{(2)} K_{22} - \omega_{31}^{(1)} K_{51} - \frac{1}{2} W K_{21} +$$

$$\frac{1}{2} C_F \left[ \lambda_6(1) - K_{63} \right] ,$$

$$\dot{K}_{22} = \omega_{21}^{(2)} K_{21} - \omega_{31}^{(1)} K_{52} - \frac{1}{2} W K_{22} +$$

$$\frac{1}{6} C_F \left[ 3\lambda_3(1) + 3\lambda_7(1) + \right.$$

$$\left. 3\sqrt{3}\lambda_8(1) - 4\lambda_3(2) - 3K_{33} + 3K_{73} - \sqrt{3} K_{83} \right] , \qquad (3.656)$$

$$\dot{K}_{23} = -\omega_{31}^{(1)} K_{53} - W\lambda_2(1) - W K_{23} +$$

$$\frac{1}{6} C_F \left[ 4\lambda_2(2) + 3K_{32} + 3K_{61} - 3K_{72} + \sqrt{3} K_{82} \right] ,$$

$$\dot{K}_{31} = -\omega_{21}^{(2)} K_{32} - \left( \omega_{31}^{(1)} - \omega_{21}^{(1)} \right) K_{61} - \frac{1}{2} W K_{31} +$$

$$\frac{1}{2} C_F \left[ \lambda_4(1) + \lambda_5(1) + K_{43} + K_{53} \right] ,$$

$$\dot{K}_{32} = \omega_{21}^{(2)} K_{31} - \frac{1}{2} W K_{32} - \left( \omega_{31}^{(1)} - \omega_{21}^{(1)} \right) K_{62} -$$

$$\frac{1}{2} C_F \left[ \lambda_1(1) + \lambda_2(1) + K_{13} + K_{23} \right] ,$$

$$\dot{K}_{33} = - \left( \omega_{31}^{(1)} - \omega_{21}^{(1)} \right) K_{63} - W\lambda_3(1) - W K_{33} +$$

$$\frac{1}{2} C_F \left[ K_{12} + K_{22} - K_{41} - K_{51} \right] ,$$

$$\dot{K}_{41} = -\omega_{21}^{(2)} K_{42} + \omega_{21}^{(1)} K_{11} - \frac{1}{2} W K_{41} +$$
$$\frac{1}{6} C_{\mathrm{F}} \left[ -3\lambda_3(1) - 6\lambda_7(1) + 4\lambda_3(2) + 3K_{33} - 2\sqrt{3} K_{83} \right] ,$$

$$\dot{K}_{42} = \omega_{21}^{(1)} K_{12} + \omega_{21}^{(2)} K_{41} - \frac{1}{2} W K_{42} +$$
$$\frac{1}{2} C_{\mathrm{F}} \left[ -\lambda_6(1) + K_{63} \right] ,$$

$$\dot{K}_{43} = \omega_{21}^{(1)} K_{13} - W \lambda_4(1) - W K_{43} +$$
$$\frac{1}{6} C_{\mathrm{F}} \left[ -4\lambda_1(2) - 3K_{31} - 3K_{62} + 2\sqrt{3} K_{81} \right] ,$$

$$\dot{K}_{51} = \omega_{31}^{(1)} K_{21} - \frac{1}{2} W K_{51} - \omega_{21}^{(2)} K_{52} +$$
$$\frac{1}{6} C_{\mathrm{F}} \left[ -3\lambda_3(1) - 3\lambda_7(1) - 3\sqrt{3} \lambda_8(1) + 4\lambda_3(2) + \right.$$
$$\left. 3K_{33} - 3K_{73} + \sqrt{3} K_{83} \right] ,$$

$$\dot{K}_{52} = -\omega_{31}^{(1)} K_{22} + \omega_{21}^{(2)} K_{51} - \frac{1}{2} W K_{52} + \tag{3.657}$$
$$\frac{1}{2} C_{\mathrm{F}} \left[ \lambda_6(1) - K_{63} \right] ,$$

$$\dot{K}_{53} = \omega_{31}^{(1)} K_{23} - W \lambda_5(1) - W K_{53} +$$
$$\frac{1}{6} C_{\mathrm{F}} \left[ -4\lambda_1(2) - 3K_{31} + 3K_{62} + 3K_{71} - \sqrt{3} K_{81} \right] ,$$

$$\dot{K}_{61} = -\left( -\omega_{31}^{(1)} + \omega_{21}^{(1)} \right) K_{31} - \omega_{21}^{(2)} K_{62} - \frac{1}{2} W K_{61} +$$
$$\frac{1}{2} C_{\mathrm{F}} \left[ \lambda_1(1) - \lambda_2(1) + K_{13} - K_{23} \right] ,$$

$$\dot{K}_{62} = -\left( -\omega_{31}^{(1)} + \omega_{21}^{(1)} \right) K_{32} - \omega_{21}^{(2)} K_{61} - \frac{1}{2} W K_{62} +$$
$$\frac{1}{2} C_{\mathrm{F}} \left[ -\lambda_4(1) + \lambda_5(1) - K_{43} + K_{53} \right] ,$$

$$\dot{K}_{63} = -\left( -\omega_{31}^{(1)} + \omega_{21}^{(1)} \right) K_{33} - W \lambda_6(1) - W K_{63} +$$
$$\frac{1}{2} C_{\mathrm{F}} \left[ -K_{11} + K_{21} - K_{42} + K_{52} \right] ,$$

$$\dot{K}_{71} = -\frac{1}{2}WK_{71} - \omega_{21}^{(2)}K_{72} +$$

$$\frac{1}{2}C_{\mathrm{F}}\left[2\lambda_4(1) + \lambda_5(1) - K_{53}\right] ,$$

$$\dot{K}_{72} = \omega_{21}^{(2)}K_{71} - \frac{1}{2}WK_{72} +$$

$$\frac{1}{2}C_{\mathrm{F}}\left[-2\lambda_1(1) - \lambda_2(1) + K_{23}\right] ,$$

$$\dot{K}_{73} = -W\lambda_7(1) - WK_{73} -$$

$$\frac{1}{2}C_{\mathrm{F}}\left[K_{22} - K_{51}\right] ,$$

$$\dot{K}_{81} = -\frac{1}{2}WK_{81} - \omega_{21}^{(2)}K_{82} +$$

$$\frac{1}{6}C_{\mathrm{F}}\left[3\lambda_5(1) - 2\sqrt{3}K_{43} + \sqrt{3}K_{53}\right] ,$$

$$\dot{K}_{82} = \omega_{21}^{(2)}K_{81} - \frac{1}{2}WK_{82} +$$

$$\frac{\sqrt{3}}{6}C_{\mathrm{F}}\left[-3\lambda_2(1) + 2K_{13} - K_{23}\right] ,$$

$$\dot{K}_{83} = -W\lambda_8(1) - WK_{83} -$$

$$\frac{\sqrt{3}}{6}C_{\mathrm{F}}\left[2K_{12} - K_{22} - 2K_{41} + K_{51}\right] .$$

$$(3.658)$$

These equations exemplify how with increasing $n$ (local state dimension) and increasing $N$ (number of nodes) the dynamics tends to be dominated by non-local (correlation) terms. The damping parameter $W$ enters all equations involving subsystem 2.

### 3.3.7.5 Quantum Beats in $SU(3) \otimes SU(2)$.
After excitation of an ensemble of quantum systems, relaxation processes like spontaneous emission

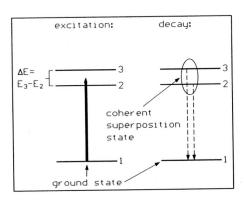

Fig. 3.55. Excitation and decay of a coherent superposition state in a 3-level system

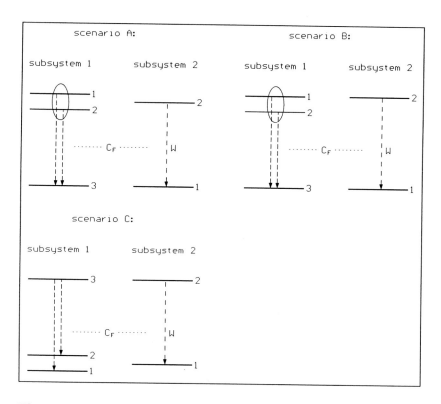

**Fig. 3.56.** Decay of a coherent superposition in an $SU(3) \otimes SU(2)$ network with symmetric (A) and non-symmetric (B) coupling. The two decay channels are indistinguishable but less so in case (B). Note the inverse numbering of states: the doublet is always 1, 2. Scenario C: decay of a "common" state via two distinguishable decay channels

cause the decay of the excitation state. As already discussed in Sect. 3.3.6.2, such a decay may be non-exponential, i.e modulated by oscillations (called *quantum beats*). They come into being if at least two indistinguishable decay channels exist so that *interference* occurs. Quantum beats are observable in ensembles of 3-level systems: after optical excitation by a light pulse with duration $\Delta t$. If the energy uncertainty $\hbar/\Delta t$ is larger than the energy difference $\Delta E = E_3 - E_2$ of the two excitation levels, the excitation causes a coherent superposition state which may decay non-exponentially. In Fig. 3.55 this situation is sketched.

In the following, quantum beats will be studied in more detail. However, instead of a single 3-level node coupled to a bath including non-secular terms (overlapping channels, see Sect. 3.3.4.4), an $SU(3) \otimes SU(2)$ network (3-level node coupled to a damped 2-level node via Förster interaction) will be considered: such an extended environment allows indistinguishability to be modelled and varied. In Fig. 3.56 these scenarios are depicted. The transitions of

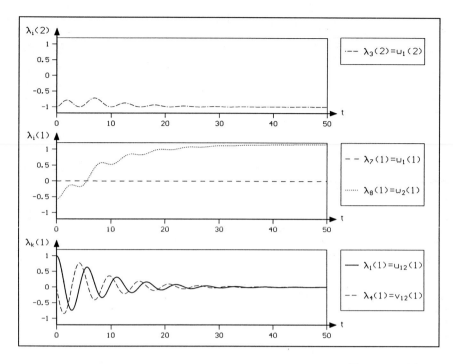

**Fig. 3.57.** Decay in an $SU(3) \otimes SU(2)$ system (scenario A). $\omega_{31}^{(1)} = -7$, $\omega_{21}^{(1)} = -6$, $\omega_{21}^{(2)} = 6.5$, $W = 1$, $C_F = 0.25$. Simulation by R. Wawer (cf. [149])

subsystem 1 will be symmetric (scenario A) or non-symmetric (scenario B) with respect to the transition in subsystem 2. For comparison, the decay of a "common" state via distinguishable decay channels will be studied (scenario C). It will be assumed that at time $t = 0$ (initial state) subsystem 1 is in the highest energy state, and subsystem 2 is in the ground state. The inter-node coupling constant is $C_F$; $W$ describes the damping in subsystem 2. We discuss these scenarios based on the corresponding network equations derived in Sect. 3.3.7.4.

In Figs. 3.57, 3.58, 3.59 numerical solutions for the different decay scenarios A, B, C are shown. $W$, $C_F$, and $\omega_{21}^{(1)}$, $\omega_{31}^{(1)}$, $\omega_{21}^{(2)}$ are the control parameters. In Fig. 3.57 the scenario A is taken as a basis for the numerical simulation. The initial state ($t = 0$) is the coherent state (subsystem 1) and the ground state (subsystem 2). The coherent state decays non-exponentially: this is shown by the behaviour of $\lambda_8(1) = w_2(1)$, which is a measure for the occupation difference between the two upper levels and the ground state. Due to the symmetrically coupled subsystem 2, the vector component $\lambda_7(1)$ remains zero. It describes the occupation difference between the two upper levels (1 and 2). The components $\lambda_1(1) = u_{12}(1)$, $\lambda_4(1) = v_{12}(1)$ oscillate periodically and vanish in the course of time.

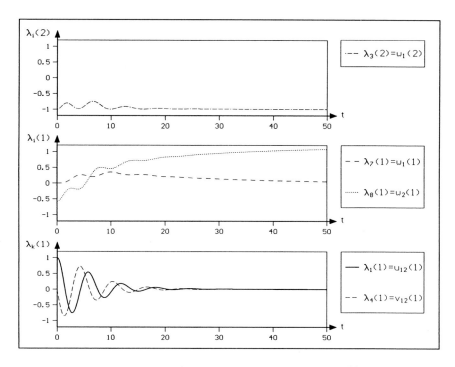

**Fig. 3.58.** Decay in an $SU(3) \otimes SU(2)$ system (scenario B). $\omega_{31}^{(1)} = -7$, $\omega_{21}^{(1)} = -6$, $\omega_{21}^{(2)} = 7$, $W = 1$, $C_{\mathrm{F}} = 0.25$. Simulation by R. Wawer (cf. [149])

In Fig. 3.58 a non-symmetrically coupled subsystem 2 is taken as a basis (scenario B). The initial state is the same as before. In this case the component $\lambda_7(1) = w_1(1)$ does not remain zero, but shows quantum beats, too. The beating of $\lambda_8$ is attenuated; it would disappear completely if the asymmetry was further enlarged.

Fig. 3.59 represents the decay process for the scenario C. In this scenario, at time $t = 0$, subsystem 1 is in the state 3 and subsystem 2 is in the ground state. In contrast to the scenarios A, B, the decay process in scenario C uses two distinguishable channels. In such a case an exponential decay process without quantum beats comes into being: $\lambda_8(1) = w_2(1)$ decays exponentially. The final state is a coherent superposition of the two lower states (no damping included). A "which-way measurement" could now be performed to distinguish between these two.

**3.3.7.6 Network in $k$ Space: Semiconductor Bloch Equations.** Condensed matter consists of many, often strongly interacting subunits (such as molecules). This interaction gives rise to approximate eigenstates (modes) which can no longer be ascribed to any individual local subunit. In translationally invariant systems these modes are conveniently specified by a wave vector $k$ (compare Fig. 3.60). If the local ("atomic") states $|j\rangle$ at position $R_\nu$

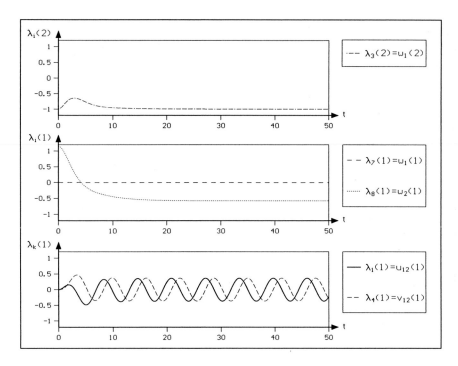

**Fig. 3.59.** Decay in an $SU(3) \otimes SU(2)$ system (scenario C). $\omega_{31}^{(1)} = 7$, $\omega_{21}^{(1)} = 1$, $\omega_{21}^{(2)} = 6.5$, $W = 1$, $C_F = 0.25$. Simulation by R. Wawer (cf. [149])

are denoted by $|j(R_\nu)\rangle$, the wave functions local in $k$ space read (cf. [103])

$$|j(k_s)\rangle \sim \sum_\nu e^{-ik_s R_\nu} |j(R_\nu)\rangle , \qquad (3.659)$$

$$|j(R_\nu)\rangle \sim \sum_s e^{-ik_s R_\nu} |j(k_s)\rangle . \qquad (3.660)$$

If discrete, the set of $k$ vector modes may now be considered to form a network in $k$ space rather than in real space. Formally, all the details discussed so far also apply to this (more abstract) situation. Let us start with an optically driven $SU(2)$ network without node (= mode) interactions. Then, in density matrix form, we have for a given *dispersion relation* $E_{21}(k)$:

$$\frac{d}{dt} R_{11}(k) = -\frac{i}{\hbar} d_{21}(k) \left[ R_{12}(k)\epsilon(t) - R_{21}(k)\epsilon^*(t) \right] , \qquad (3.661)$$

$$\frac{d}{dt} R_{22}(k) = \frac{i}{\hbar} d_{21}(k) \left[ R_{12}(k)\epsilon(t) - R_{21}(k)\epsilon^*(t) \right] , \qquad (3.662)$$

$$\frac{d}{dt} R_{12}(k) = \frac{1}{i\hbar} E_{21}(k) R_{12}(k) + \frac{1}{i\hbar} d_{21}(k)\epsilon(t) \left[ R_{11}(k) - R_{22}(k) \right] \qquad (3.663)$$

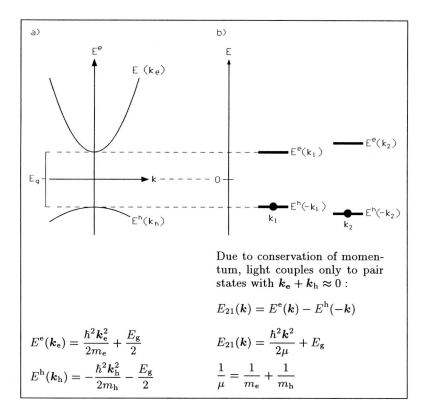

**Fig. 3.60.** Electron–hole picture, part a), and pair–picture, part b) (the spin degrees of freedom are neglected)

(cf. (3.530)). Assuming an electromagnetic field

$$\epsilon(t) = \epsilon_0 e^{-i\omega_L t} = u\epsilon_0 e^{-i\omega_L t} , \qquad (3.664)$$

introducing the abbreviation

$$G_k(t) = \frac{i}{\hbar} g(k) \left[ R_{12}(k) e^{-i\omega_L t} - R_{21}(k) e^{i\omega_L t} \right] \qquad (3.665)$$

with

$$g(k) = M(k)\epsilon_0 = d_{21}(k)\epsilon_0 , \qquad (3.666)$$

and transforming into the *electron–hole picture*,

$$
\boxed{
\begin{aligned}
R_{11}(k) &= 1 - f^h_{-k} , \\
R_{22}(k) &= f^e_k , \\
R_{12}(k) &= P_k , \\
R_{21}(k) &= P^*_k
\end{aligned}
}
\qquad (3.667)
$$

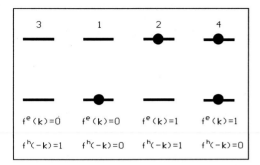

$f^e(k)=0$    $f^e(k)=0$    $f^e(k)=1$    $f^e(k)=1$

$f^h(-k)=1$    $f^h(-k)=0$    $f^h(-k)=1$    $f^h(-k)=0$

**Fig. 3.61.** "Neutral" (1, 2) and "charged" states (3, 4) in $k$ space

(cf. Fig. 3.60), one obtains the evolution equations

$$
\begin{aligned}
\frac{d}{dt} f^h_{-k} &= G_k(t) , \\
\frac{d}{dt} f^e_k &= G_k(t) , \\
\frac{d}{dt} P_k &= \frac{1}{i\hbar} E_{21}(k) P_k + \frac{1}{i\hbar} d_{21}(k) \epsilon_0 e^{-i\omega_L t} \left(1 - f^e_k - f^h_{-k}\right) .
\end{aligned}
$$

(3.668)

From

$$ R_{11} + R_{22} = 1 $$

(3.669)

it follows that

$$ f^e_k = f^h_{-k} . $$

(3.670)

For an incoherent light field, the polarization (i.e. the off-diagonal density matrix element) can be eliminated: Equation (3.505) reads for $n = 2$

$$ \frac{d}{dt} R_{22} = W^{(\text{eff})}_{21} (R_{11} - R_{22}) , $$

(3.671)

or, in the electron–hole picture,

$$ \frac{d}{dt} f^e_k = W^{(\text{eff})}_{21} \left(1 - f^e_k - f^h_{-k}\right) := G^{cl}_k(k) , $$

(3.672)

where $G^{cl}_k(k)$ is called *classical generation rate*. We thus end up with

$$
\begin{aligned}
\frac{d}{dt} f^h_{-k} &= G^{cl}_k(k) , \\
\frac{d}{dt} f^e_k &= G^{cl}_k(k) .
\end{aligned}
$$

(3.673)

Let us now turn to node–node interactions. As it turns out, the energy–transfer interaction (Förster mechanism), would be far too restrictive here. In fact, we have to enlarge the state space to accommodate the most dominant type of coupling, charge transfer. Consider, in addition to the "neutral" states

$$ f^e_k = f^h_{-k} = 0 $$

(3.674)

and

$$f^{\mathrm{e}}_{k} = f^{\mathrm{h}}_{-k} = 1 \; , \qquad\qquad\qquad\qquad (3.675)$$

the "charged" states (note that this is not in real space) as sketched in Fig. 3.61 (cf. Sect. 2.2.5.6). The charged states $\nu = 3, 4$ are optically inactive and can thus not be reached from the neutral states 1 and 2 by optical coupling. Such transitions can be modelled as incoherent single-particle scattering processes (then $f^{\mathrm{e}}_{k}$, $f^{\mathrm{h}}_{-k}$ become independent parameters):

$$I^{\mathrm{e,h}}_{k}(t) = \sum_{q} \left[ W^{\mathrm{e,h}}_{k-q,k} f^{\mathrm{e,h}}_{k} \left( 1 - f^{\mathrm{e,h}}_{k-q} \right) - W^{\mathrm{e,h}}_{k,k-q} f^{\mathrm{e,h}}_{k-q} \left( 1 - f^{\mathrm{e,h}}_{k} \right) \right] \; . (3.676)$$

$W^{\mathrm{e,h}}$ could, for example, be due to electron–(hole)–phonon scattering. The additional factors $1 - f^{\mathrm{e,h}}$ guarantee the Pauli principle for fermions: no scattering into an already occupied state. When combined with (3.673), one gets the *semiclassical Boltzmann equations* with source.

Adding the $I^{\mathrm{e,h}}$ terms to the original Bloch equations then leads to a variant of the *semiconductor Bloch equations*:

$$\boxed{\begin{aligned}
\frac{\mathrm{d}}{\mathrm{d}t} f^{\mathrm{h}}_{-k} &= G^{\mathrm{cl}}_{k}(k) - I^{\mathrm{h}}_{k}(t) \; , \\
\frac{\mathrm{d}}{\mathrm{d}t} f^{\mathrm{e}}_{k} &= G^{\mathrm{cl}}_{k}(k) - I^{\mathrm{e}}_{k}(t) \; , \\
\frac{\mathrm{d}}{\mathrm{d}t} P_{k} &= \frac{1}{\mathrm{i}\hbar} E_{21}(k) P_{k} + \frac{1}{\mathrm{i}\hbar} d_{21}(k) \epsilon_{0} \mathrm{e}^{-\mathrm{i}\omega_{\mathrm{L}}t} \left( 1 - f^{\mathrm{e}}_{k} - f^{\mathrm{h}}_{-k} \right) - \\
&\quad \gamma_{k} P_{k} \; .
\end{aligned}} \qquad (3.677)$$

($\gamma_{k}$ is a phenomenological damping parameter.) Though they indeed look like modified Bloch equations for 2-state systems labelled by $k$, they are actually defined in a 4-dimensional state space with coherence constrained to the states $\nu = 1, 2$ within single node $k$. (Coherence between state 3 and 1 (2) would include hole (electron) wave packets localized in real space.) Extended versions may include 2-node coherence, e.g. between electron–hole pairs and phonons. There are many additional complications, all being traced back to many-particle effects. Also the single-particle energies become renormalized.

**3.3.7.7 Incoherent Network with Local Damping.** Due to the interactions with its environment, an incoherent network is constrained to move in the "classical sector", where single-node as well as many-node coherence is completely suppressed (with respect to its measurement basis, cf. Sect. 2.6). In this basis the total density matrix is diagonal; or, in the $SU(n)$ language, any local node-coherence vector remains in its $w_{l}(\nu)$ subspace and all correlation tensors M are zero up to any order. This is a tremendous simplification. The dynamics is then described by the *Pauli master equation* (compare (3.403) and (3.407))

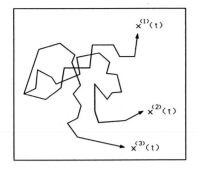

**Fig. 3.62.** Random walk in real space. The paths can be identified, for example, with paths of a Brownian particle in a liquid

$$\dot{R}_{ss}(t) = \sum_{m,m\neq s} W_{sm} R_{mm}(t) - R_{ss}(t) \sum_{m,m\neq s} W_{ms} \,. \tag{3.678}$$

The Hamilton model will enter here at most in an indirect way via the rates $W_{sm}$.

These equations may be interpreted to result from a stochastic process generating paths $s^{(\alpha)}$ ($\alpha = 1, 2, \ldots$) in (discrete) state space such that

$$\lim_{Z\to\infty} \frac{1}{Z} \sum_{\alpha=1}^{Z} \delta_{s(t),s^{(\alpha)}(t)} = R_{ss}(t) \,, \tag{3.679}$$

where the trajectories $s^{(\alpha)}$ all start from the same initial state. This is reminiscent of Brownian motion (see Fig. 3.62) for which the trajectory is in real space. Discrete state space trajectories are sketched in Fig. 3.63.

If a network of $N$ interacting nodes (with $n$ states each) is considered, the index $s$ can be decomposed into

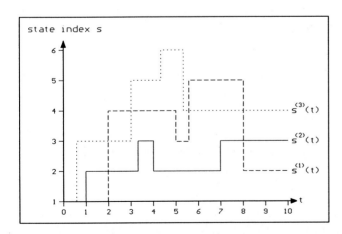

**Fig. 3.63.** Random walk in discrete state space. Energy states are indicated by $s$. Three paths are shown starting from the same initial state at $t = 0$

$$s \to \{l\} = \{l_1, l_2, l_3, \ldots, l_\nu, \ldots, l_N\} \ , \tag{3.680}$$

where each index $l_\nu$ can have the values

$$l_\nu = 1, 2, 3, \ldots, n \ . \tag{3.681}$$

If there is only local damping (cf. Sect. 3.3.7.1), the allowed elementary transitions connect only network states which differ in the state of just one node $\nu$:

$$\{l\} = \{l_1, l_2, l_3, \ldots, l_\nu, \ldots, l_N\} \to \{l'\} \tag{3.682}$$

with

$$\{l'\} = \{l_1, l_2, l_3, \ldots, l'_\nu, \ldots, l_N\} \ . \tag{3.683}$$

The corresponding transition rate will be denoted by $W_{l_\nu l'_\nu}^{(\nu)}$ (this notation simplifies bookkeeping: we only specify the changing states) so that the *network rate equation* then reads

$$\boxed{\dot{R}_{\{l\}\{l\}}(t) = \sum_{\nu=1}^{N} \sum_{\substack{l'_\nu = 1 \\ l'_\nu \neq l_\nu}}^{n} \left[ W_{l_\nu l'_\nu}^{(\nu)} R_{l'_\nu l'_\nu}^{(\nu)}(t) - W_{l'_\nu l_\nu}^{(\nu)} R_{l_\nu l_\nu}^{(\nu)}(t) \right] \ .} \tag{3.684}$$

$R_{\{l\}\{l\}}(t)$ is a diagonal density matrix element in full network space. All indices on the r.h.s. of this equation (other than those referring to node $\nu$) are the same as appearing on the l.h.s. of this equation.

## 3.4 Summary

In this chapter dynamic properties of networks with $N$ nodes ($N \leq 3$) have been studied. Ways to extend the approach to $N > 3$ have been shown.

Both *closed systems* (systems without coupling to their environment) and *open systems* (systems which are influenced by their environment ("bath")) have been discussed. Closed systems are described by *unitary dynamics* (and the unitary group), open systems by master equations like the *Markovian master equation* which corresponds to the so-called quantum dynamical semigroup.

Basic equations defining unitary motion (the time-dependent Schrödinger equation, Liouville equation, Heisenberg's equation of motion) have been studied, where – as the simplest example – a 2-level-1-node system driven by *optical forces* was introduced. The transition to an $SU(2)$ representation then led to *Bloch equations* without damping. The extension to 3-level systems is then straightforward, including various *optical driving scenarios*. A generalization to *2-node networks* in $SU(2) \otimes SU(2)$ has been considered, where the dynamics of *non-local correlations* (2-node correlations) has been discussed.

Allowing for local damping effects (due to a coupled bath) then led to the *Markovian master equation*. Passing over to an $SU(2)$ representation, the *damped Bloch equations* have been obtained, while the transition to an $SU(n)$ representation led to *generalized Bloch equations*. Such a representation describes the evolution of expectation values of the respective generating operators represented by the components of the *coherence vector*.

In the last section, *network equations* in $SU(2) \otimes SU(2)$ and $SU(3) \otimes SU(2)$ including damping have been introduced. In the latter scenario the damped $SU(2)$ subsystem coupled to a coherently excited 3-level system was used to model various damping scenarios: if the damping channel cannot differentiate between decay paths, *quantum beats* are shown to result.

# 4. Quantum Stochastics

## 4.1 Introduction

The theory of *stochastic processes* represents an extension of conventional probability theory to time-dependent variables (cf. [146]). Examples abound in classical physics. Due to lack of control, such processes are described as a stochastic rather than a *deterministic process*. (For a deterministic evolution the dependence on the time parameter would be specified by a function.) The origin of limited control may be different from example to example; in the case of a *Brownian particle* it is the *thermal motion* of the surrounding fluid which is uncontrolled.

In the quantum regime the origin of *indeterminism* is believed to be of fundamental nature. However, to date there is no general agreement on the extent to which non-classical indeterminism might also be traced back to stochastic processes in the conventional sense. In too broad a sense this is certainly not true: *Heisenberg's uncertainty relation* deals with indeterminism, but one should refrain from naively assigning a stochastic process to the position variable of a quantum object prepared in a momentum eigenstate. The same applies, for example, to the so-called *zero-point fluctuations* of a harmonic oscillator prepared in its ground state. Historical names can be misleading here: in quantum mechanics the term "fluctuations" arises since observables are typically expressed as ensemble averages, and the variances of such quantities in general do not vanish.

*Quantum noise* discussed, for example, in the context of charge transport is essentially due to charge quantization and – possibly – correlations between the carriers; in any way, this noise is related to a concrete measurement scenario, i.e. stochasticity occurs (and becomes observable) only after the quantum system has been embedded into an appropriate ("classical") environment. A similar situation occurs in optics, where a stochastic process clearly implies *dissipation*, e.g., due to *spontaneous emission*. *Quantum stochastics* is thus intimately related to open systems, as discussed in the last chapter. Though fundamental (and as such unavoidable), this behaviour is not inherent to the quantum system in the strict sense: it is too obvious that stochastic rules also depend on the coupling to, and the type of, surrounding.

Before we start with a brief introduction into *stochastic modelling of quantum processes*, we review some other lines of thought and their respective motivation.

### 4.1.1 Quantum Noise and Langevin Equations

Classical models of noise are often based on a Langevin-type equation, i.e. an equation of motion for an observable $A$ to which random forces $f(t)$ are added, typically requiring the correlation property

$$\langle f(t)f(t')\rangle = F\delta(t-t') .\tag{4.1}$$

Statistically equivalent to this model is the formalism of the *Fokker-Planck equation* for the probability density $P(A,t)$ (e.g. see [150]).

Quantum dynamics of open systems can be specified by the master equation

$$\dot{R}_{nm} = \sum_{i,j}\left(\frac{i}{\hbar}D_{nmij} + R_{nmij}\right)R_{ij} ,\tag{4.2}$$

where the matrix $\tilde{D}$ (containing all matrix elements $D_{nmij}$) specifies the coherent, $\tilde{R}$ (containing all matrix elements $R_{nmij}$) the incoherent part of the evolution. In general, this equation is complicated and lacks intuitive feeling and appropriate ways of approximations. This has been one dominant motivation for the development of the *quasi-probability representation* (P representation) in optics (cf. [53, 126]). Under restrictive conditions, the resulting equation of motion for this P function is, again, a Fokker-Planck equation and thus defines a classical stochastic process. The use of the quasi-probability concept has the advantage that it uses a language familiar from classical statistical physics. However, this concept breaks down when the quantum fluctuations show their typical non-classical behaviour: due to negative regions in the distribution function, variances become negative, and a corresponding *Langevin equation* does not exist. The so-called *positive P representation* defined in a further enlarged space avoids these difficulties, though this function is no longer unique. This function is able to account for non-classical fluctuations, but at the price of losing intuitive appeal. The method has been very successful, though, in laser physics.

### 4.1.2 Self-reduction

The intuitive expectation that "large" systems should inherently become classical is formalized by the concept of *self-reduction*. Here one searches for a general law of propagation of the wave function, which would prevent the development of far-away superpositions in the case of macrosystems (cf. [51]). The cure proposed for the Schrödinger equation consists of additional but *intrinsic* stochastic terms implying *self-reductions* of the wave function: the total dynamics would result in a kind of "breathing behaviour", the Hamiltonian spreading would be stopped by stochastic reductions. In this way (i.e. by choosing appropriate *ad hoc parameters*) macroscopic bodies would possess classical properties even if they were isolated. As one goes towards smaller masses, the self-reductions become less frequent, so that in the microscopic domain proper, the (additional) intrinsic stochasticity would become unobservable (to avoid conflict with experiments).

The so-called *GRW model* (Ghirardi, Rimini, Weber, cf. [56]) does not give any hypothesis concerning the origin of the breakdown of the superposition principle. The *K model* (Károlyházy) assumes that *fluctuations of the metric* might cause loss of phases. Both models differ in quantitative aspects. The *quantum state diffusion model* (Gisin, Percival, cf. [57]) is another elaboration based on similar ideas. Though of principal theoretical interest, these schemes face the severe problem of giving experimental evidence for their assertions, i.e. locating weak additional decoherence within the abundance of environment-induced reductions.

### 4.1.3 Stochastics as a Source of Information

Stochastic models have become increasingly popular after experiments on *quantum jumps* had been brought to the attention of the physics community (cf. [15, 102]). *Stochastic modelling* tries to find (microscopic) *stochastic processes* consistent with the *ensemble limit* described by the density matrix. However, this requirement does not define a unique solution and a *model process* which leads to the correct ensemble limit does not necessarily represent the "real" microscopic process in all quantum mechanical details. Thus, the motivation may be, at least, two-fold. On the one hand, one can use such a stochastic approach as a kind of *quantum Monte Carlo method* (cf. [31, 100]) for the calculation of ensemble properties including *multi-time correlations*. This method has advantages over other techniques. On the other hand, stochastic modelling can give an intuitive approach to quantum fluctuations, down to the "trajectories" of single quantum objects (cf. [67]). These quantum trajectories can be considered "real" in so far as they most efficiently reproduce and predict typical experimental results (measurement protocols). Stochastics thus becomes intimately related with information dynamics. This is contrary to ensemble behaviour for which no additional information is needed. Stochastic simulation should thus be able to link information dynamics with the system dynamics (which is not directly observable). This requirement is more restrictive than it might appear at first sight: the rules of the game are based on quantum measurement.

Quantum measurement theory has long since suffered from ambiguities and thus often been considered to be of academic interest only. However, this theory deals with *information retrieval,* and the language of information has often been used very successfully as a pragmatic tool to understand quantum experiments. For example, the *which-way information* (cf. [129]) is said to destroy the interference in Young's double-slit experiment, a set of excited atoms decays differently if the light-field cannot "distinguish" from which atom the photon is emitted, beat phenomena occur if the damping channel cannot distinguish between two alternative decay paths, etc. It should thus not be too surprising if a kind of *information dynamics* might supply us with well-defined stochastic processes even for definitely non-classical behaviour. In this context it is essential to note that the notion of information is contextual, i.e. we have to define which kind of alternative events are at disposal. Information, however, does not require actual observation by any human being.

In this book, stochastic modelling is based on the concept of *continuous measurement* (cf. [31, 67, 77, 84, 100, 111]) which supplies a process interpretation of the *damping channels*. A comparison with the state diffusion model has been made in [54]. The spontaneously occurring *measurement projections* do not stop the motion due to other types of forces in the equation of motion, but they modify it, even if "nothing happens". (Note that "nothing happens" is also an information historically called zero measurement). A continuous and complete measurement – as an idealization – will render the *individual network* to remain in a state of (almost) zero entropy, despite dissipation. Just like a classical Brownian particle, the network exercises a (discrete) *random walk* in its respective state space. This state space, however, will, in general, be very non-classical and also the type of events will often be at odds with classical intuition.

The discussion of continuous measurement scenarios will be the starting point of this chapter. Stochastics, supposed to underlie the ensemble behaviour of open systems, must explain the emerging properties of the damping channels: what do the rules look like which are able to reproduce phenomena like memory effects, coherence by dissipation, and non-locality?

## 4.2 Continuous Measurement

Contents: measurement scenarios, subensemble approach, ensemble dynamics and stochastic trajectories, random hopping, implementation of stochastic behaviour.

### 4.2.1 Basics

One may distinguish two types of stochastics: those contributing uncontrolled *fluctuations* (and contributing to the entropy of the system), and those related to (possible) information retrieval by the environment. We will mainly focus here on the latter.

In the previous chapters of this book we have studied measurements based on the projection postulate and the dynamics of open systems modified by damping channels. While the former provides information (decisions) without saying "how and when", the latter considers the averaged back-action of the environment on the quantum system in question without exploiting this interaction for information retrieval. Quantum stochastics tries to bridge this gap. It gives dynamical rules for the measurement projections such that the ensemble average over the resulting "quantum trajectories" leads back to the solution of the respective master equation.

Continuous measurement scenarios and their underlying stochastic trajectories will be considered in this section. However, we will restrict ourselves to well-controlled scenarios which provide the basis for a clear (binary) decision. Detections in the optical regime are here particularly advantageous as the zero-temperature limit can easily be reached to a very good approximation. (As in everyday life, there are often situations in which a specific event does not logically imply a unique interpretation: a wet street does not necessarily mean that it has rained, detecting a photon of specific frequency will signal the relaxation of an atomic probe only if considerable care is taken to shield away other light sources.)

The time-evolution of the state specified, for example by a coherence vector $\boldsymbol{\lambda}(t)$, can be characterized as a continuous motion. However, a more detailed investigation shows that the dynamics can be decomposed into a set of stochastic trajectories illustrating underlying microscopic processes. The ensemble behaviour then is obtained by averaging over the stochastic ("discontinuous") trajectories with the same initial state:

$$\boldsymbol{\lambda}(t) = \lim_{Z \to \infty} \frac{1}{Z} \sum_{\alpha=1}^{Z} \boldsymbol{\lambda}^{(\alpha)}(t) . \tag{4.3}$$

Here, $\boldsymbol{\lambda}^{(\alpha)}(t)$ represents the stochastic quantum trajectory $\alpha$.

### 4.2.2 A Simple Measurement Scenario

Consider a 2-level node in contact with a photon bath at temperature $T \approx 0$. Restricting ourselves to transitions from the upper state into the ground state, see Fig. 4.1 (l.h.s.), the ensemble dynamics of this node is described by the master equations

$$\dot{R}_{22}(t) = -W_{12} R_{22}(t) ,$$

$$\dot{R}_{11}(t) = W_{12} R_{22}(t) ,$$

$$\dot{R}_{21}(t) = -\frac{1}{2} W_{12} R_{21}(t) + i\omega_{21} R_{21}(t) , \tag{4.4}$$

$$\dot{R}_{12}(t) = -\frac{1}{2} W_{12} R_{12}(t) - i\omega_{21} R_{12}(t) .$$

Here we have applied (3.509) with $D_{12}(t) = i\omega_{21} R_{12}(t)$, $D_{21}(t) = -i\omega_{21} R_{21}(t)$, $D_{11}(t) = D_{22}(t) = 0$, and the damping parameters $W_{21}$, $\mathrm{Re}\{\tilde{\gamma}_{ij}\}$, $\Delta\omega_{ij}$ identically zero. The corresponding Bloch equations are

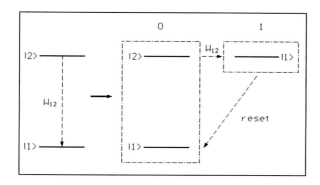

**Fig. 4.1.** The original transition process (l.h.s) and decomposition into two subensembles (r.h.s.) indicated by the numbers 0 and 1

$$\dot{\lambda}_1 = -\frac{1}{2}W_{12}\lambda_1 - \omega_{21}\lambda_2 ,$$

$$\dot{\lambda}_2 = -\frac{1}{2}W_{12}\lambda_2 + \omega_{21}\lambda_1 , \qquad (4.5)$$

$$\dot{\lambda}_3 = -W_{12}\lambda_3 - W_{12}$$

(compare (3.525) for $\tilde{\delta} = \omega_{21}$).

**4.2.2.1 Subensemble Approach.** The relaxation process $|2\rangle \rightarrow |1\rangle$ as sketched in Fig. 4.1 (l.h.s.) is accompanied by the change of the environment from state 0 to state 1 (e.g. 0 photon $\rightarrow$ 1 photon). We thus distinguish a 0-photon subensemble (indicated by the index (0)) and a 1-photon subensemble (indicated by the index (1)). The relaxation process is then decomposed into the following 2 steps (cf. the r.h.s. of Fig. 4.1):

1. The (higher) initial node state $|2\rangle$ in subensemble 0 turns into the lower node state (with a finite probability determined by the transition rate $W_{12}$) in subensemble 1.
2. The photon is registered ("reset") so that the final state $|1\rangle$ in subensemble 0 results.

Due to this decomposition, two occupation probabilities $R_{11}^{(0)}$ and $R_{11}^{(1)}$ have to be taken into account. The upper index 1 then denotes a node state with a photon, and 0 denotes a node state without a photon. This *subensemble approach* is thus described by

$$R_{11}(t) = R_{11}^{(0)}(t) + R_{11}^{(1)}(t) ,$$

$$R_{22}(t) = R_{22}^{(0)}(t) ,$$

$$R_{21}(t) = R_{21}^{(0)}(t) , \qquad (4.6)$$

$$R_{12}(t) = R_{12}^{(0)}(t) .$$

Inserting this decomposition into (4.4), one obtains the corresponding systems of ensemble equations:

$$\dot{R}_{22}^{(0)}(t) = -W_{12}R_{22}^{(0)}(t) \, ,$$

$$\dot{R}_{11}^{(0)}(t) = 0 \, ,$$

$$\dot{R}_{21}^{(0)}(t) = -\frac{1}{2}W_{12}R_{21}^{(0)}(t) + \mathrm{i}\omega_{21}R_{21}^{(0)}(t) \, ,$$

$$\dot{R}_{12}^{(0)}(t) = -\frac{1}{2}W_{12}R_{12}^{(0)}(t) - \mathrm{i}\omega_{21}R_{12}^{(0)}(t)$$

(4.7)

and

$$\dot{R}_{11}^{(1)}(t) = W_{12}R_{22}^{(0)}(t) \, .$$

(4.8)

While the $SU(2)$ representation of (4.4) is given by (4.5), the $SU(2)$ representation of (4.7) has yet to be derived: for (3.514), the components $\lambda_1 = u_{12}$, $\lambda_2 = v_{12}$, $\lambda_3 = w_1$ of a coherence vector in $SU(2)$ are related to the density matrix R by

$$\lambda_1 = R_{12}(t) + R_{21}(t) \, ,$$

$$\lambda_2 = \mathrm{i}\left[R_{12}(t) - R_{21}(t)\right] \, ,$$

$$\lambda_3 = -R_{11}(t) + R_{22}(t) \, ,$$

(4.9)

supplemented by the trace

$$\lambda_0 = R_{11} + R_{22} \, .$$

(4.10)

(Note that the trace is not conserved in the subensembles.) We thus get

$$\dot{\lambda}_0^{(0)} = -\frac{1}{2}W_{12}\left(\lambda_0^{(0)} + \lambda_3^{(0)}\right) \, ,$$

$$\dot{\lambda}_1^{(0)} = -\frac{1}{2}W_{12}\lambda_1^{(0)} - \omega_{21}\lambda_2^{(0)} \, ,$$

$$\dot{\lambda}_2^{(0)} = -\frac{1}{2}W_{12}\lambda_2^{(0)} + \omega_{21}\lambda_1^{(0)} \, ,$$

$$\dot{\lambda}_3^{(0)} = -\frac{1}{2}W_{12}\left(\lambda_0^{(0)} + \lambda_3^{(0)}\right)$$

(4.11)

and

$$\dot{\lambda}_0^{(1)} = \frac{1}{2}W_{12}\left(\lambda_0^{(0)} + \lambda_3^{(0)}\right) \, ,$$

$$\dot{\lambda}_1^{(1)} = 0 \, ,$$

$$\dot{\lambda}_2^{(1)} = 0 \, ,$$

$$\dot{\lambda}_3^{(1)} = -\frac{1}{2}W_{12}\left(\lambda_0^{(0)} + \lambda_3^{(0)}\right) \, .$$

(4.12)

Both systems of equations (4.7)–(4.8) and (4.11)–(4.12) describe the dynamics of the ensemble without the subsequent reset (= photon detection). A

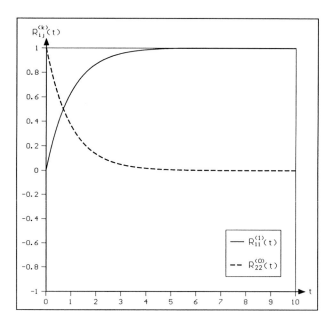

**Fig. 4.2.** Subensemble dynamics. The time-evolution of the reduced density matrix. Transition rate: $W_{12} = 1$. Initial state: $R_{22}^{(0)} = 1$, $R_{11}^{(0)} = R_{21}^{(0)} = R_{12}^{(0)} = 0$, $R_{11}^{(1)} = 0$. The matrix elements $R_{11}^{(0)}$, $R_{21}^{(0)}$, $R_{12}^{(0)}$ remain zero here (incoherent process)

numerical example described by the ensemble equations (4.7)–(4.8) is illustrated in Fig. 4.2. Fig. 4.3 illustrates the corresponding motion of the coherence vector described by (4.11)–(4.12).

**4.2.2.2 Single Node Processes.** In contrast to the continuous ensemble dynamics, the dynamics of a single node can be characterized as *a discontinuous hopping* in state space: consider the initial state (time $t = t_1$)

$$\hat{R}^{(0)}(t_1) = |2\rangle\langle 2| \, , \tag{4.13}$$

i.e.

$$R_{22}^{(0)}(t_1) = 1 \, . \tag{4.14}$$

(All other matrix elements at time $t_1$ are identically zero. Due to (4.7), the off-diagonal matrix elements remain zero.)

At time $t_2 = t_1 + \Delta t$, a transition of the single node into the state $\hat{R}^{(0)}(t_2)/p^{(0)}$ or $\hat{R}^{(1)}(t_2) = |1\rangle\langle 1|$ (controlled by transition probabilities $p^{(0)}$, $p^{(1)}$) occurs:

$$\hat{R}^{(0)}(t_2) \overset{p^{(0)}}{\underset{p^{(1)}}{\lessgtr}} \begin{array}{l} \hat{R}^{(0)}(t_2)/p^{(0)} \\ \hat{R}^{(1)}(t_2) = |1\rangle\langle 1| \end{array} \, . \tag{4.15}$$

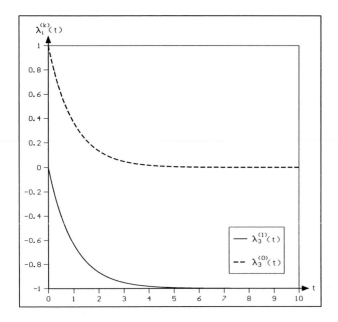

**Fig. 4.3.** Subensemble dynamics. The time-evolution of the coherence vector. Transition rate: $W_{12} = 1$. Initial state: $\lambda_0^{(0)} = 1$, $\lambda_3^{(0)} = 1$, $\lambda_1^{(0)} = \lambda_2^{(0)} = 0$, $\lambda_0^{(1)} = 0$, $\lambda_3^{(1)} = 0$, $\lambda_1^{(1)} = \lambda_2^{(1)} = 0$. $\lambda_1^{(0)}$, $\lambda_2^{(0)}$, $\lambda_1^{(1)}$, $\lambda_2^{(1)}$ remain zero here

The transition probabilities $p^{(0)}$, $p^{(1)}$ are given by

$$p^{(0)} = \text{Tr}\left\{\hat{R}^{(0)}(t_2)\right\} , \tag{4.16}$$

$$p^{(1)} = 1 - \text{Tr}\left\{\hat{R}^{(0)}(t_2)\right\} . \tag{4.17}$$

In Fig. 4.4 the resulting one step *hopping process* is sketched. In this example, at time $t_5$ the node switches to the ground state in subensemble 1, and is then reset to subensemble 0.

**4.2.2.3 The Measurement Protocol.** With the final "reset-click" a measurable signal is associated (photon detection) so that this *decay event* can be registered by measurement (see Fig. 4.4). In contrast to a continuously decaying signal produced by the whole ensemble, a decay process of a single node is represented by a single *counter signal*. This reset process is supposed to happen almost immediately (time scale ≪ any other inherent time scale) so that the system will never generate a substantial average population within subensemble 1, let alone subensembles with larger photon numbers. The latter condition becomes important if the 2-level node is driven and will thus produce a sequence of photons. In $SU(2)$ this process may be summarized by

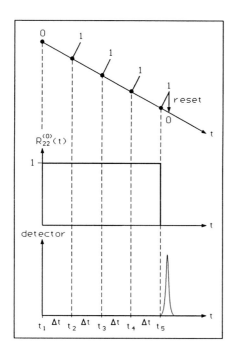

**Fig. 4.4.** Single node process: decision tree (subensemble 0 and 1), trajectory $R_{22}^{(0)}(t)$, and measurement protocol are shown

$$\lambda_0^{(0)}, \lambda^{(0)} \quad \begin{array}{l} \lambda_0^{(0)'} = 1, \lambda^{(0)'} = \lambda^{(0)}/\lambda_0^{(0)} \; : \quad p^{(0)} = \lambda_0^{(0)} \\[2mm] \lambda_0^{(1)'} = 1, \lambda^{(1)'} = (0,0,-1) \; : \quad p^{(1)} = 1 - p^{(0)} \end{array}$$

$$\downarrow$$

$$\lambda_0^{(0)'} = 1, \lambda^{(0)'} = (0,0,-1) \tag{4.18}$$

Only the reset is equivalent to a measurement projection (cf. (2.433)).

One should note that this measurement scenario is based on prior-knowledge and a well-defined logic of inference: we know there is one atom in its excited state, no other photon source, and the detector has an efficiency of 100 %. Then, as long as no click has occurred, the atom is in state $|2\rangle$, after the click it is in state $|1\rangle$. We assume, to be sure, that this type of behaviour results from the type of embedding, irrespective of whether or not we actually

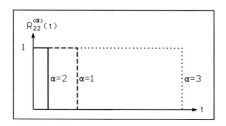

**Fig. 4.5.** Set of stochastic trajectories (as we restrict ourselves to the subensemble 0 we have suppressed the index 0)

have access to the information: a detector efficiency of less than 100 % would limit our control, though.

We may repeat such a simulation starting from the same initial condition and generate a set of trajectories $R_{ii}^{(\alpha)}$ (compare Fig. 4.5). The ensemble limit then is recovered from

$$R_{ii}(t) = \lim_{Z \to \infty} \frac{1}{Z} \sum_{\alpha=1}^{Z} \delta_{R_{ii}(t),R_{ii}^{(\alpha)}(t)} \cdot \tag{4.19}$$

As long as the dynamics is completely incoherent, the introduction of the 1-photon subensemble is merely a detour, which does not change the overall stochastics: in fact, as the 1-photon subensemble is immediately emptied by reset, this state can adiabatically be eliminated, remembering that the decay is associated with a "detector-click".

### 4.2.3 Decomposition of the Lindblad Operator: Stochastic Algorithm

One easily convinces oneself that (4.7) is an example of the truncated master equation of Sect. 3.3.3,

$$\boxed{\dot{\hat{\rho}}(t) = -\frac{i}{\hbar} \left[ \hat{H}, \hat{\rho}(t) \right]_- + \mathcal{L}_{\text{inc}}^{(2)} \hat{\rho}(t) ,} \tag{4.20}$$

for

$$\hat{\mathcal{L}}_{\text{inc}}^{(2)} \hat{\rho} = -\frac{1}{2} W_{12} \left( \hat{\rho} \hat{P}_{22} + \hat{P}_{22} \hat{\rho} \right) , \tag{4.21}$$

$$\hat{P}_{22} = \hat{\sigma}_{12}^+ \hat{\sigma}_{12}^- , \tag{4.22}$$

$$\hat{\sigma}_{12}^\pm = \hat{u}_{12} \pm i \hat{v}_{12} . \tag{4.23}$$

As we have seen, this equation does not preserve $\text{Tr}\,\{\hat{\rho}(t)\}$: consider the above evolution equation in the form

$$\hat{\rho}(t + dt) = \hat{\rho}(t) - \frac{i}{\hbar} \left[ \hat{H}, \hat{\rho}(t) \right]_- dt + \mathcal{L}_{\text{inc}}^{(2)} \hat{\rho}(t) dt . \tag{4.24}$$

Given $\hat{\rho}(t = 0)$ with $\text{Tr}\,\{\hat{\rho}(t = 0)\} = 1$, one defines

$$\text{Tr}\,\{\hat{\rho}(t + dt)\} = 1 - dp , \tag{4.25}$$

where

$$0 < dp \leq 1 . \tag{4.26}$$

For a continuous measurement on a system with a single dissipative channel, two alternative events may occur at time $t + dt$:

1. No measurement projection occurs (i.e. in the previous example, no photon is emitted and detected).

2. A measurement projection occurs (in the previous example, a photon is detected).

These two possibilities are described by

$$\hat{\rho}'(t + dt) = \frac{\hat{\rho}(t + dt)}{\text{Tr}\left\{\hat{\rho}(t + dt)\right\}} \quad \text{if } r \geq dp, \tag{4.27}$$

$$\hat{\rho}'(t + dt) = \frac{\mathfrak{L}^{(1)}\hat{\rho}(t + dt)}{\text{Tr}\left\{\mathfrak{L}^{(1)}\hat{\rho}(t + dt)\right\}} \quad \text{if } r \leq dp, \tag{4.28}$$

where (for a single decay channel $|n\rangle \rightarrow |m\rangle$)

$$\mathfrak{L}^{(1)}\hat{\rho} = W_{mn}\hat{P}_{mn}\hat{\rho}\hat{P}_{nm} = W_{mn}\rho_{nn}\hat{P}_{mm}, \tag{4.29}$$

$$\text{Tr}\left\{\mathfrak{L}^{(1)}\hat{\rho}\right\} = W_{mn}\rho_{nn}. \tag{4.30}$$

$r$ is a random number (with $0 \leq r \leq 1$) generated to decide between the two possible actions. After updating, the dynamics continues starting with $\hat{\rho}'(t + dt)$ and the updated trace $\text{Tr}\left\{\hat{\rho}'(t + dt)\right\} = 1$. This rule can be generalized to include several decay channels: the projection alternative (4.28) then has to be qualified further: the decay variants are weighted by their respective prefactors $W_{mn}\rho_{nn}$.

This algorithm allows the numerical simulation of stochastic trajectories associated with measurement projections. The procedure is based on the Markovian master equation and thus applies also to simplified variants like the Pauli master equation. With respect to networks, the present scheme represents a subsystem measurement. The effect of a projection in subsystem 2 on subsystem 1, say, can then be calculated along the lines of Sects. 2.4.5, 2.5.4. The resulting measurement protocol makes the observed system remain in a zero-entropy state (i.e. in a pure state in total system space). However, as measurements are typically due to local damping channels, any total wave

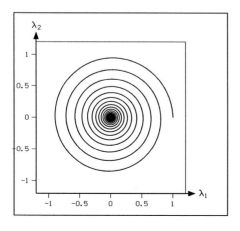

**Fig. 4.6.** Ensemble description: the coherence vector reaches a simple attractor state, here shown in the $\lambda_1, \lambda_2$ plane. $W_{12} = 3$, $\omega_{21} = 30$

vector description would completely lack intuitive appeal. Local descriptions, on the other hand, have to make use of density matrix descriptions.

It is clear that $\hat{\mathcal{L}}_{\text{inc}}\rho$ could include transitions $|n\rangle \to |m\rangle$ and $|m\rangle \to |n\rangle$. Only for a zero-temperature bath can we restrict ourselves to a one-way relaxation: only this zero-temperature bath defines a "zero-photon" reference state basic to our introduction of the subensembles 0 and 1. Information retrieval from photon counting becomes very limited if this reference state has no well-defined photon number to begin with.

### 4.2.4 Decay of a Coherent Superposition

In Sect. 4.2.2 an eigenstate $|2\rangle$ was taken as an initial state of the decay process. A more interesting measurement scenario is provided by a coherent superposition as an initial state: consider

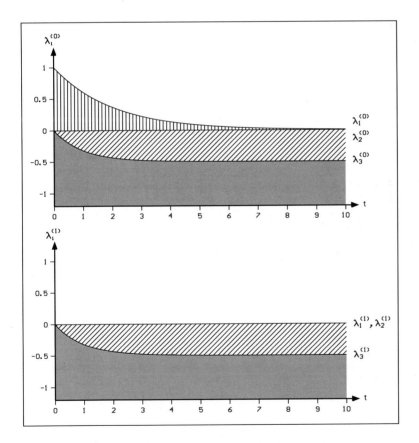

**Fig. 4.7.** Decay of the coherence vectors in the subensembles 0 and 1. $W_{12} = 1$. Initial state $(t = 0)$: $\lambda_0^{(0)} = 1$, $\lambda_1^{(0)} = 1$, $\lambda_2^{(0)} = \lambda_3^{(0)} = \lambda_0^{(1)}\lambda_1^{(1)} = \lambda_2^{(1)} = \lambda_3^{(1)} = 0$, $\omega_{21} = 0$ (rotating frame)

$$|\psi(t_1)\rangle = \frac{1}{\sqrt{2}} \left(|1\rangle + |2\rangle\right) , \qquad (4.31)$$

i.e.

$$R(t_1) = \begin{pmatrix} 1/2 & 1/2 \\ 1/2 & 1/2 \end{pmatrix} = R^{(0)}(t_1) \qquad (4.32)$$

or

$$\boldsymbol{\lambda}(t_1) = (1,0,0) = \boldsymbol{\lambda}^{(0)}(t_1) . \qquad (4.33)$$

According to (4.5), the system is for $t \rightarrow \infty$ in the ground state $|1\rangle$, i.e. $\boldsymbol{\lambda}(t) = (0,0,-1)$.

The resulting motion of the coherence vector $\boldsymbol{\lambda}(t)$ in the $\lambda_1, \lambda_2$ plane (described by (4.5)) is depicted in Fig. 4.6. In Fig. 4.7 the respective subensem-

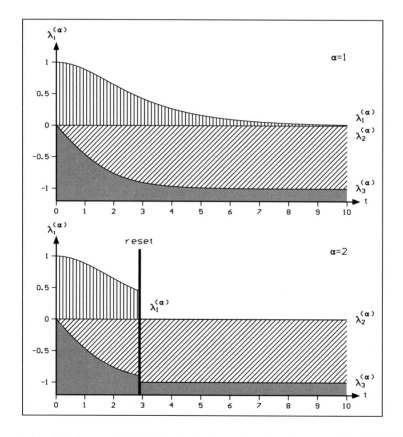

**Fig. 4.8.** Stochastic trajectories in $SU(2)$. $W_{12} = 1$, $\omega_{21} = 0$. The initial state is a coherent state: $\lambda_1^{(\alpha)} = 1$, $\lambda_2^{(\alpha)} = \lambda_3^{(\alpha)} = 0$. Top: a stochastic trajectory ($\alpha = 1$) which represents the motion of the coherence vector without photon emission. Below: a stochastic trajectory ($\alpha = 2$) with emission of a photon. Simulation by M. Keller (cf. [76])

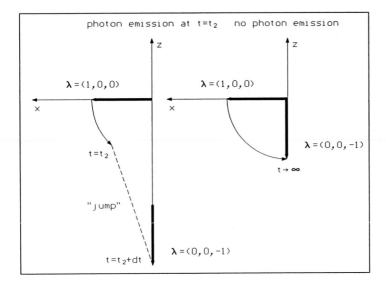

**Fig. 4.9.** The possible transition subprocesses depicted in the corresponding Bloch sphere. $\omega_{21} = 0$. $W_{12} = 1$

ble dynamics is shown for $\omega_{21} = 0$ (defining a corresponding rotating frame). These graphs represent a solution of the equations (4.11)–(4.12). The asymptotic state of the (total ensemble) coherence vector, i.e.

$$\boldsymbol{\lambda}(t_1) = (1,0,0) \rightarrow \boldsymbol{\lambda}(t)|_{t\to\infty} = (0,0,-1) , \tag{4.34}$$

is now split according to

$$\boldsymbol{\lambda}(t_1) = (1,0,0) \begin{array}{l} \boldsymbol{\lambda}^{(0)}(t)|_{t\to\infty} = (0,0,-1/2) \quad (1) \\ \boldsymbol{\lambda}^{(1)}(t)|_{t\to\infty} = (0,0,-1/2) \quad (2) \end{array} . \tag{4.35}$$

Subprocess (2) represents a transition into subensemble 1, i.e. a transition into the node state $|1\rangle$ + photon (with subsequent detection of the photon). Subprocess (1) represents a transition without photon. On average, either subprocess will happen in 50 % of the cases (cf. Fig. 4.8). If we measure a photon, we know the system is in state $|1\rangle$; but if we do not measure a photon, we cannot conclude for sure that the system is still in its initial state. In fact, if we have not seen a photon for a long time, it is likely that the subprocess (1) is realized.

Due to formula (4.3), the time-dependent coherence vector $\boldsymbol{\lambda}(t)$ can be decomposed into an infinite set of stochastic trajectories $\boldsymbol{\lambda}^{(\alpha)}(t)$. These trajectories represent realizations of the two possible subprocesses. Based on (4.5) and the algorithm introduced in Sect. 4.2.3, such stochastic trajectories can be calculated. In Fig. 4.8 the result is shown. In Fig. 4.9, the corresponding motion of the coherence vector in the Bloch sphere is indicated.

## 4.3 Incoherent Networks

Contents: rate equations, incoherent optical driving, random telegraph signals (RTSs) (normal RTSs, anomalous RTSs, minimal model for complex RTSs, large amplitude RTSs, cooperative jumps), ferro and antiferro coupling, macro-variables, 2-level defects, Ising model, $SU(2)$ and $SU(3)$ chains, state space compression, adiabatic elimination, classical versus quantum correlation.

### 4.3.1 Basics

In this section, we apply the stochastic approach to incoherent quantum networks. In this case the quantum trajectories are confined to the measurement basis (in the $SU(n)$ description: to the $w_l$ subspace). Absence of coherence makes them look classical in the sense of Sect. 2.6.3 even though the alternate states may result from a quantum model. The corresponding models consist of localized few-level nodes that interact among one another. Coupling to the environment (which represents the uncontrolled environment as well as the, at least partially, controlled measurement scenario) leads to a large repertoire of correlated and uncorrelated quantum jumps giving rise to *random telegraph signals* (RTSs).

The ensemble dynamics is described by a master equation or a network equation in $SU(n) \otimes SU(n) \ldots \otimes SU(n)$. In particular, we consider a network of interacting *2-level traps* in a heat bath and a charge transfer *quantum-dot array* in an incoherent electromagnetic field as prototype for *synthetic quantum networks*. The dynamical repertoire of the second model is larger, as transition rates can easily be adjusted by the spectral density of the light field. Under certain restrictions it can be mapped onto the first model.

The stochastic behaviour of such quantum systems can be observed only in rather refined experiments. One reason is that usually ensembles of small networks or individual nodes are observed, in which case the quantum noise is "washed out". Pertinent examples are an atom beam or *elementary excitations* within a single crystal. On a mathematical level, such systems are described by sums of identical Hamiltonians without interaction. A second reason is that incoherent transitions, lacking a clear environmental reference state, do not (by themselves) define measurement decisions. The fluctuating state may, however, modulate other physical parameters.

Stochastic dynamics (*quantum jumping*) is typically observed in relatively small systems with only few ensemble members, or in larger systems with inter-node interactions so that cooperative dynamics prevents the system from approaching an ensemble of non-interacting systems. Examples for the first case are single ions in a *Paul trap* (cf. [15, 102, 124]) or *single dopant molecules* in a host crystal (in the wings of the *inhomogeneously broadened line*, cf. [3, 99]). The second case can be realized by a small *tunnel junction* for which large *amplitude current fluctuations* are triggered by correlated *defect stochastics* (cf. [5, 46, 118, 128]) indicating that the respective nodes "jump" between discrete quantum levels (cf. [31, 67, 138]).

**4.3.1.1 Rate Equations.** The network rate equation has been shown to be (cf. Sect. 3.3.7.7)

$$\dot{R}_{\{l\}\{l\}}(t) = \sum_{\nu=1}^{N} \sum_{\substack{l'_{\nu}=1 \\ l'_{\nu} \neq l_{\nu}}}^{n} \left[ W^{(\nu)}_{l_{\nu}l'_{\nu}} R^{(\nu)}_{l'_{\nu}l'_{\nu}}(t) - W^{(\nu)}_{l'_{\nu}l_{\nu}} R^{(\nu)}_{l_{\nu}l_{\nu}}(t) \right] . \qquad (4.36)$$

For incoherent networks the Hamiltonian does not play the dominant role we are used to. As a matter of fact, parameters of the Hamiltonian only enter the dynamics of the network in so far as they influence the transition rates. This can happen with respect to external driving forces and via internal interactions. Both leave their mark on the resulting processes.

**4.3.1.2 Hamiltonian.** The Hamiltonian of an interacting quantum network consisting of $N$ nodes $\nu$ with $n$ energy levels is given by

$$\boxed{\hat{H} = \sum_{\nu} \hat{H}_0(\nu) + \frac{1}{2} \sum_{\nu,\nu',\nu \neq \nu'} \hat{H}(\nu,\nu') ,} \qquad (4.37)$$

where

$$\boxed{\hat{H}_0(\nu) = \sum_{l_{\nu}=1}^{n} E^{(\nu)}_{l_{\nu}} \hat{P}_{l_{\nu}l_{\nu}}(\nu) = \sum_{l_{\nu}=1}^{n} E^{(\nu)}_{l_{\nu}} \left| E^{(\nu)}_{l_{\nu}} \right\rangle \left\langle E^{(\nu)}_{l_{\nu}} \right|} \qquad (4.38)$$

represents the Hamiltonian of a node $\nu$ with $n$ local states, and

$$\boxed{\hat{H}(\nu,\nu') = \sum_{l,l'=1}^{n} h^{\nu\nu'}_{ll'} \hat{P}_{ll}(\nu) \otimes \hat{P}_{l'l'}(\nu')} \qquad (4.39)$$

$(\nu \neq \nu')$ describes diagonal 2-node interactions (Coulomb coupling, cf. Sect. 2.4.2.1) between two different nodes. $h^{\nu\nu'}_{ll'}$ is defined by (2.847),

$$h^{\nu\nu'}_{ll'} = \frac{1}{4\pi\epsilon_r\epsilon_0 \left| \boldsymbol{R}_{\nu\nu'} \right|^3} \left\{ d_{l_{\nu}}(\nu) d_{l_{\nu'}}(\nu') - \right.$$
$$\left. \frac{3}{\left| \boldsymbol{R}_{\nu\nu'} \right|^2} \left[ \boldsymbol{R}_{\nu\nu'} d_{l_{\nu}}(\nu) \right] \left[ \boldsymbol{R}_{\nu\nu'} d_{l_{\nu'}}(\nu') \right] \right\} , \qquad (4.40)$$

where we take

$$h^{\nu\nu'}_{l_{\nu}l_{\nu'}} = h^{\nu'\nu}_{l_{\nu'}l_{\nu}} . \qquad (4.41)$$

The single-node states $\left| E^{(\nu)}_{l} \right\rangle$ are determined by the eigenvalue equation

$$\hat{H}_0(\nu)\left|E_{l_\nu}^{(\nu)}\right\rangle = E_{l_\nu}^{(\nu)}\left|E_{l_\nu}^{(\nu)}\right\rangle , \tag{4.42}$$

and the $n^N$ product states

$$\left|E_{l_1}^{(1)}\dots E_{l_N}^{(N)}\right\rangle = \left|E_{l_1}^{(1)}\right\rangle \otimes \dots \otimes \left|E_{l_N}^{(N)}\right\rangle \tag{4.43}$$

are the eigenstates of the network:

$$\hat{H}\left|E_{l_1}^{(1)}\dots E_{l_N}^{(N)}\right\rangle = E_{l_1\dots l_N}\left|E_{l_1}^{(1)}\dots E_{l_N}^{(N)}\right\rangle \tag{4.44}$$

with the energy eigenvalues given by

$$E_{l_1\dots l_N} = \sum_{\nu=1}^{N} E_{l_\nu}^{(\nu)} + \frac{1}{2}\sum_{\substack{\nu,\nu'=1\\\nu\neq\nu'}}^{N} h_{l_\nu l_{\nu'}}^{(\nu\nu')} . \tag{4.45}$$

Transition energies within node $\nu$ thus depend, in general, on all other node states:

$$E_{l_1,l_2,\dots,l_{\nu-1},m_\nu,l_{\nu+1},\dots,l_N} - E_{l_1,l_2,\dots,l_{\nu-1},l_\nu,l_{\nu+1},\dots,l_N}$$
$$= E_{m_\nu}^{(\nu)} - E_{l_\nu}^{(\nu)} + \sum_{\substack{\nu'=1\\\nu\neq\nu'}}^{N}\left(h_{m_\nu l_{\nu'}}^{(\nu\nu')} - h_{l_\nu m_{\nu'}}^{(\nu\nu')}\right)$$
$$:= \tilde{E}_{ml}^{(\nu)} = \hbar\tilde{\omega}_{ml}^{(\nu)} , \tag{4.46}$$

while the transition energy of the isolated subsystem is given by

$$E_{ml}^{(\nu)} = E_{m_\nu}^{(\nu)} - E_{l_\nu}^{(\nu)} = \hbar\omega_{ml}^{(\nu)} . \tag{4.47}$$

If the transition rate in the network rate equation (4.36) depends on these transition energies $\tilde{E}_{ml}^{(\nu)}$, the rate equations between different nodes become coupled.

### 4.3.2 Single 2-Level Node: Random Telegraph Signals

For a single node the network rate equation (4.36) reduces to

$$\dot{R}_{s,s}(t) = \left[W_{s,-s}R_{s,-s}(t) - W_{-s,s}R_{s,s}(t)\right] . \tag{4.48}$$

For the damping we may apply either the thermal model (cf. (3.490))

$$W_{s,-s} = \frac{1}{2\tau_0}\left[1 - \tanh\left(\frac{E_{1,-1}}{2k_{\mathrm{B}}T}s\right)\right] , \tag{4.49}$$

or the optical model based on incoherent optical driving forces (cf. (3.504))

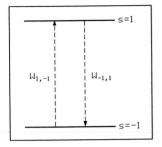

**Fig. 4.10.** Energy spectrum of an isolated 2-level-1-node network with $s = \pm 1$

$$\boxed{W_{1,-1} = \frac{\pi}{\epsilon_0 \hbar^2} |d_{12}|^2 U(|\omega_{12}|) ,} \tag{4.50}$$

where

$$W_{-1,1} = W_{1,-1} + S_{-1,1} \tag{4.51}$$

($S_{-1,1}$ = spontaneous decay). For small temperatures there is no incoherent absorption. In Fig. 4.10 the energy spectrum of a 2-level-1-node system pertinent to both models is sketched.

The corresponding Markov process is a random hopping between the two states $s = \pm 1$ with hopping rate $W_{s,-s}$. In Fig. 4.11 an example is shown. Only the spontaneous decay would be observable as a luminescence signal (photons other than contained in the driving field). In the stationary case, this 2-level random telegraph signal is characterized by an exponential decay of the 2-time-correlation function

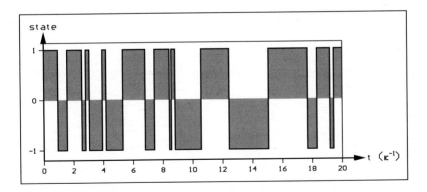

**Fig. 4.11.** Random hopping in a 2-dimensional state space. The random telegraph signal $s(t)$ of an isolated node. Computer simulation with parameters $W_{1,-1} = W_{-1,1} = \kappa$

$$\langle s(t)s(t+\tau)\rangle$$
$$= \left(\frac{W_{1,-1} - W_{-1,1}}{W_{1,-1} + W_{-1,1}}\right)^2 +$$
$$\left(\frac{4W_{1,-1}W_{-1,1}}{W_{1,-1} + W_{-1,1}}\right) \exp\left[-(W_{1,-1} + W_{-1,1})|\tau|\right] .$$

(4.52)

This (stationary) correlation function does not depend on $t$ but on the time difference $\tau$. Random telegraph signals described by such an exponential correlation function are called *normal*. These telegraph signals for a strong incoherent (optical) driving might be contrasted with the overdamped but coherent field case. Under these conditions the 2-level system virtually remains always in the ground state: tiny excursions are reset by spontaneous emission events. Though observable, these "clicks" are not really relevant for the reconstruction of the atomic state.

### 4.3.3 Interacting 2-Level Pair: Classical Correlations

A minimal model for *anomalous stochastic dynamics* is given by two coupled 2-level nodes with

$$s_1, s_2 = \pm 1$$

(4.53)

the dynamics of which proceeds in a 4-dimensional state space described by $(s_1, s_2)$.

For this minimal model eight transition rates appear:

$$\left\{W^{(1)}_{1,s_2,-1,s_2}, W^{(1)}_{-1,s_2,1,s_2}\right\} , \quad s_2 = \pm 1$$
$$\left\{W^{(2)}_{s_1,1,s_1,-1}, W^{(2)}_{s_1,-1,s_1,1}\right\} , \quad s_1 = \pm 1 .$$

(4.54)

The corresponding eight processes

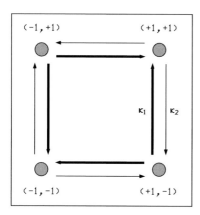

Fig. 4.12. Stochastic dynamics in the 4-dimensional state space. The thickness of the arrows indicates the magnitude of the transition rates in the case of ferro coupling ($\kappa_1 > \kappa_2$)

$$(s_1, s_2) \rightarrow (s_1', s_2') \tag{4.55}$$

are sketched in Fig. 4.12.

Assuming identical nodes, these 8 transition rates must be equal in pairs leaving the 4 parameters

$$W_{s_1,s_2,s_1,-s_2} = \{W_{1,1,1,-1}, W_{-1,-1,-1,1}, W_{-1,1,-1,-1}, W_{1,-1,1,1}\} . \tag{4.56}$$

In the following we restrict ourselves to

$$W_{-1,-1,-1,1} = W_{1,1,1,-1} := \kappa_1 , \quad W_{-1,1,-1,-1} = W_{1,-1,1,1} := \kappa_2 . \tag{4.57}$$

The case $\kappa_1 > \kappa_2$ defines so-called *ferro coupling* of the two nodes: in Fig. 4.12 the thickness of the arrows indicates the magnitude of the transition rates. In this case, the two nodes tend to stay in the same state. The diagonal interaction, according to Fig. 2.28, implies for degenerate local energy levels:

$$E_{-1,-1} = E_{1,1} = \Delta E , \quad E_{-1,1} = E_{1,-1} = -\Delta E . \tag{4.58}$$

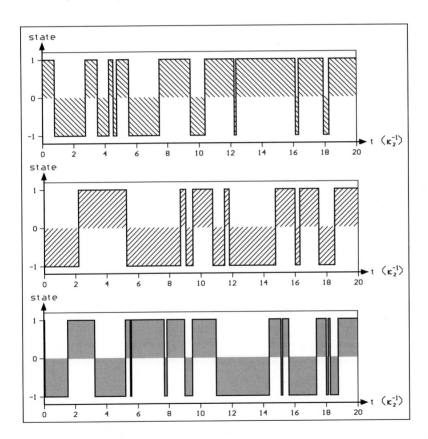

**Fig. 4.13.** Random telegraph signals in two coupled subsystems. Top: the random telegraph signal $s_1(t)$, $\kappa_1 = \kappa_2$. Middle: the signal $s_2(t)$, $\kappa_1 = \kappa_2$. Below: $\kappa_1 = 1000\kappa_2$. Apart from the starting point, the signals $s_1(t)$, $s_2(t)$ are virtually identical

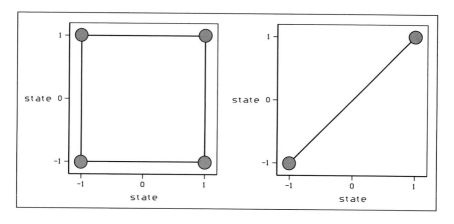

**Fig. 4.14.** The stochastic trajectories $s_1(t)$ and $s_2(t)$ in the $s_1, s_2$ plane. The shaded circels indicate the states of the 4-dimensional state space. L.h.s.: uncoupled nodes ($\kappa_1 = \kappa_2$). A stochastic hopping between the four states $(-1, -1)$, $(+1, -1)$, $(+1, +1)$, $(-1, +1)$ occurs. R.h.s.: strong coupled nodes ($\kappa_1 = 1000\kappa_2$). The hopping takes place between the two states $(-1, -1)$, $(+1, +1)$

The first pair of transitions of (4.57) is thus related to the transition energy $-2\Delta E$, the second pair to $+2\Delta E$. In the thermal model one thus gets

$$
\begin{aligned}
\kappa_1 &= \frac{1}{2\tau_0}\left[1 + \tanh\left(\frac{\Delta E}{k_B T}\right)\right] , \\
\kappa_2 &= \frac{1}{2\tau_0}\left[1 - \tanh\left(\frac{\Delta E}{k_B T}\right)\right] .
\end{aligned}
\tag{4.59}
$$

Using these expressions, ferro coupling is expressed by the familiar condition $\Delta E > 0$. An effective model based on optical control will be discussed later (cf. Sect. 4.3.6.1).

Stochastic trajectories $s_1(t)$ and $s_2(t)$ are depicted in Fig. 4.13: for $\kappa_1 = \kappa_2$ the two subsystems are not coupled so that they jump independently. However, if one increases the ratio $\kappa_1/\kappa_2$, a strong correlation develops so that both subsystems are synchronized: as this synchronization happens in the classical sector of the dynamical state space, we associate this behaviour with a "classical correlation" (note that the quantum correlation M is always zero). In Fig. 4.14 these stochastic modes are illustrated in the $s_1, s_2$ plane.

Averaging over the stochastic signal, one obtains the corresponding correlation functions,

$$
\begin{aligned}
\langle s_1(t)s_1(t+\tau)\rangle &= \langle s_2(t)s_2(t+\tau)\rangle \\
&= \frac{\kappa_1}{\kappa_1 + \kappa_2}\exp\left(-2\kappa_2\,|\tau|\right) + \\
&\quad \frac{\kappa_2}{\kappa_1 + \kappa_2}\exp\left(-2\kappa_1\,|\tau|\right) ,
\end{aligned}
\tag{4.60}
$$

as well as the functions of the cross-correlation,

$$\langle s_1(t)s_2(t+\tau)\rangle = \frac{\kappa_1}{\kappa_1 + \kappa_2} \exp\left(-2\kappa_2 |\tau|\right) - \frac{\kappa_2}{\kappa_1 + \kappa_2} \exp\left(-2\kappa_1 |\tau|\right) . \tag{4.61}$$

For $\kappa_1 \neq \kappa_2$ the decay in either subsystem is non-exponential, the random telegraph signals are *anomalous*. For $\kappa_1 > \kappa_2$ (ferro coupling) the cross-correlation is positive and for $\kappa_1 < \kappa_2$ (antiferro coupling) it is negative. Maximal correlation

$$\langle s_1(t)s_2(t)\rangle = 1 \tag{4.62}$$

results for $\tau = 0$ and in the limit $\kappa_1 \gg \kappa_2$, maximal anticorrelation

$$\langle s_1(t)s_2(t)\rangle = -1 \tag{4.63}$$

for $\kappa_1 \ll \kappa_2$.

For the *macro-variable*

$$S(t) = s_1(t) + s_2(t) \tag{4.64}$$

the correlation is given by

$$\langle S(t)S(t+\tau)\rangle = \frac{4\kappa_1}{\kappa_1 + \kappa_2} \exp\left(-2\kappa_2 |\tau|\right) . \tag{4.65}$$

### 4.3.4 $SU(2)$ Chain

"Natural" quantum networks may exist in various forms. As an example, a chain composed of interacting defects (*2-level traps*) is investigated in this section.

**4.3.4.1 Effective Ising Model.** Consider *defects* described by the index $\nu$, where every defect can be represented by a 2-level system with states $s_\nu = \pm 1$ and energy spacing $\Delta E$. With an additional linear coupling to a static external field $F$, the local energies $E_{l_\nu}^{(\nu)}$ then have the form

$$E_{l_\nu}^{(\nu)} := E_{s_\nu} = \left(\frac{\Delta E}{2} - \chi F\right) s_\nu , \tag{4.66}$$

where $\chi$ denotes the coupling constant to the external field. (Compare Fig. 4.15, part a).)

Considering a linear *chain of identical defects* realized by dipoles with two states $s_\nu = \pm 1$, and restricting ourselves to *nearest-neighbour interaction*, the only non-vanishing matrix elements $h_{l_\nu l_{\nu'}}^{\nu\nu'}$ are given by

$$h_{s_\nu,s_{\nu+1}}^{\nu,\nu+1} = h_{s_{\nu+1},s_\nu}^{\nu+1,\nu} = -J s_\nu s_{\nu+1} , \tag{4.67}$$

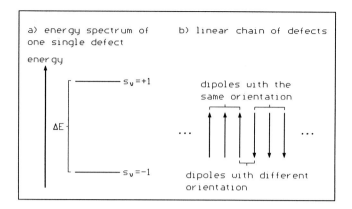

a) energy spectrum of one single defect

b) linear chain of defects

energy

$s_\nu = +1$

$\Delta E$

$s_\nu = -1$

dipoles with the same orientation

. . .    . . .

dipoles with different orientation

**Fig. 4.15.** Interacting 2-level defects in a heat bath. The energy spectrum of a single defect (l.h.s.), and an example of an arrangement of defects in real space (r.h.s.)

where $J$ is the energy decrease (increase) if the neighbours are in the same (opposite) state. In Fig. 4.15, part b), this scenario is sketched. A *heat bath* may then induce transitions between the network states

$$\{s(\nu)\} = (s_1, s_2, \ldots, s_{\nu-1}, s_\nu, s_{\nu+1}, \ldots, s_N) \tag{4.68}$$

and

$$\{-s(\nu)\} = (s_1, s_2, \ldots, s_{\nu-1}, -s_\nu, s_{\nu+1}, \ldots, s_N) \ . \tag{4.69}$$

Each single transition represents a "flip" from a state $\{s(\nu)\}$ into the state $\{-s(\nu)\}$ (affecting node $\nu$).

This model is equivalent to the 1-dimensional *Ising model* with the magnetic field

$$\boxed{H = \chi F - \Delta E/2} \tag{4.70}$$

and *exchange interaction* $J$. Using (4.70), the eigenenergies $E_{s_1 \ldots s_N}$ of the whole dipole chain ("spin chain") can be represented by

$$E_{s_1 \ldots s_N} = -H \sum_{\nu=1}^{N} s_\nu - J \sum_{\nu=1}^{N} s_\nu s_{\nu+1} \ , \tag{4.71}$$

and the transition energy of a single environment-induced transition can be written in the form

$$\begin{aligned}
\tilde{E}^{(\nu)}_{-s_\nu, s_\nu} &= E_{s_1, s_2, \ldots, s_{\nu-1}, -s_\nu, s_{\nu+1}, \ldots, s_N} - E_{s_1, s_2, \ldots, s_{\nu-1}, s_\nu, s_{\nu+1}, \ldots, s_N} \\
&= 2H s_\nu + 2J s_\nu \left( s_{\nu+1} + s_{\nu-1} \right) \\
&= -\tilde{E}^{(\nu)}_{s_\nu, -s_\nu} \ .
\end{aligned} \tag{4.72}$$

**4.3.4.2 Ensemble Dynamics.** The ensemble dynamics of this model – which can be considered as *kinetic Ising model* – is governed by the *rate equation*

$$\dot{R}_{\{s\},\{s\}}(t) = \sum_{\nu=1}^{N} \left[ W^{(\nu)}_{s_\nu,-s_\nu} R^{(\nu)}_{-s_\nu,-s_\nu}(t) - W^{(\nu)}_{-s_\nu,s_\nu} R^{(\nu)}_{s_\nu,s_\nu}(t) \right] \qquad (4.73)$$

(compare (4.36)). Restricting ourselves to states in thermal equilibrium (i.e. stationary states), the l.h.s. of this equation is zero:

$$\sum_{\nu=1}^{N} \left[ W^{(\nu)}_{s_\nu,-s_\nu} R^{(\nu)}_{-s_\nu,-s_\nu}(t) - W^{(\nu)}_{-s_\nu,s_\nu} R^{(\nu)}_{s_\nu,s_\nu}(t) \right] = 0 . \qquad (4.74)$$

Due to the *principle of detailed balance*, for systems in thermal equilibrium, the terms in brackets vanish individually. Thus, this stationary condition can be replaced by

$$W^{(\nu)}_{s_\nu,-s_\nu} R^{(\nu)}_{-s_\nu,-s_\nu} = W^{(\nu)}_{-s_\nu,s_\nu} R^{(\nu)}_{s_\nu,s_\nu} \qquad (4.75)$$

or by

$$\frac{W^{(\nu)}_{s_\nu,-s_\nu}}{W^{(\nu)}_{-s_\nu,s_\nu}} = \frac{R^{(\nu)}_{s_\nu,s_\nu}}{R^{(\nu)}_{-s_\nu,-s_\nu}} . \qquad (4.76)$$

With (4.72) we get from the Glauber model (cf. (3.490)):

$$\begin{aligned} W^{(\nu)}_{s_\nu,-s_\nu} \\ = \frac{1}{2\tau_0} \left( 1 - \tanh \left\{ -\frac{1}{k_B T} s_\nu \left[ H + J \left( s_{\nu-1} + s_{\nu+1} \right) \right] \right\} \right) . \end{aligned} \qquad (4.77)$$

Replacing in (4.77) $s_\nu$ by $-s_\nu$, one obtains the transition rate $W^{(\nu)}_{-s_\nu,s_\nu}$. In Fig. 4.16 the transition rate $W^{(\nu)}_{s_\nu,-s_\nu}$ is illustrated.

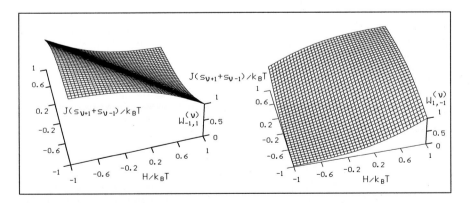

**Fig. 4.16.** The transition rate $W^{(\nu)}_{s_\nu,-s_\nu}$, $(s_\nu = \pm 1)$ in the Glauber model. $1/2\tau_0 := 1$

### 4.3.5 Single 3-Level Node: Random Telegraph Signals

In the last few sections, random telegraph signals in 2-level node systems have been discussed. In this section the considerations will be extended to 3-level nodes.

**4.3.5.1 Rate Equations.** Consider a 3-level node characterized by the energy spectrum shown in Fig. 4.17 ($\Lambda$ scenario): the spontaneous and induced transitions are caused by the bath or the pump light field, respectively. Both transitions ($1 \rightarrow 3$ and $2 \rightarrow 3$) are optically driven. A direct optically induced charge transfer via coupling of the light field to the transition $1 \rightarrow 2$ can be disregarded due to the missing overlap of both states and/or detuning.

The rate equations for this 3-level node are (cf. (3.505) and (4.36))

$$\frac{\mathrm{d}}{\mathrm{d}t}R_{11}(t) = W_{12}R_{22}(t) + W_{13}R_{33}(t) - (W_{21} + W_{31})\,R_{11}(t)\,,$$

$$\frac{\mathrm{d}}{\mathrm{d}t}R_{22}(t) = W_{21}R_{11}(t) + W_{23}R_{33}(t) - (W_{12} + W_{32})\,R_{22}(t)\,, \qquad (4.78)$$

$$\frac{\mathrm{d}}{\mathrm{d}t}R_{33}(t) = W_{31}R_{11}(t) + W_{32}R_{22}(t) - (W_{13} + W_{23})\,R_{33}(t)$$

with

$$\begin{aligned}
W_{12} &= S_{12}\,, \\
W_{13} &= S_{13} + B_{13}U\left(|\omega_{13}|\right)\,, \\
W_{21} &= 0\,, \\
W_{23} &= S_{23} + B_{23}U\left(|\omega_{23}|\right)\,, \\
W_{31} &= B_{13}U\left(|\omega_{13}|\right)\,, \\
W_{32} &= B_{23}U\left(|\omega_{23}|\right)\,,
\end{aligned} \qquad (4.79)$$

where $B_{ll'}U\left(|\omega_{ll'}|\right)$ represents transition rates $I_{ll'}$ of induced transitions (cf. (3.504)):

$$I_{13} = B_{13}U\left(|\omega_{13}|\right)\,, \quad I_{23} = B_{23}U\left(|\omega_{23}|\right)\,. \qquad (4.80)$$

The rates $S_{ll'}$ represent spontaneous emission (bath coupling for low temperatures).

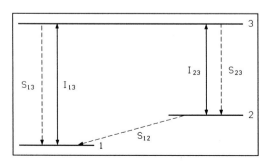

**Fig. 4.17.** Energy states of a driven 3-level system (minimal model for optically controllable charge transfer). Basic state 1, metastable state 2, and transient state 3. The spontaneous transition rates are denoted by $S_{ll'}$, and the transition rates induced by the pump light field by $I_{ll'} = B_{ll'}U\left(|\omega_{ll'}|\right)$

**4.3.5.2 Stationary Solutions.** Let

$$S_{12} = 0 \,. \tag{4.81}$$

In this case stationary solutions $(\dot{R}_{ll'}(t) = 0)$ are determined by

$$0 = W_{13}R_{33} - W_{31}R_{11} \,,$$
$$0 = W_{23}R_{33} - W_{32}R_{22} \,, \tag{4.82}$$
$$0 = W_{31}R_{11} + W_{32}R_{22} - (W_{13} + W_{23})R_{33} \,.$$

For

$$I_{13} = I_{23} \,, \quad S_{13} = S_{23} \tag{4.83}$$

this stationary system of rate equations reduces to

$$I_{13}R_{11} = (I_{13} - S_{13})R_{33} \,,$$
$$I_{13}R_{22} = (I_{13} + S_{13})R_{33} \tag{4.84}$$

so that

$$R_{11} = R_{22} = R_{33}\frac{I_{13} + S_{13}}{I_{13}} \,. \tag{4.85}$$

Observing

$$2R_{11} + R_{33} = 1 \,, \tag{4.86}$$

we readily find the solution

$$\boxed{\begin{aligned} R_{11} &= \frac{I_{13} + S_{13}}{3I_{13} + 2S_{13}} = R_{22} \,, \\ R_{33} &= \frac{I_{13}}{3I_{13} + 2S_{13}} = R_{22} \,. \end{aligned}} \tag{4.87}$$

For $I_{13} \gg S_{13}$ this solution reduces to $R_{11} \approx R_{22} \approx R_{33} \approx 1/3$, while for $I_{13} \ll S_{13}$ one obtains $R_{11} \approx R_{22} \approx 1/2$, $R_{33} \approx 0$, i.e. the occupation of the transient state 3 goes to zero (*effective 2-level dynamics*).

The rate equations above describe the ensemble dynamics of identical 3-level nodes characterized by the scenario of Fig. 4.17. However, the dynamics of the individual node is a stochastic random hopping $l(t)$ in its three-dimensional state space.

**4.3.5.3 Random Telegraph Signals.** Figure 4.18 shows the dynamics discussed in Sect. 4.3.5.2 as a stochastic process for a single 3-level node, where the abbreviations

$$\kappa := I_{13} = I_{23} \,, \quad S := S_{13} = S_{23} \tag{4.88}$$

are used.

In the case $\kappa = 5S$ both transitions are "saturated", the mean probability to find the node in a state $l$ is the same in all three states. In the case of

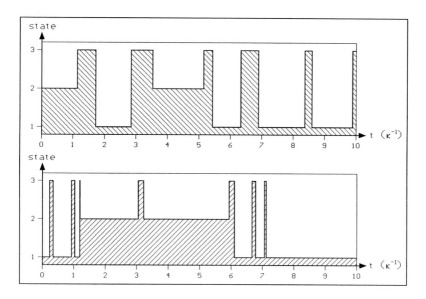

**Fig. 4.18.** Stochastic signal $l(t)$ of an isolated subsystem. Parameters: $S_{13} = S_{23} = S$, $S_{31} = S_{32} = S_{21} = S_{12} = 0$, $\kappa = I_{13} = I_{23}$. Top: $\kappa = 5S$. Below: $\kappa = 0.2S$

a weaker light field, $\kappa = 0.2S$, the occupation probability of state 3, which decays on the time scale $1/S_{13} = 1/S_{23}$, is negligibly small. The system effectively jumps only between states 1 and 2.

### 4.3.6 State Space Compression

The state space of the incoherent networks is already very restricted as compared to that of fully coherent systems. Nevertheless, this state space can quite often be further reduced observing certain hierarchies of scales. We will study two typical examples in which such a reduction is indeed possible. One possibility is already indicated in our study of the isolated $SU(3)$ model, Sect. 4.3.5. There we encountered a situation in which the occupation of the third level – though essential for any transition – becomes a very brief transient phenomenon. One may thus expect that an effective 2-level process should do almost the same job – on an appropriate time scale.

#### 4.3.6.1 $SU(3)$ Chain as an Effective $SU(2)$ Chain.

*Specification.* The rate equation is given by (cf. (4.36))

$$\dot{R}_{\{s\}\{s\}}(t) = \sum_{\nu=1}^{N} \sum_{\substack{s'=1 \\ s' \neq s}}^{n} \left[ W_{ss'}^{(\nu)} R_{s's'}^{(\nu)}(t) - W_{s's}^{(\nu)} R_{ss}^{(\nu)}(t) \right] . \tag{4.89}$$

The local transition rates are those of the optical model:

$$W_{ss'}^{(\nu)} = S_{ss'}^{(\nu)} + I_{ss'}^{(\nu)} . \qquad (4.90)$$

$I_{ss'}^{(\nu)}$ is the induced transition rate according to (3.504), and $S_{ss'}^{(\nu)}$, again, represents the spontaneous emission. At zero temperature we have

$$S_{21}^{(\nu)} = S_{31}^{(\nu)} = S_{32}^{(\nu)} = 0 . \qquad (4.91)$$

The optical driving field is that of the $\Lambda$ scenario, coupling the states 1 and 3 as well as 2 and 3. Furthermore,

$$\boxed{I_{ss'}^{(\nu)} \ll S_{13}^{(\nu)} = S_{23}^{(\nu)} \; (s,s' = 1,3 , \; s \neq s')} \qquad (4.92)$$

is assumed.

The Hamiltonian model for the network is that discussed in Sect. 4.3.1.2. The "diagonal" interaction renormalizes the transition frequencies $\omega_{31}$, $\omega_{32}$ depending on the neighbour states:

$$\boxed{\tilde{\omega}_{3,s}^{(\nu)}(\mathbf{s}_\nu) = \omega_{3,s}^{(\nu)} - \sum_{\substack{\nu'=1 \\ \nu \neq \nu'}}^{N} C_{\nu\nu'} s_\nu s_{\nu'}} \qquad (4.93)$$

with

$$C_{\nu\nu'} = \frac{d^2}{4\pi\hbar\epsilon_r\epsilon_0 \left|\mathbf{R}_{\nu\nu'}\right|^3} \left\{ e_z(\nu)e_z(\nu') - \right.$$

$$\left. \frac{3}{\left|\mathbf{R}_{\nu\nu'}\right|^2} \left[\mathbf{R}_{\nu\nu'}e_z(\nu)\right]\left[\mathbf{R}_{\nu\nu'}e_z(\nu')\right] \right\} . \qquad (4.94)$$

This frequency relation defines for each of the two values $s_\nu = \pm 1$ three non-overlapping frequency bands: the central frequencies result from next nearest neighbour interactions and are separated by

$$\Delta\omega_1 = 2C_0 , \qquad (4.95)$$

where $C_0$ represents the coupling constant of dipole–dipole interactions between nearest neighbours. All other more distant subsystems induce the *inhomogeneous bandwidth*

$$\Delta\omega_2 = 4C_0 \sum_{\nu=2}^{N} \frac{1}{\nu^3} \approx 0.808C_0 . \qquad (4.96)$$

In Fig. 4.19 the corresponding energy spectrum is sketched. As the induced transitions $S_{ss'}^{(\nu)}$ are controlled by the spectral density of the light field, they thus depend on the states of the neighbours around node $\nu$.

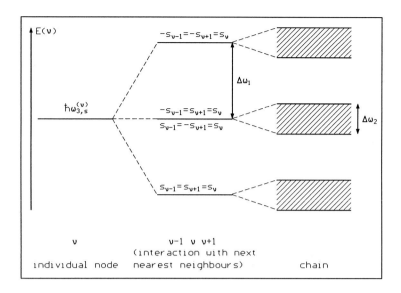

**Fig. 4.19.** Local transition energy $E(\nu)$ of a node $\nu$ in a linear chain: dipole–dipole interactions with the two nearest neighbours lead to three distinguishable state environments (splitting $\Delta\omega_1$) which broaden to three bands if all possible neighbour states are included (width $\Delta\omega_2 < \Delta\omega_1$).

*Adiabatic Elimination.* Due to (4.92), the occupation probabilities $R_{ss}^{(\nu)}$ obey the inequalities

$$R_{p_1,p_2,\ldots,p_{\nu-1},p_\nu,p_{\nu+1},\ldots,p_N}^{(\nu)}(t)$$
$$\gg R_{p_1,p_2,\ldots,p_{\nu-1},3,p_{\nu+1},\ldots,p_N}^{(\nu)}$$
$$\gg R_{p_1,p_2,\ldots,p_{\nu'-1},3,p_{\nu'+1},\ldots,p_{\nu-1},3,p_{\nu+1},\ldots,p_N}^{(\nu)}$$
$$\gg \ldots \tag{4.97}$$

with

$$p_\nu = 1,2 , \tag{4.98}$$

i.e. for each occupied state 3 in the network, the occupation probability is reduced by one order. Comparing orders of magnitude, the rate equation (4.89) can be approximated in first order by

$$\dot{R}_{p_1\ldots p_N}(t) = \sum_{\nu=1}^{N} \left[ S_{p_\nu 3}^{(\nu)} R_{p_1,p_2,\ldots,p_{\nu-1},3,p_{\nu+1},\ldots,p_N}^{(\nu)}(t) - \right.$$
$$\left. B_{p_\nu 3}^{(\nu)} U\left(\omega_{3,p_\nu}^{(1\ldots N)}\right) R_{p_1,p_2,\ldots,p_{\nu-1},p_\nu,p_{\nu+1},\ldots,p_N}^{(\nu)}(t) \right] \tag{4.99}$$

for occupation probabilities with no occupied level 3, and

$$\dot{R}_{p_1,p_2,\ldots,p_{\nu-1},3,p_{\nu+1},\ldots,p_N}(t)$$

$$= \sum_{m_\nu=1}^{2} \left[ B_{m_\nu 3}^{(\nu)} U\left(\omega_{3,m_\nu}^{(1\ldots N)}\right) R_{p_1,p_2,\ldots,p_{\nu-1},m_\nu,p_{\nu+1},\ldots,p_N}^{(\nu)}(t) - S_{m_\nu 3}^{(\nu)} R_{p_1,p_2,\ldots,p_{\nu-1},3,p_{\nu+1},\ldots,p_N}^{(\nu)}(t) \right] \quad (4.100)$$

for probabilities with exactly one occupied level 3.

Now, the hierarchy (4.92) allows an *adiabatic elimination* of the variable $R_{p_1,p_2,\ldots,p_{\nu-1},3,p_{\nu+1},\ldots,p_N}(t)$ in the system of equations (4.99)-(4.100): the time-derivative in (4.100) can be set to zero:

$$\dot{R}_{p_1,p_2,\ldots,p_{\nu-1},3,p_{\nu+1},\ldots,p_N}(t) = 0 . \quad (4.101)$$

Then combining both equations, one obtains the rate equation in $p_\nu$ subspace:

$$\dot{R}_{p_1\ldots p_N}(t) = \sum_{\nu=1}^{N} \left[ \tilde{W}_{p_\nu,3-p_\nu}^{(\nu)} R_{p_1,p_2,\ldots,p_{\nu-1},3-p_\nu,p_{\nu+1},\ldots,p_N}^{(\nu)}(t) - \tilde{W}_{3-p_\nu,p_\nu}^{(\nu)} R_{p_1,p_2,\ldots,p_{\nu-1},p_\nu,p_{\nu+1},\ldots,p_N}^{(\nu)}(t) \right] \quad (4.102)$$

with

$$\tilde{W}_{p_\nu,3-p_\nu}^{(\nu)} = \frac{S_{p_\nu 3}^{(\nu)}}{S_{13}^{(\nu)} + S_{23}^{(\nu)}} B_{3-p_\nu 3}^{(\nu)} U(\tilde{\omega}_{3,3-p_\nu}) . \quad (4.103)$$

**4.3.6.2 $SU(2)$ Chain as an Effective $SU(2)$ Node.** (See [80].) In this section, a linear chain of localized 2-level systems with energy splitting $\Delta E$ is considered, where only interactions between the nearest neighbours are assumed. Furthermore, the thermal model defined by the transition rates (4.77) with $H = -\Delta E/2$ will be taken as a basis ($s_\nu = \pm 1$):

$$W_{s_\nu,-s_\nu}^{(\nu)} = \frac{1}{2\tau_0}\left(1 - \tanh\left\{-\frac{1}{k_B T} s_\nu \left[-\frac{\Delta E}{2} + J\left(s_{\nu-1} + s_{\nu+1}\right)\right]\right\}\right)$$

$$:= W_{-s_\nu}^{(\nu)}(s_{\nu-1}, s_{\nu+1}) . \quad (4.104)$$

We define the macro-variable

$$S(t) = \sum_{\nu=1}^{N} s_\nu(t) . \quad (4.105)$$

Figure 4.20 illustrates for $\Delta E = 0$ the stochastic behaviour of $S(t)$. For low values of the control parameter $J/k_B T$ the transition rate (4.104) does not depend on the state of the two neighbours; each node changes independently. Increasing the value of the control parameter, *cooperative behaviour* emerges so that, eventually, only 2-state fluctuations between the extrema $S = \pm N$ remain; the resulting *large amplitude random telegraph signal* is a *2-level signal*. This signal is shown in Fig. 4.21 on a finer time scale: the large amplitude

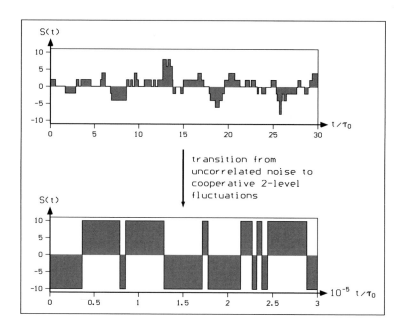

**Fig. 4.20.** Random telegraph signals of a linear chain of localized 2-level systems: the stochastic behaviour of the macro-variable $S(t)$. The change of the control parameter $J/K_BT$ generates a transition into a cooperative fluctuation trace. Parameters: $N = 10$, $\Delta E = 0$. Top: $J/k_BT = 0$. Below: $J/k_BT = 2.4$

signal emerges on a coarse-grained scale as a consequence of correlated jumps of the individual nodes. In Fig. 4.22 this large amplitude signal in the case $\Delta E \neq 0$ is depicted. As can be seen, the signal becomes asymmetric with respect to upper and lower state occupation.

*Hierarchy of Transition Rates.* For $\Delta E = 0$, two stationary collective states described by $S = Ns$ ($s = \pm1$) corresponding to the occupation probabilities

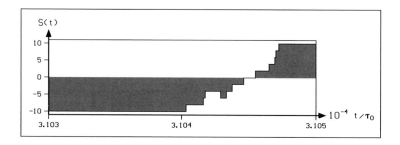

**Fig. 4.21.** The first large amplitude jump (of the lower graph in the preceding figure) on a finer time scale

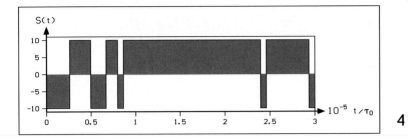

**Fig. 4.22.** Large amplitude random telegraph signal in the case $\Delta E \neq 0$. Parameters: $\Delta E/J = -0.04$, $J/k_B T = 2.4$

$R_{s\ldots s}$ $(s = \pm 1)$ are possible, where the case of cooperative 2-level signals is restricted by the condition

$$1 - (R_{1\ldots 1} + R_{-1\ldots -1}) \ll 1 . \tag{4.106}$$

As the stationary occupation probabilities are given by

$$
\begin{aligned}
R_{1\ldots 1} &= R_{-1\ldots -1} \\
&= \frac{\exp\left(NJ/k_B T\right)}{\left[2\cosh\left(J/k_B T\right)\right]^N + \left[2\sinh\left(J/k_B T\right)\right]^N} \\
&:= R_{Ns} ,
\end{aligned}
\tag{4.107}
$$

(4.106) implies

$$\boxed{N \ll -\frac{1}{\ln\tanh\left(J/k_B T\right)} := L .} \tag{4.108}$$

In the thermodynamic limit $N \to \infty$, $L$ describes the *correlation length*, i.e. the decay of 2-point correlations:

$$\langle s_\nu s_{\nu+r} \rangle = \exp\left(-\frac{r}{L}\right) . \tag{4.109}$$

The above inequality shows that cooperative 2-level signals will exist in finite systems if the correlation length is larger than the *system size* $N$.

The above inequality implies a hierarchy in the transition rates $W^{(\nu)}_{s_\nu, -s_\nu}$ of the considered thermal model. Due to the fact that these transition rates only depend on the next neighbours (compare (4.77)), the more explicit notation $W^{(\nu)}_{s_\nu}(s_{\nu-1}, s_{\nu+1})$ can be used, i.e.

$$W^{(\nu)}_{s_\nu, -s_\nu} = W^{(\nu)}_{-s_\nu}(s_{\nu-1}, s_{\nu+1}) \tag{4.110}$$

with

$$s_\nu, s_{\nu-1}, s_{\nu+1} = s = \pm 1 . \tag{4.111}$$

Using this notation, the hierarchy implied by the above inequality reads

$$\boxed{NW_{-s}^{(\nu)}(-s,-s) \ll W_{-s}^{(\nu)}(-s,s), W_{-s}^{(\nu)}(s,s) \quad (s = \pm 1).} \tag{4.112}$$

This hierarchy describes the fact that it is very unlikely to have a flipping inside a cluster (cluster = connected area in which the subsystems are all in the same state). Due to this hierarchy, again, an adiabatic elimination procedure is possible to compress the state space of the network.

*Adiabatic Elimination.* The hierarchy (4.112) implies, in turn, a hierarchy of the occupation probabilities (we assume periodic boundary conditions):

$$R_{1\ldots1}^{(\nu)}(t), R_{-1\ldots-1}^{(\nu)}(t)$$
$$\gg R_{1,1,\ldots,1|-1,-1,\ldots,-1|1,1,\ldots,1}^{(\nu)}(t), R_{-1,-1,\ldots,-1|1,1,\ldots,1|-1,-1,\ldots,-1}^{(\nu)}(t)$$
$$\gg R_{1,1,\ldots,1|-1,-1,\ldots,-1|1,1,\ldots,1|-1,-1,\ldots,-1|1,1,\ldots,1}^{(\nu)}(t),$$
$$R_{-1,-1,\ldots,-1|1,1,\ldots,1|-1,-1,\ldots,-1|1,1,\ldots,1|-1,-1,\ldots,-1}^{(\nu)}(t)$$
$$\gg \ldots \tag{4.113}$$

Due to this hierarchy, the appearance of an additional cluster reduces the order of magnitude of an occupation probability. So, considering the orders of magnitude of the transition rates (4.112), the set of rate equations defined by (4.73) can – in first order approximation – be replaced by a simple set of equations: using the notation (4.110), the basic rate equation (4.73) turns into

$$\dot{R}_{s_1\ldots s_N}(t)$$
$$= \sum_{\nu=1}^{N} \left[ W_{-s_\nu}^{(\nu)}(s_{\nu-1}, s_{\nu+1}) R_{s_1,s_2,\ldots,s_{\nu-1},-s_\nu,s_{\nu+1},\ldots,s_N}^{(\nu)}(t) - \right.$$
$$\left. W_{s_\nu}^{(\nu)}(s_{\nu-1}, s_{\nu+1}) R_{s_1,s_2,\ldots,s_{\nu-1},s_\nu,s_{\nu+1},\ldots,s_N}^{(\nu)}(t) \right]. \tag{4.114}$$

This differential equation describes the time-evolution of network states represented by the occupation probabilities $R_{s_1\ldots s_N}(t)$. For a collective state $R_{s\ldots s}(t)$, this rate equation reduces to

$$\dot{R}_{s\ldots s}(t) = \sum_{\nu=1}^{N} \left[ W_{-s}^{(\nu)}(s,s) R_{s,s,\ldots,s,-s,s,\ldots,s}^{(\nu)}(t) - \right.$$
$$\left. W_{s}^{(\nu)}(s,s) R_{s,s,\ldots,s,s,s,\ldots,s}^{(\nu)}(t) \right]. \tag{4.115}$$

Here one should take into account that the transition rates $W_s^{(\nu)}(s,s)$ do not depend on the special node $\nu$, so that

$$W_{\pm s}^{(\nu)}(s,s) = W_{\pm s}(s,s), \tag{4.116}$$

and introduce a more comprehensive cluster notation: the single cluster state will be specified by

$$R^{(\nu)}_{s,s,\ldots,s,s,s,\ldots,s}(t) = R^{(\nu)}_{Ns}(t) \,, \tag{4.117}$$

and 2-cluster states by

$$R^{(\nu)}_{s,s,\ldots,s,-s,s,\ldots,s}(t) = R^{(\nu)}_{(N-1)s}(t) \,,$$
$$R^{(\nu)}_{s,s,\ldots,s,-s,-s,\ldots,s,s,s,\ldots,s}(t) = R^{(\nu)}_{(N-2)s}(t) \,, \tag{4.118}$$

etc. .

The above rate equation then turns into

$$\dot{R}_{Ns}(t) = W_{-s}(s,s) \sum_{\nu=1}^{N} R^{(\nu)}_{(N-1)s}(t) - W_s(s,s) \sum_{\nu=1}^{N} R^{(\nu)}_{Ns}(t) \,. \tag{4.119}$$

Due to the fact that there is only one state to gain from

$$R_{(N-1)s}(t) = \sum_{\nu=1}^{N} R^{(\nu)}_{(N-1)s}(t) \,, \tag{4.120}$$

but $N$ states to lose into,

$$\sum_{\nu=1}^{N} R^{(\nu)}_{Ns}(t) = N R_{Ns}(t) \,, \tag{4.121}$$

one obtains the rate equation

$$\dot{R}_{Ns}(t) = W_{-s}(s,s) R_{(N-1)s}(t) - N W_s(s,s) R_{Ns}(t) \,. \tag{4.122}$$

In order to eliminate $R_{(N-1)s}(t)$ we need to consider the whole hierarchy:

$$\dot{R}_{(N-1)s}(t) = 2 W_{-s}(-s,s) R_{(N-2)s}(t) + N W_s(s,s) R_{Ns}(t) -$$
$$[W_{-s}(s,s) + 2 W_s(-s,s)] R_{(N-1)s}(t) \quad (s = \pm 1) \,, \tag{4.123}$$

and then

$$\dot{R}_{S_1}(t) = 2 W_1(-1,1) R_{S_1+1}(t) + 2 W_{-1}(-1,1) R_{S_1-1}(t) -$$
$$2 [W_{-1}(-1,1) + W_1(-1,1)] R_{S_1}(t) \tag{4.124}$$

for

$$S_1 = N - 2, N - 3, \ldots, -N + 3, -N + 2 \,. \tag{4.125}$$

In adiabatic approximation,

$$\dot{R}_{S_2}(t) = 0 \,, \tag{4.126}$$

$$S_2 = N - 1, N - 2, N - 3, \ldots, -N + 3, -N + 2, -N + 1 \,, \tag{4.127}$$

we obtain from (4.124) and (4.123):

$$\boxed{R_{S_2}(t) = \alpha(t) \left[ \frac{W_{-1}(-1,1)}{W_1(-1,1)} \right]^{S_2/2} + \beta(t)} \tag{4.128}$$

with

$$
\begin{aligned}
\alpha(t) &= \Lambda_0 \Lambda_1(-1) W_1(1,1) R_N(t) - \\
&\quad \Lambda_0 \Lambda_1(1) W_{-1}(-1,-1) R_{-N}(t) \, , \\
\beta(t) &= \Lambda_0 \Lambda_1(-1) W_{-1}(1,1) W_{-1}(-1,-1) R_{-N}(t) - \\
&\quad \Lambda_0 \Lambda_1(1) W_1(1,1) W_1(-1,-1) R_N(t)
\end{aligned}
\tag{4.129}
$$

with

$$
\Lambda_0 = \frac{N}{\Lambda_1(-1)\Lambda_2(-1)W_{-1}(1,1) - \Lambda_1(1)\Lambda_2(1)W_1(-1,-1)} \, ,
\tag{4.130}
$$

$$
\Lambda_1(s) = W_{-s}(s,s) + 2\left[W_s(-s,s) - W_{-s}(-s,s)\right] \, ,
\tag{4.131}
$$

$$
\Lambda_2(s) = \left(\frac{W_s(-s,s)}{W_{-s}(-s,s)}\right)^{N/2-1} .
\tag{4.132}
$$

With the help of (4.128), the probability $R_{(N-1)s}(t)$ occurring in the first rate equation (4.122) can be eliminated: one finally obtains the *effective rate equation*

$$
\dot{R}_{Ns}(t) = W^{(\mathrm{eff})}_{-Ns} R_{-Ns}(t) - W^{(\mathrm{eff})}_{Ns} R_{Ns}(t) \, ,
\tag{4.133}
$$

where the *effective transition rates* are defined by

$$
\begin{aligned}
&W^{(\mathrm{eff})}_{Ns} \\
&= \frac{2N\Lambda_1(s)W_s(s,s)W_s(-s,-s)\left[W_s(-s,s) - W_{-s}(-s,s)\right]}{\Lambda_1(s)\Lambda_2(s)W_s(-s,-s) - \Lambda_1(-s)\Lambda_2(-s)W_{-s}(s,s)} .
\end{aligned}
\tag{4.134}
$$

This effective rate equation determines the time-evolution of a collective network state represented by the collective variable $S = Ns$. The corresponding stationary solution is defined by (4.107).

The mean time $\tau_{Ns}$ spent in the network state $S = Ns$ is given by the reciprocal of the effective transition rate $W^{(\mathrm{eff})}_{Ns}$:

$$
\tau_{Ns} = \frac{1}{W^{(\mathrm{eff})}_{Ns}} \, .
\tag{4.135}
$$

Inserting the explicit expression for the transition rates $W^{(\mathrm{eff})}_{Ns}$ (cf. (4.134)), and making use of the Glauber function (cf. (4.77)), for $|N\Delta E/k_B T|^2 \ll 1$ this mean time reads

$$
\tau_{Ns} = \tau_0 \exp\left(\frac{4J - Ns\Delta E/2}{k_B T}\right) .
\tag{4.136}
$$

In Fig. 4.23 an example of this mean time is depicted as a function of its pertinent parameters. This model can be applied to explain 2-level fluctuations observed in small tunnel junctions (cf. [5, 46, 118, 128]).

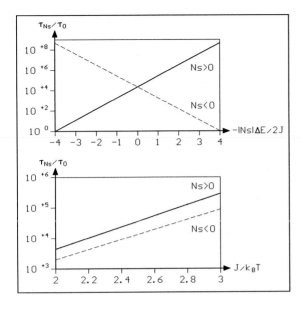

**Fig. 4.23.** The mean time spent in the network state $S = Ns$. Top: $J/k_\mathrm{B}T = 2.5$. Below: $Ns\Delta E/J = -0.4$

## 4.4 Partly Coherent Networks

Contents: reacting and non-reacting environment, antibunching, quantum Zenon effect, superradiance.

### 4.4.1 Preliminary Remarks

Partly coherent systems are the most general; they thus deserve special attention. On the single-object level, the dynamics is a combination of continuous rotations in $SU(n)$ and jumps induced by the damping channels. Typically,

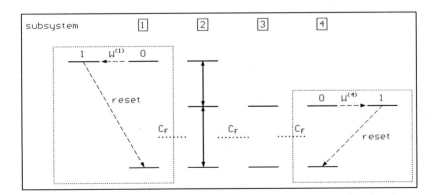

**Fig. 4.24.** 4-node network with 2 local damping channels (in subsystem 1 and 4, respectively)

these jumps will not all be registered by the experimentalist so that the entropy of the system state will increase. This applies, in particular, to systems coupled to a high-temperature bath. In an idealized situation, however, dissipation is connected with information retrieval so that, in the extreme case, the entropy may even remain zero.

The model we use now is an extension of the continuous measurement model discussed before (cf. Fig. 4.24). Each reset is a subsystem measurement according to the rules of (2.831), (2.705) and generalizations thereof. Absence of a reset ("no-click") is a continue command, adjusting the trace in the no-photon subspace (indicated by the index 0) back to the value 1. A typical network of that kind is shown in Fig. 4.24: a 4-node system coupled by Förster interactions is locally damped in node 1 ($W^{(1)}$) and 4 ($W^{(4)}$), respectively. These two (distinguishable) decay channels are each interpreted via zero-photon and 1-photon subensembles with subsequent reset actions degenerating the measurement protocol.

### 4.4.2 Parameter Fluctuations: A Model for $W_{s's}^{\mathrm{ad}}$

Parameters enter any Hamiltonian model and must be specified from the start. However, they may also fluctuate in time due to interactions with the environment. The effect of these fluctuations on the dynamics will be discussed now within a simple model. We start with a network consisting of an undamped 2-level subsystem coupled to a fluctuating environment, modelled as another 2-level system exposed to *incoherent thermal transitions*.

Consider an $SU(2) \otimes SU(2)$ scenario as depicted in Fig. 4.25. We assume that subsystem 1 is prepared in the superposition state (4.33), i.e.

$$\boldsymbol{\lambda} = (1, 0, 0) , \tag{4.137}$$

while subsystem 2 is subject to thermal transitions

$$E_2(2) - E_1(2) \ll k_{\mathrm{B}}T \tag{4.138}$$

with the thermal transition rates

$$W^+ = W^- . \tag{4.139}$$

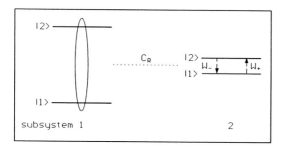

**Fig. 4.25.** An $SU(2) \otimes SU(2)$ network: node 1 is prepared in a superposition state (l.h.s.) and node 2 is subject to thermal transitions (r.h.s.)

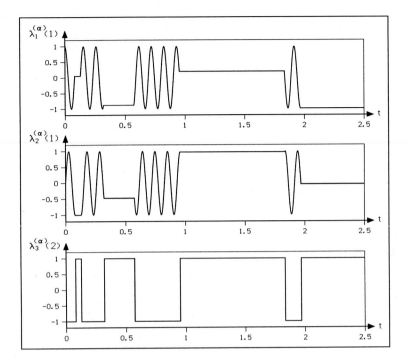

**Fig. 4.26.** Stochastic dynamics. Top: stochastic motion of the coherence vector of subsystem 1. Initial state: $\lambda_1(1) = 1$, $\lambda_2(1) = \lambda_3(1) = 0$. Transition frequency: $\omega_{21} = 30$. $C_R = -30$. Below: random hopping of $\lambda_3(2)$. Initial state: $\lambda_1(2) = \lambda_2(2) = 0$, $\lambda_3(2) = -1$. $W^+ = W^- = 5$

The autonomous dynamics of subsystem 2 is described by a simple rate equation (cf. 3.678):

$$\dot{\lambda}_3(2) = (W^+ - W^-) - (W^+ + W^-)\lambda_3(2) \,. \tag{4.140}$$

The time-dependent motion of this stochastic variable can be characterized as *random hopping*. The effective transition frequency of subsystem 1 is influenced by the state of subsystem 2 by means of the Coulomb interaction $C_R$. This is described by

$$
\begin{aligned}
\dot{\lambda}_1(1) &= -\lambda_2(1)\left[\omega_{21} + C_R\lambda_3(2)\right], \\
\dot{\lambda}_2(1) &= \lambda_1(1)\left[\omega_{21} + C_R\lambda_3(2)\right], \\
\dot{\lambda}_3(1) &= 0 \,.
\end{aligned}
\tag{4.141}
$$

A special stochastic trajectory $\lambda_3^{(\alpha)}(2)$ then generates a fluctuating angular frequency of $\boldsymbol{\lambda}^{(\alpha)}(1)$. When averaged over many trajectories $\alpha$, phase memory is lost, describing a $T_2$-damping. This interaction is "adiabatic", i.e. it

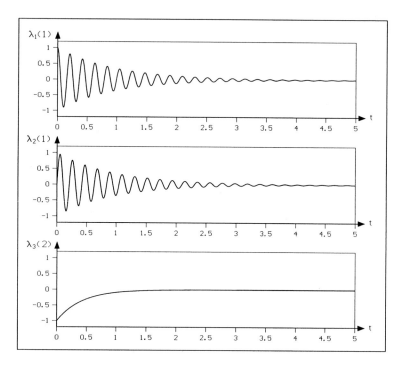

**Fig. 4.27.** Ensemble dynamics. Motion of the coherence vectors of subsystems 1 and 2. Initial state: $\lambda_1(1) = 1$, $\lambda_2(1) = \lambda_3(1) = 0$, $\lambda_1(2) = \lambda_2(2) = 0$, $\lambda_3(2) = -1$.

represents a model for $W_{s's}^{\text{ad}}$ according to (3.396) (all other parameters being zero).

The resulting stochastic dynamics based on (4.140), (4.141) is shown in Fig. 4.26. We are in a rotating frame in which the coherence vector $\boldsymbol{\lambda}(1)$ is at rest if $\boldsymbol{\lambda}(2) = (0,0,1)$. The motion of the corresponding ensemble is illustrated in Fig. 4.27. We realize that in the ensemble description both subsystems end up in a completely mixed state. The *destruction of coherence* in the ensemble results from averaging over the randomized phases. Any individual system, however, remains in a pure state in the $\lambda_1, \lambda_2$ plane for all times. This is similar to the inhomogeneous ensemble discussed in Sect. 3.3.5.8 for $t \ll T_2$. (Here, however, the decoherence of the ensemble cannot be undone: the fluctuation source is uncontrollable.) Though the simulation of this network proceeds in the manner outlined in Sect. 4.4.1, it is not a measurement scenario proper. The thermal bath inducing up and down transitions does not allow the introduction of two alternative subensembles 0 and 1; there is no information retrieval. In this sense, the parameter fluctuations (here, of $\omega_{21}$) do not destroy coherence of single quantum objects; rather we lose control of the current phases. This statement can be generalized to an individual network and its possible entanglement: those fluctuations act like a kind of time-dependent Hamiltonian.

### 4.4.3 Reacting and Non-reacting Environment

The "cut" between system and environment may often appear ambiguous. It should therefore be worthwhile considering a model in which criteria for the validity of a specific cut can easily be found.

For this purpose we consider a 2-level system coupled to a *reacting environment*, consisting of a second (damped) 2-level system. The coupling is via the Förster-type Coulomb interaction. We will demonstrate that for a given coupling strength $C_F$ but increasing damping the role played by the environment changes. For small damping the dynamics occurs in the whole $SU(2) \otimes SU(2)$ state space, while for large damping, the environment (including its 2-level subsystem) acts more and more like a rigid boundary condition: the dynamics is essentially confined to the system's $SU(2)$ state space.

**4.4.3.1 Evolution Equations in $SU(2) \otimes SU(2)$.** Consider a 2-level-2-node system described by (3.639) and (3.640),

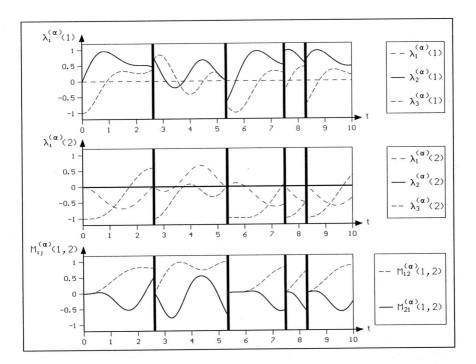

**Fig. 4.28.** Stochastic dynamics in $SU(2) \otimes SU(2)$. Non-zero parameters: $W_{12}^{(2)} = 1$, $g^{(1)} = 2$, $C_F = 1$. Vertical lines: reset events. Simulation by M. Keller (cf. [76])

$$\dot{\lambda}_1(v) - \dot{\lambda}_1^c(v) = -\frac{1}{2}W_{12}^{(v)}\lambda_1(v) \,,$$

$$\dot{\lambda}_2(v) - \dot{\lambda}_2^c(v) = -\frac{1}{2}W_{12}^{(v)}\lambda_2(v) \,, \tag{4.142}$$

$$\dot{\lambda}_3(v) - \dot{\lambda}_3^c(v) = -W_{12}^{(v)}\lambda_3(v) - W_{12}^{(v)}$$

and

$$\dot{K}_{kl}(1,2) - \dot{K}_{kl}^c(1,2) = -\frac{1}{2}\left(W_{12}^{(1)} + W_{12}^{(2)}\right)K_{kl}(1,2) \,,$$

$$\dot{K}_{k3}(1,2) - \dot{K}_{k3}^c(1,2) = -\left(\frac{1}{2}W_{12}^{(1)} + W_{12}^{(2)}\right)K_{k3}(1,2)$$
$$- W_{12}^{(2)}\lambda_k(1) \,,$$

$$\dot{K}_{3l}(1,2) - \dot{K}_{3l}^c(1,2) = -\left(W_{12}^{(1)} + \frac{1}{2}W_{12}^{(2)}\right)K_{3l}(1,2) \tag{4.143}$$
$$- W_{12}^{(1)}\lambda_l(2) \,,$$

$$\dot{K}_{33}(1,2) - \dot{K}_{33}^c(1,2) = -\left(W_{12}^{(1)} + W_{12}^{(2)}\right)K_{33}(1,2)$$
$$- W_{12}^{(1)}\lambda_3(2) - W_{12}^{(2)}\lambda_3(1)$$

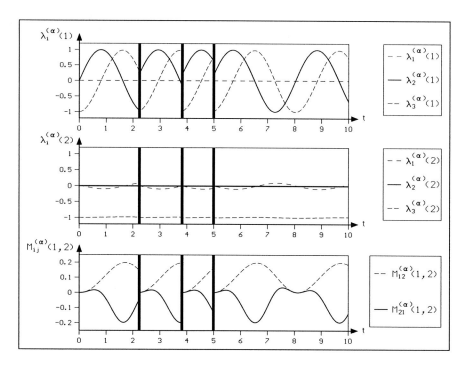

**Fig. 4.29.** Stochastic dynamics in $SU(2)\otimes SU(2)$. Non-zero parameters: $W_{12}^{(2)} = 20$, $g^{(1)} = 2$, $C_{\mathrm{F}} = 1$. Vertical lines: reset events. Simulation by M. Keller (cf. [76])

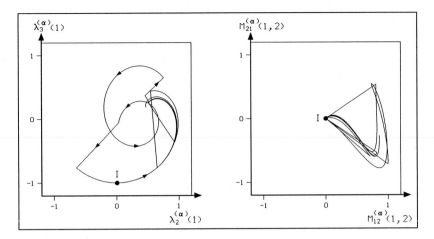

**Fig. 4.30.** Stochastic portrait in the $\lambda_2^{(\alpha)}(1), \lambda_3^{(\alpha)}(1)$ and $M_{12}^{(\alpha)}(1,2), M_{21}^{(\alpha)}(1,2)$ plane. $W_{12}^{(2)} = 1$. The initial state (I) is represented by the black dots. The straight lines represent the reset events. The arrows indicate the motion up to the second reset event

with

$$k, l = 1, 2 . \tag{4.144}$$

The subsystem 2 is now modelled by a stochastic trajectory as discussed before, i.e. after each finite time step the subsystem is updated to either

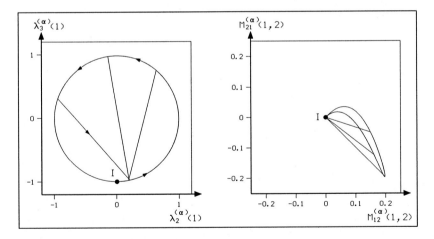

**Fig. 4.31.** Stochastic portrait in the $\lambda_2^{(\alpha)}(1), \lambda_3^{(\alpha)}(1)$ and $M_{12}^{(\alpha)}(1,2), M_{21}^{(\alpha)}(1,2)$ plane. $W_{12}^{(2)} = 20$. The initial state (I) is represented by the black dots. The straight lines represent the reset events. The arrows indicate the motion up to the first reset event

remain in its subensemble 0 or to change to subensemble 1 with immediate reset. This updating changes the values of all other state parameters $\lambda_i(1)$, $M_{ij}(1,2)$ accordingly. The reset is equivalent to an indirect measurement already studied in Sect. 2.4.5. Note that also the dynamics between reset events is non-unitarian.

Figures 4.28–4.31 show stochastic trajectories based on (4.142)–(4.143) and the algorithm presented in Sect. 4.2.3. In Fig. 4.28 the weak damping case is shown, in Fig. 4.29 the strong damping case is depicted. In the weak damping case, the dynamics of the coherence vector $\lambda(2)$ is fully developed; after reset (indicated by the vertical lines) acting on subsystem 2, subsystem 1 is kicked to almost any state. After each reset, $M(1,2) = 0$, as required. In the strong damping case, the dynamics of $\lambda(2)$ is largely suppressed, while now the reset projects subsystem 1 onto the $\lambda(1) = (0,0,-1)$ state. This dynamics is very close to the scenario with just one single damped 2-level system. This example illustrates that the network approach is not restricted to considering the network as the actual system: part of it can be used to simulate a reacting environment. It further demonstrates that the "correct" positioning of the "cut" can be tested: as it is most convenient to keep the quantum system to be treated in detail as small as possible, overdamped parts can safely be included as parts of bath boundary conditions.

Figures 4.30, 4.31 illustrate the stochastic portrait in coherence vector space and entanglement space. This complexity reminds us of chaotic motions observed in macroscopic (nonlinear) systems like liquids or laser systems (however, without the typical properties of deterministic chaos, compare Fig. 1.2). We note in passing that for (correlated) unitary mapping sequences in $SU(2)$ a type of chaotic behaviour has already been demonstrated (cf. [19]).

### 4.4.4 Antibunching

The trajectories $\lambda^{(\alpha)}(1)$ depicted in Fig. 4.29 give an intuitive account of the well-known phenomenon of *photon antibunching*: after each reset event (connected with the spontaneous emission of a photon), the 2-level atom first has to restart its venturing out into the non-classical (coherent) state space, before the next emission event is possible (cf. Fig. 4.32). This becomes most transparent in the *waiting time distribution* $f(\tau)$ (between subsequent luminescence photons), normalized according to

$$\int_0^\infty f(\tau)\mathrm{d}\tau = 1 . \tag{4.145}$$

A numerical analysis is shown in Fig. 4.33. It can be directly compared with the analytically calculated correlation function $g(\tau)$ shown in Fig. 4.34. This function is derived as follows: consider the master equations (3.530). For zero-detuning, $\tilde{\delta} = 0$, the master equations valid for the subensemble 0 (no photon) read

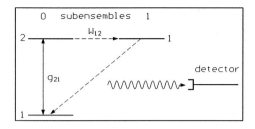

**Fig. 4.32.** Measurement of a driven 2-level system

$$\dot{R}_{11}^{(0)}(t) = -\frac{i}{2}g_{21}\left[R_{12}^{(0)}(t) - R_{21}^{(0)}(t)\right] \;,$$

$$\dot{R}_{22}^{(0)}(t) = -W_{12}R_{22}^{(0)}(t) + \frac{i}{2}g_{21}\left[R_{12}^{(0)}(t) - R_{21}^{(0)}(t)\right] \;, \tag{4.146}$$

$$\dot{R}_{12}^{(0)}(t) = -\frac{1}{2}W_{12}R_{12}(t) + \frac{i}{2}g_{21}\left[R_{22}^{(0)}(t) - R_{11}^{(0)}(t)\right] \;,$$

where $W_{12}$ denotes the decay probability of the node in the excited state. Starting from the ground state as the initial state (prepared by the photon emission at time $t = 0$), one obtains the analytical solution

$$R_{22}^{(0)}(t) = \begin{cases} \dfrac{g_{21}^2}{(|W_{12}^2 - 4g_{21}^2|)}L_{22}^{(1)}(t) & \text{for} \quad g_{21} < \dfrac{W_{12}}{2} \\[2ex] \dfrac{W_{12}^2}{16}t^2 L_{22}^{(2)}(t) & \text{for} \quad g_{21} = \dfrac{W_{12}}{2} \\[2ex] \dfrac{2g_{21}^2}{(|W_{12}^2 - 4g_{21}^2|)}L_{22}^{(3)}(t) & \text{for} \quad g_{21} > \dfrac{W_{12}}{2} \end{cases} \tag{4.147}$$

with

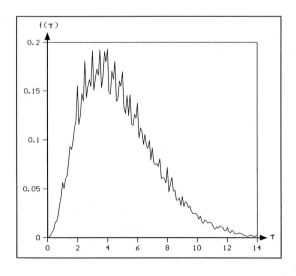

**Fig. 4.33.** Distribution of waiting times between two subsequent photon counts. Numerical simulation (cf. [149]). $W_{12} = 1$. $g_{21} = 0.6$

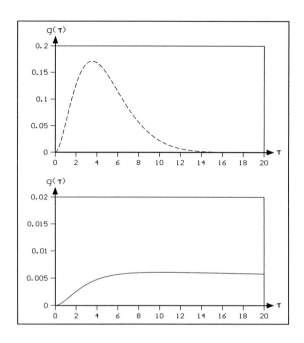

**Fig. 4.34.** Correlation function $g(\tau)$ (analytical solution). $W_{12} = 1$. Top: $g_{21} = 0.6$. Below: $g_{21} = 0.08$

$$L_{22}^{(1)}(t) = e^{\left(-\frac{1}{2}W_{12}+\frac{1}{2}\sqrt{|W_{12}^2-4g_{21}^2|}\right)t} +$$
$$e^{\left(-\frac{1}{2}W_{12}-\frac{1}{2}\sqrt{|W_{12}^2-4g_{21}^2|}\right)t} -$$
$$2e^{-\frac{1}{2}W_{12}t} , \tag{4.148}$$

$$L_{22}^{(2)}(t) = e^{-\frac{1}{2}W_{12}t} , \tag{4.149}$$

$$L_{22}^{(3)}(t) = e^{-\frac{1}{2}W_{12}t}\left[1 - \cos\left(\frac{\sqrt{|W_{12}^2 - 4g_{21}^2|}}{2}t\right)\right] . \tag{4.150}$$

The expected probability for the next photon emission in the time interval $\tau \leq t \leq \tau + d\tau$ is then given by

$$dp = W_{12}R_{22}^{(0)}(\tau)d\tau . \tag{4.151}$$

Using the analytical solution (4.147), the 2-time-photon-correlation function

$$g(\tau) = \frac{dp}{d\tau} \tag{4.152}$$

can be calculated:

$$g(\tau) = W_{12}R_{22}^{(0)}(\tau) . \tag{4.153}$$

In Fig. 4.34 two cases are depicted. Any of these go to zero for $\tau \to 0$, the antibunching effect proper. If the driving field intensity is small ($g_{21} \ll$

$W_{12}$), emissions become a very rare event (compare both graphs), and the 2-level system hardly ever leaves its ground state (coherence can eventually be neglected).

### 4.4.5 Quantum Zenon Effect

The *quantum Zenon effect*, perhaps more appropriately known as the "watchdog effect", was originally designed to challenge the predictions of quantum mechanics (cf. [98]): if a resonantly driven 2-level system is observed, it will be found either in state $|2\rangle$ or $|1\rangle$, its measurement basis. If one now repeats such measurements at times $t_i = iT/N$ after the system has started in state $|1\rangle$ at time $t = 0$, the probability to find the system in state $|2\rangle$ after a $\pi$ pulse (defined by $g_{21}T = \pi$) decreases. In fact for $N \to \infty$ their probability goes to zero, an apparently paradoxical situation.

*Proof.* Consider a driven 2-level system (cf. (3.94)),

$$\boldsymbol{\Gamma} = (g_{21}, 0, 0) ,$$ (4.154)

with initial state

$$\boldsymbol{P} = (0, 0, -1) .$$ (4.155)

The solution of the corresponding Bloch equations is (cf. (3.113))

$$P_x(t) = 0 , \quad P_y(t) = \sin(g_{21}t) , \quad P_z(t) = -\cos(g_{21}t)$$ (4.156)

so that after time $T$ with $g_{21}T = \pi$ ($\pi$ pulse)

$$\boldsymbol{P}(T) = (0, 0, 1) .$$ (4.157)

Suppose now that for a given integer $N > 1$ and times

$$t_i = \frac{T}{N}i \quad (i = 1, 2, 3, \ldots, N)$$ (4.158)

there are ideal measurement projections on

$$\boldsymbol{P}'(t_i) = (0, 0, \pm 1)$$ (4.159)

with probability

$$p_{\mp 1} = \frac{1}{2}[1 \mp p_z(t_i)]$$ (4.160)

(cf. (2.434)). For $t = t_1^-$ just before measurement one then obtains

$$\boldsymbol{P}(t_1^-) = (0, \sin(\pi/N), -\cos(\pi/N))$$ (4.161)

so that after ensemble measurement we have

$$\boldsymbol{P}(t_1^+) = (0, 0, -\cos(\pi/N)) .$$ (4.162)

This is now taken as the initial state for the next period of coherent evolution until $t_2$. If this procedure is repeated $N$ times, we find at time $t_N = T$:

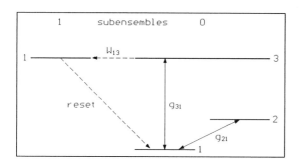

**Fig. 4.35.** The $\nu$ scenario of the quantum Zenon effect

$$\boldsymbol{P}(T) = \left[0, 0, -\cos^N\left(\pi/N\right)\right] \ . \tag{4.163}$$

Taking into account that

$$\cos^N\left(\pi/N\right) \simeq \left(1 - \frac{\pi^2}{2N^2} + \ldots\right)^N \simeq 1 - \frac{\pi^2}{2N^2} \ , \tag{4.164}$$

we conclude that

$$\boxed{\boldsymbol{P}(T) = (0, 0, -1) \ .} \tag{4.165}$$

Such a "suppression of motion" has meanwhile been observed experimentally with ensembles of 3-level atoms, all prepared in the ground state at same initial time $t = 0$ (cf. [73]). A second light field connecting the ground state 1 to a damped level 3 (which can be switched on and off) serves as a measurement channel. As shown by the appropriate master equations, these measurements indeed reduce the population of level 2 at time $g_{21}T = \pi$ (cf. [52]).

However, the suppression of motion is only one feature: more details emerge in the stochastic simulation of a single 3-level atom: consider the $\nu$ scenario sketched in Fig. 4.35 with

$$\delta_{21} = \delta_{31} = 0 \tag{4.166}$$

(i.e. resonant transitions) and

$$W_{12} = W_{21} = W_{23} = W_{32} = W_{31} = 0 \tag{4.167}$$

(i.e. level 2 – at least approximately – represents a stable state). The respective equations of motion are (cf. (3.596)):

$$\dot{\lambda}_1 = -\frac{1}{2}g_{31}\lambda_6 \ , \tag{4.168}$$

$$\dot{\lambda}_2 = \frac{1}{2}g_{21}\lambda_6 - \frac{1}{2}W_{13}\lambda_2 \ , \tag{4.169}$$

$$\dot{\lambda}_3 = \frac{1}{2}g_{31}\lambda_4 + \frac{1}{2}g_{21}\lambda_5 - \frac{1}{2}W_{13}\lambda_3 \ , \tag{4.170}$$

$$\dot{\lambda}_4 = -\frac{1}{2}g_{31}\lambda_3 - g_{21}\lambda_7 \;, \tag{4.171}$$

$$\dot{\lambda}_5 = -\frac{1}{2}g_{21}\lambda_3 - \frac{1}{2}g_{31}\lambda_7 - \frac{1}{2}\sqrt{3}g_{31}\lambda_8 - \frac{1}{2}W_{13}\lambda_5 \;, \tag{4.172}$$

$$\dot{\lambda}_6 = \frac{1}{2}g_{31}\lambda_1 - \frac{1}{2}g_{21}\lambda_2 - \frac{1}{2}W_{13}\lambda_6 \;, \tag{4.173}$$

$$\dot{\lambda}_7 = g_{21}\lambda_4 + \frac{1}{2}g_{31}\lambda_5 - \frac{1}{\sqrt{3}}W_{13}\lambda_8 - \frac{1}{3}W_{13} \;, \tag{4.174}$$

$$\dot{\lambda}_8 = \frac{1}{2}\sqrt{3}g_{31}\lambda_5 - W_{13}\lambda_8 - \frac{1}{\sqrt{3}}W_{13} \;. \tag{4.175}$$

This system is now simulated stochastically. A light field of frequency $\omega^{(21)} = \omega_{21} = (E_2 - E_1)/\hbar$ and coupling constant $g_{21} = 0.002$ induces coherent oscillations between the states 1 and 2. In Fig. 4.36, part a) this dynamics is shown. This behaviour is completely coherent but inaccessible to an outside observer. If the light field $g_{31}$ is switched on (trace b)), dissipation becomes active which qualitatively modifies the smooth oscillation of case a): projections to the ground state are connected with the emission of luminescence photons. These luminescence photons may be detected; here we assume that the detector efficiency is 100 %. Comparing the protocol c) with the system dynamics b), it is fairly obvious that the following "logic" applies: if we see a photon, the system is in the ground state. For higher intensity (trace d)) the coherent motion has completely given way to a stochastic telegraph signal. The corresponding photon counting protocol shows light and dark sections (trace e)). The watchdog effect is clearly seen: during the light period the system is "frozen" in state $\lambda_7 = -1$. However, this does not last forever: even for $g_{31} \gg W_{13}$ (saturation) the system will eventually flip to $\lambda_7 = 1$ and "freeze" again. There are actually three time scales: on the time scale $T = \pi/g_{21}$, motion is suppressed; on a longer time scale, we see a flipping between light and dark periods; on a still larger time scale, we are back to an averaged continuous light emission.

The watchdog effect is but one aspect in the transition from coherent to stochastic motion. Contrary to axiomatic measurement projections (which are taken to be local in time), real measurements require a finite interaction time: therefore, an "infinite number" of measurements within a finite time period is impossible; motion cannot completely be suppressed for finite damping rates.

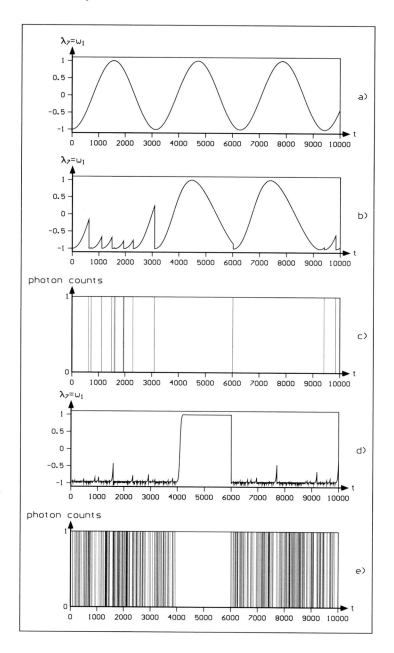

**Fig. 4.36.** Continuous quantum Zenon effect. a) $g_{31} = 0$: coherent dynamics, no damping, no information retrieval. b) $g_{31} = 0.03$: interrupted oscillations, rare detection events. c) Photon counting protocol corresponding to b). d) $g_{31} = 0.2$: stochastic motion between state 1 and 2. e) Photon counting protocol of d) ([149])

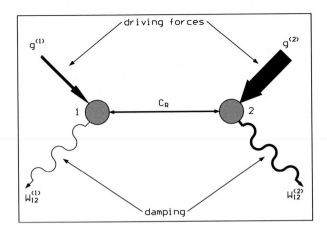

**Fig. 4.37.** 2-node Zenon scenario. Two physically different 2-level systems coupled by Coulomb interaction $C_R$, with local driving $g^{(\nu)}$ and damping $W_{12}^{(\nu)}$

### 4.4.6 Zenon Effect in $SU(2) \otimes SU(2)$

The Zenon effect may also be realized within 3-level subspaces of composite systems. Let us consider two 2-level systems, coupled via the renormalization interaction $C_R$ (cf. Figs. 4.37 and 2.28). Both subsystems are driven by light fields with coupling constant $g^{(i)}$ ($g^{(1)} \ll g^{(2)}$). The non-zero damping parameters are $W_{12}^{(1)} \ll W_{12}^{(2)}$. The respective frequencies are selected to match resonance as sketched in Fig. 4.38. Without interaction, $C_R = 0$, the system 1 – weakly coupled to its environment – would hardly be seen: system 2 would dominate the luminescence. With $C_R \neq 0$ system 1 induces light and dark periods, while (at the same time) it is forced into the "classical sector" $\lambda_3(1) = \pm 1$ by system 2. The stochastic simulation is shown in Fig.

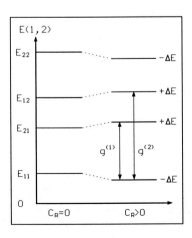

**Fig. 4.38.** 2-node Zenon: the level scheme. Shown are the two resonant transitions

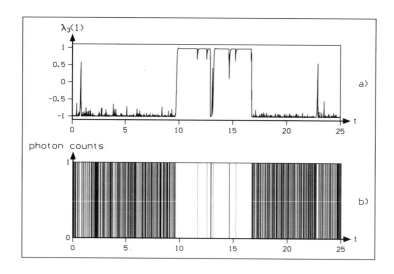

**Fig. 4.39.** 2-node Zenon effect. Stochastic simulation of a single 2-level pair. a) Coherent dynamics. b) The number of emitted photons: every line represents a single photon ([76])

4.39. Such effects may explain the fluctuating luminescence intensity traces observed in [4].

### 4.4.7 Non-local Damping: Superradiance

So far, we have restricted ourselves to local damping channels, i.e. dissipative transitions induced within individual nodes. As a result, these transitions produce local decisions and local information if registered. As has been discussed in Sect. 2.4.5, these subsystem measurements may, nevertheless, produce non-local updating effects in the whole network.

Non-local damping channels are expected to address subgroups of $N \geq 2$ nodes in such a way that within such a subgroup entanglement will not necessarily be destroyed: one may say that the damping cannot "distinguish" between the nodes within this subgroup (the volume occupied by this subgroup is then often called *coherence volume*).

As the simplest example we consider an $SU(2) \otimes SU(2)$ network without mutual interactions:

$$\hat{H} = \hat{H}_1 + \hat{H}_2 , \tag{4.176}$$

$$\hat{H}_1 = \hat{H}(1) \otimes \hat{1}(2) , \tag{4.177}$$

$$\hat{H}_2 = \hat{1}(1) \otimes \hat{H}(2) , \tag{4.178}$$

and

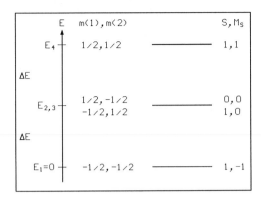

Fig. 4.40. Superradiance: the level scheme

$$\hat{H}(\nu) = \frac{1}{2}\Delta E \hat{w}_1(\nu) , \tag{4.179}$$

$$\Delta E = E_2 - E_1 . \tag{4.180}$$

(Note that $\Delta E$ is the same for each subsystem $\nu = 1, 2$.) The eigenenergies and eigenstates of the total system are

$$
\begin{aligned}
E_1 &= 0 : & |\psi_1\rangle &= \left|-\tfrac{1}{2}, -\tfrac{1}{2}\right\rangle , \\
E_{2,3} &= \Delta E : & |\psi_2\rangle &= \left|-\tfrac{1}{2}, \tfrac{1}{2}\right\rangle , \quad |\psi_3\rangle = \left|\tfrac{1}{2}, -\tfrac{1}{2}\right\rangle , \\
E_4 &= 2\Delta E : & |\psi_4\rangle &= \left|\tfrac{1}{2}, \tfrac{1}{2}\right\rangle .
\end{aligned}
\tag{4.181}
$$

We recall that the spin angular momenta are defined as

$$\hat{s}_z(\nu) = \frac{\hbar}{2}\hat{w}_1(\nu) , \tag{4.182}$$

and the total spin vector is given by

$$\hat{\boldsymbol{S}} = \hat{\boldsymbol{s}}(1) + \hat{\boldsymbol{s}}(2) . \tag{4.183}$$

The four eigenstates can now alternatively be written as eigenstates of the total spin: $|S, M_S\rangle$. The singlet state

$$|0,0\rangle = \frac{1}{\sqrt{2}}\left(\left|-\tfrac{1}{2}, \tfrac{1}{2}\right\rangle - \left|\tfrac{1}{2}, -\tfrac{1}{2}\right\rangle\right) := |\psi_-\rangle \tag{4.184}$$

is degenerate with the triplet state

$$|1,0\rangle = \frac{1}{\sqrt{2}}\left(\left|-\tfrac{1}{2}, \tfrac{1}{2}\right\rangle + \left|\tfrac{1}{2}, -\tfrac{1}{2}\right\rangle\right) := |\psi_+\rangle . \tag{4.185}$$

(See Fig. 4.40) It is the type of coupling which implies a preference for one representation over the other. Assume that each 2-level subsystem is coupled to a photon bath inducing transitions independent of $\nu$:

$$\left\langle -\tfrac{1}{2}(\nu) \left| \hat{H}'(\nu) \right| \tfrac{1}{2}(\nu) \right\rangle = \frac{\hbar}{2}M_\mp , \tag{4.186}$$

$$\left\langle \tfrac{1}{2}(\nu) \left| \hat{H}'(\nu) \right| -\tfrac{1}{2}(\nu) \right\rangle = \frac{\hbar}{2}M_\pm \tag{4.187}$$

with the interaction operator

$$\hat{H}'(\nu) = M_\mp \hat{\sigma}_-(\nu) + M_\pm \hat{\sigma}_+(\nu) \ . \tag{4.188}$$

Thus the total Hamiltonian

$$\hat{H} = \sum_\nu \hat{H}(\nu) + \sum_\nu \hat{H}'(\nu) \tag{4.189}$$

can be written as

$$\hat{H} = \frac{1}{\hbar} \Delta E \hat{S}_z + M_\mp \hat{S}_- + M_\pm \hat{S}_+ \tag{4.190}$$

with

$$\hat{S}_\pm = \frac{\hbar}{2} \sum_\nu \hat{\sigma}_\pm(\nu) \ , \tag{4.191}$$

$$\hat{S}_z = \sum_\nu \hat{s}_z(\nu) \ . \tag{4.192}$$

One shows that

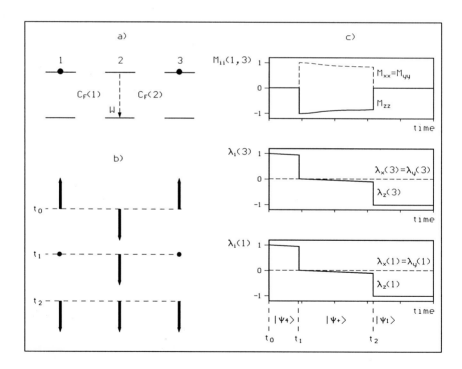

**Fig. 4.41.** Superradiance: decay scenario (part a)), flip of the coherence vectors (part b)), and time-evolution of coherence vectors and correlation tensor (part c)). $C_F(1) = C_F(2) = 2$. $W = 20$

$$\left\langle SM_S \left| \hat{S}_\mp \right| S'M'_S \right\rangle = \frac{\hbar}{2}\sqrt{(S+1)S - M(M-1)}\delta_{SS'}\delta_{M'M\pm 1} , \quad (4.193)$$

i.e. $S$ is conserved in these transitions; the singlet state cannot be reached from the triplet states. Furthermore, if the transition matrix element within the individual atom is proportional to

$$\left\langle -\frac{1}{2}(\nu) \left| \hat{S}_-(\nu) \right| \frac{1}{2}(\nu) \right\rangle = \frac{\hbar}{2} \quad (\nu = 1,2) , \quad (4.194)$$

the transition from the entangled state $|1,0\rangle$ is enhanced:

$$\left\langle 1,-1 \left| \hat{S}_-(\nu) \right| 1,0 \right\rangle = \hbar \quad (4.195)$$

(*superradiance*). This scheme is readily generalized to $N > 2$ atoms (cf. [37]).

The decay of an $SU(2) \otimes SU(2)$ system prepared in the initial state $|\psi\rangle = \left|\frac{1}{2},\frac{1}{2}\right\rangle$ can easily be simulated. Rather than modifying directly the damping model, we can enlarge the system as shown in Fig. 4.41. The 2-level system 2 now stands for an overdamped photon mode to be measured. With a symmetric Förster coupling to its excited neighbours (subsystems 1, 3) the observed system cannot distinguish between the left and the right source: with the first photon click the 2-atom system is projected into the entangled state with $\lambda(1) = \lambda(2) = 0$, $M_{xx} = M_{yy} = 1$, $M_{zz} = -1$. (The interactions $C_F(1)$, $C_F(2)$ slightly change the eigenenergies not included in the rotating frame representations: this is why the state parameters are not exactly constant.) The entanglement is reduced if the coupling becomes asymmetric (i.e. $C_F(1) \neq C_F(2)$) or if there is asymmetric detuning. This is the simplest example in which entanglement is generated by dissipation. However, this entangled state is only transient and decays to the ground state.

### 4.4.8 Driven 3-Node System: Relaxation into Entanglement

Dissipation plays a constructive role in non-equilibrium phase transitions (cf. [64]). Dephasing-induced phenomena have also long since been noted in nonlinear optics: coherent optical mixing resonances that would not occur without damping have been studied experimentally with atom vapours. Here, collisions are the source of incoherent perturbation.

Interacting quantum networks show such a behaviour in a much more pronounced way. For this purpose we consider a chain of four 2-level systems with nearest-neighbour coupling $C_R$ (see Fig. 4.42). Systems E and D are damped, systems 1 and 2 are undamped. In a first step ($t = t_1$), 1 and 2 are each prepared in a local superposition state $\lambda_y(1) = 1$, $\lambda_y(2) = -1$. Then, in a second step, system E is optically driven at a frequency that would be resonant for E with the neighbours in state $\lambda_z(1) = -1$, $\lambda_z(2) = 1$, or $\lambda_z(1) = 1$, $\lambda_z(2) = -1$ (these 2 environments are indistinguishable in energy space). If system E is found to be in resonance (high luminescence signal) its environment consisting of system 1 and 2 is transferred into an entangled state

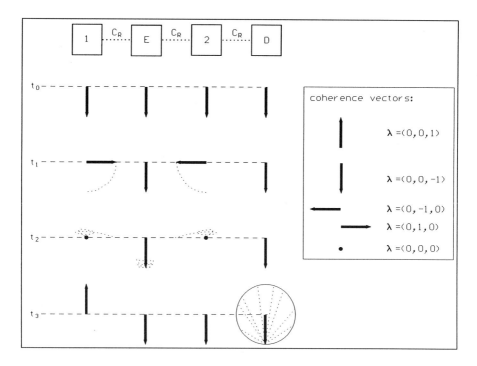

**Fig. 4.42.** Entanglement in a dissipative chain. The last step (time $t_3$) only serves for confirmation of the entanglement. The inset defines the type of states used

with $M_{xx}(1,2) = M_{yy}(1,2) = M_{zz}(1,2) = -1$ (EPR state): local properties have disappeared. This is indicated by the two dots representing $\lambda = 0$ states. Then the driving is switched off. (The entangled state is an attractor state of the chain.) The last step consists of a local measurement performed on system 2. For this purpose system D is driven in such a way that resonance is obtained if system 2 is in the ground state $\lambda_z(2) = -1$. The broken lines in the Bloch circle for system D indicate the jumping to the ground state. In the present simulation (cf. Fig. 4.43) system 2 happens to be projected into the ground state; as a consequence, system 1 is kicked into the excited state. $M = 0$ after this subsystem measurement.

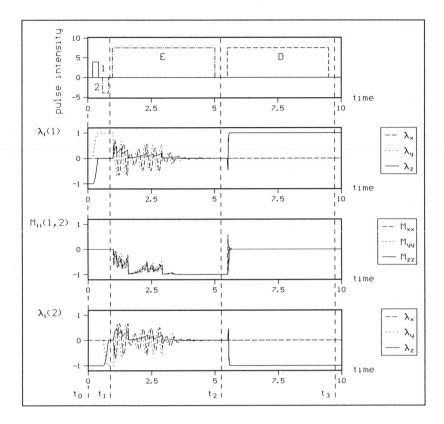

**Fig. 4.43.** Motion of coherence vectors and correlation tensor, numerical simulation (by M. Keller, cf. [76] )

## 4.5 Summary

In this chapter we have studied stochastic processes associated with continuous (idealized) *measurement events* (= counting events). The stochastic rules, based on a decomposition of the *Lindblad operator* (defining the most general form of a Markovian master equation), allow to generate "quantum trajectories". In general, these trajectories consist of a continuous (non-unitarian) evolution, interrupted by *discontinuous measurement projections*. Local damping implies local subsystem projections, which – due to non-local correlations (entanglement) – tend to update the whole network. To avoid misunderstandings; the quantum trajectories are not "ontological" properties of the respective quantum systems proper, rather they result from the embedding, like "footprints in the snow".

In the incoherent (i.e. "classical") limit the quantum trajectories reduce to *random telegraph signals*: various examples have been discussed for illustration, including *cooperative quantum jumps*. Cooperation results here from "classical" correlations: the rate of one node depends on the actual state of the other. Even in the incoherent limit, further *state space compression* is possible if certain *parameter hierarchies* are fulfilled.

The partly coherent case is the most general. We have restricted ourselves to a discussion of well-known effects (*superradiance, quantum Zenon effect*) and scenarios in which dissipation gives rise to entangled states as attractors. The network approach allows the simulation of effectively non-Markovian behaviour: subsystem measurements combined with entanglement generate memory effects. The network approach also allows an intuitive account of non-local damping to be given.

# 5. Summary

## 5.1 Background

We started our journey through quantum networks with the *electron* and the *photon spin*, both described by the $SU(2)$ *algebra*. We then generalized to pseudospins realized by 2-level systems or restricted *number states* (for example of *cavity photon modes*) and to 3- or 4-level systems implementing $SU(3)$ or $SU(4)$ algebras. Both static and dynamic properties have been investigated. In particular, the coupling to a *measurement scenario* has been considered, where the most interesting feature of such *1-node networks* is that direct (ideal) measurements will always find the system in one of the $n$ *measurement states*, which – by the way – resembles throwing dice in the classical world.

In general, however, *quantum objects* live in a large $(n^2 - 1)$-dimensional space. Furthermore, if the system consists of $N$ interacting nodes, many-node coherence ("entanglement") has to be taken into account, in addition to single-node coherence. The theoretical description of coherence in the case of 2- and 3-node networks has been studied. The discussion of typical *2-node coherence* apparently started with *Aspect's experiments* on *entangled photon states* and the *Kocher-Cummings experiment*. *Greenberger-Horne-Zeilinger states* (GHZ states), i.e. special 3- and 4-node coherent states which are leading to new experimental schemes, and their interesting theoretical properties have been examined.

Access to this fascinating world was severely limited for a long time. Classical *atomic spectroscopy* has been concerned with the observation of *occupation levels* and corresponding *rate equations*; single node (single atom) coherence surfaced with the technology of *coherent optics*, and – on the theoretical level of description – with, e.g., the *Bloch equations*. Such coherence effects have also been under investigation in spin resonance experiments; *cavity electrodynamics* and *tunneling systems* provide other examples. Such networks have been considered in this book. For example, ensembles of non-interacting nodes in optics and their *coherent dynamics* have been studied.

However, more and more experimentalists are now gaining access to the world of interacting microsystems: the advances in *nanotechnology* represent examples. On the basis of such systems, *quantum computation* requiring many-node coherence may become possible. The theoretical scheme covered in this book deals with methods allowing the description of the complex behaviour of such *nanostuctures*. In particular, the concept of *density matrix* allows consideration of quantum networks coupled to a *measurement environment* which represents a special kind of *bath*. Such an environment, which very often can be treated as a kind of *boundary condition*, then leads to *incoherent damping effects*.

Starting from a general equation of motion defining the *unitary dynamics* of a *closed quantum system*, the so-called *Liouville equation*, and implementing terms describing non-unitary damping effects occurring in *open quantum systems*, one obtains *evolution equations*. In particular, the *Markovian master equation* was introduced to include incoherent damping effects caused by the environment (bath). The transcription into $SU(n)$ then led to *damped Bloch equations* and their generalized form in higher dimensions. In contrast to rate equations, which determine the time-development of *occupation probabilities* of energy states, the *generalized Bloch equations* characterize the dynamics of measurement quantities like *magnetization* or *electric polarization*. Such generalized Bloch equations describe the motion of a *coherence vector* determining 1-node coherence.

General *network equations* in $SU(n) \otimes \ldots \otimes SU(n)$ which include the generalized Bloch equations have been derived. These network equations also allow the description of the evolution of *correlations* occurring in *composite systems*, characterized by *correlation tensors*. *Numerical simulations* then led to new insight into the complex dynamics of composite systems.

The influence of *damping channels* causes *stochastic behaviour* on the level of the individual network nodes. This behaviour can be modelled by *stochastic quantum trajectories*; in the overdamped case (virtually no coherence) these trajectories reduce to *random hopping processes* in a corresponding state space. Concepts of stochastic modelling have been considered. In this context, in particular, the occurrence of *collective dynamics* (which occurs when specific *control parameters* reach critical values) has been investigated. This shows that the concept of *synergetics* may also be applicable in quantum network theory if the particular nature of quantum systems is taken into account.

The central aim of this book has been to introduce concepts for the theoretical description of quantum networks such as nanostructures and to illustrate the considerations by experimentally and technologically relevant examples. Only networks with few energy levels and few nodes have been studied in more detail. In these cases, basic *design principles* based on *hierarchical structures* (which – on a mathematical level – may lead to special *reduction procedures* like *adiabatic elimination*) have been considered. However, an extension of the concepts presented in this book to systems of higher complexity is possible. The main challenging features are related to non-locality in space and time.

In the following section, the key topics treated in this book will be summarized.

## 5.2 Key Topics

### 5.2.1 Quantum Networks and Nanostructures

In this book, systems of interacting (or non-interacting) *quantum objects* have been called *quantum networks*. Thus, *quantum networks* are composed of individual subsystems ("nodes") with specific interactions ("edges"). A "network" consisting of only 1 node then can be considered a limiting case. A network of many non-interacting identical nodes (or groups of nodes) approximates a homogeneous ensemble.

Special realizations of quantum networks are so-called *semiconductor-nanostructures*. Such nanostructures can be characterized as *assemblies of quantum objects* in the nanometer region. Arrays of single quantum objects called *quantum dots* are pertinent examples. Such nanostructures allow the development of devices with a wealth of different properties. Elementary nanostructures such as coupled *2-level systems*, *quantum-dot arrays*, and *chains of defects* were studied in this book. Other realizations involve Fock states of cavity photon modes.

### 5.2.2 Coherence and Correlation

Pure states are states which are completely described by a single quantum mechanical state vector. Thus, a pure state of a network node can be expressed as a superposition of local basis states (*coherent state*). Describing the quantum object on the basis of generating operators spanning an $SU(n)$ algebra, this *1-node coherence* is represented by $n^2 - 1$ expectation values of the generating operators. These expectation values form the so-called *coherence vector*, which can be considered as a generalized Bloch vector.

In composite systems, i.e. systems composed of more than 1 node, relations between the individual nodes occur. These correlations (entanglement) typically derive from physical interactions (like Coulomb interactions), which, however, can also be built up between nodes without any direct physical interaction (and can survive after all interactions have died away). The description of these entanglements on the basis of *correlation tensors* has been studied in detail. In contrast to *local correlations* described by a coherence vector, correlation tensors characterize *non-local correlations*.

### 5.2.3 Closed and Open Systems

Both coherence vectors of individual nodes and correlation tensors can be derived from the *density operator* which allows the description of pure and mixed states of quantum networks. (Mixed state = quantum state which cannot be described be a single state vector. Mixed states have to be described by more than one state vector, where each vector occurs with a special probability. Such mixed states result, for example, for subsystems if correlations with other subsystems are traced out.)

A *closed quantum network* is an idealization. In reality interactions with the *network environment* have to be taken into account, i.e. *open systems* provide a more realistic description. In such cases, to *coherent interactions* (for example caused by a coherent light beam) due to various kinds of *damping channels*, *incoherent interactions* with the network occur. If the environment of the network can be made to act as a kind of *boundary condition*, a closed description of the network dynamics is possible. In this book, both closed and open systems have been studied.

### 5.2.4 Network Equations

*Network equations* describe the dynamical behaviour of a quantum network. In this book, systems of *coupled Bloch equations* have been especially considered. Such network equations describe the dynamics of expectation values in closed and open networks. As a borderline case, these network equations include *generalized Bloch equations* determining the time-evolution of expectation values related to individual nodes.

Inserting a density operator in $SU(n)$ representation into the fundamental Liouville equation, and considering, in addition, *damping terms* representing the incoherent influence of the *environment*, such network equations can be shown to follow. These network equations describe the time-evolution of expectation values contained in local coherence vectors as well as the dynamics of non-local correlations represented by correlation tensors. Due to this representation of the network dynamics, such network equations are *evolution equations* in $SU(n) \otimes \ldots \otimes SU(n)$.

In contrast to Bloch equations, master equations, such as the *Pauli master equation*, describe the time-evolution of *occupation probabilities* to find a node or a network in a special energy state. Such *rate equations* have also been considered. They are a limiting case of the *damped Bloch equations*.

Evolution equations like the Liouville equation or Schrödinger's equation allow the calculation of the *coherent dynamics* of closed systems. Rate equations, such as the Pauli master equation, or overdamped network equations allow the calculation of the *incoherent dynamics* of open systems. The general (coupled) network equations cover both limiting cases and partly coherent dynamics in between.

### 5.2.5 Measurement

In order to gain information about a quantum network, *measurement equipment* is needed. Such equipment predetermines the network environment which causes changes in the observed network.

If a 1-node network is observed, and if a *direct measurement* is carried out, *measurement projections* occur which create well-defined measurement states, i.e. after measurement, observed quantum objects are found in states with well-defined properties (such as energy, momentum, etc.). The observation of spin states in a *Stern-Gerlach measurement* represents an example.

If a large network is investigated, the individual quantum objects of the network cannot be observed directly and so an *indirect measurement* is necessary. The observation of non-interacting ensembles of spins via an electron paramagnetic resonance experiment is an example in which the *absorption measurement* then changes the state of the spin ensemble.

In extended networks measurements cause damping effects via *damping channels*. The decay of the radiation intensity in an ensemble of isolated and excited gas atoms during a *luminescence measurement* is a pertinent example.

The influence of a measurement enters the specific equations of motion. For example, damping terms in evolution equations can be considered as terms describing the damping by the measurement environment.

### 5.2.6 Stochastic Dynamics and Measurement

In the case of a *continuous measurement*, the single nodes of a network show *random behaviour*. For example, observing one single node, *random hopping* in state space may become observable. The *stochastic behaviour* of one single node during a continuous measurement can be interpreted as a chain of measurement projections caused by the measurement.

This *stochastic dynamics* of single nodes can be illustrated on the basis of *stochastic quantum trajectories*. In general, these trajectories are not confined to the strictly "classical" limit (also called *Ising limit*): the indirectly observed nodes may well venture out into the coherent section of their state space. For example, an optically driven 2-level atom may continue to develop partial coherence after each jump to its ground state (registered by the emission of a luminescence photon). Only in the overdamped case is coherence virtually suppressed. The influence of a measurement on the behaviour of the individual nodes has been a main topic in this book.

# References

1. Alicki R., Lendi K.: Quantum Dynamical Semigroups and Applications (Springer, Berlin Heidelberg 1987)
2. Allen L., Eberly J.H.: Optical Resonance and Two-Level-Atoms, (Dover, New York 1987)
3. Ambrose W.P., Moerner W.E.: Fluorescence Spectroscopy and Spectral Diffusion of Single Impurity Molecules in a Crystal, Nature **349**, 225 (1991)
4. Ambrose W.P., Goodwin S.C., Martin J.C., Keller R.A.: Single Molecule Detection and Photochemistry on a Surface using Near-Field Optical Excitation, Phys. Rev. Lett. **72**, 160 (1994)
5. Andersson M.O., Xiao Z., Norrman S., Engström O.: Model based on Trap-Assisted Tunneling for 2-Level Current Fluctuations in Submicrometer Metal $SiO_2$-Si-diodes, Phys. Rev. B **41**, 9836 (1990)
6. Araki H., Lieb E.: Entropy Inequalities, Commun. Math. Phys. **18**, 160 (1970)
7. Ashburn J.R., Cline R.A., van der Burgt P.J.M., Westerweld W.B., Risley J.S.: Experimentally determined Density Matrices for H($n = 3$) formed in $H^+$-He-Collisions from 20 to 100 keV, Phys. Rev. A **41**, 2407 (1990)
8. Aspect A., Grangier P., Roger G.: Experimental Realization of Einstein-Podolski-Rosen-Bohm Gedanken Experiment, Phys. Rev. Lett. **49**, 91 (1982)
9. Awschalom D.D., DiVincenzo, Smith J.F.: Macroscopic Quantum Effects in Nanometer-Scale Magnets, Science **258**, 414 (1992)
10. Barner J.B., Ruggiero S.T.: Observation of Incremental Charging Effects of Ag-Particles by Single Electrons, Phys. Rev. Lett. **59**, 807 (1987)
11. Barnett S.M., Phoenix S.J.D.: Information Theory, Squeezing, and Correlations, Phys. Rev. A **44**, 535 (1991)
12. Barnett S.M., Phoenix S.J.D.: Information-Theoretic Limits to Quantum Cryptography, Phys. Rev. A **48**, R5 (1993)
13. Bennett C.H., Brassard G., Crepeau C., Josza R., Peres A., Wooters W.K.: Teleporting an Unknown Quantum State via Dual Classical and EPR-Channels, Phys. Rev. Lett. **70**, 1895 (1993)
14. Bergmann R., Menschig A., Lichtenstein N., Hommel J., Härle V, Scholz F., Schweizer H., Grützmacher D.: Investigation of Boundary Scattering in Dry Etched Quantum Wires, Microelectronic Engineering **23**, 429-432 (Elsevier Science B.V. 1994)
15. Bergquist J.C., Hulet R.G., Itano W.M., Wineland D.J.: Observation of Quantum Jumps in a Single Atom, Phys. Rev. Lett. **57**, 1699 (1986)
16. Bethe H.A., Jackiw R.: Intermediate Quantum Mechanics, (Benjamin, New York 1986)
17. Biafore M.: Cellular Automata for Nanometer-Scale Computation, Physica D **70**, 415 (1993)
18. Blochinzew D.I.: Grundlagen der Quantenmechanik, (Harri Deutsch, Frankfurt am Main 1966)

19. Blümel R.: Exponential Sensitivity and Chaos in Quantum Systems, Phys. Rev. Lett. **73**, 428 (1994)
20. Blum K.: Density Matrix. Theory and Applications, (Plenum Press, N.Y. 1981)
21. Bohm D., Aharonov Y.: Discussion of Experimental Proof of the Paradox of Einstein, Rosen, and Podolski, Phys. Rev. **108**, 1070 (1957)
22. Brown J.: A Quantum Revolution for Computing, New Scientist, p.24 (24. 09.1994)
23. Brune M., Haroche S., Raimond J.M., Davidovich L., Zagiry N.: Manipulations of Photons in a Cavity by Dispersive Atom-Field Coupling: QND-Measurements and Generation of Schrödinger-Cat States, Phys. Rev. A **45**, 5193 (1992)
24. Brunner K., Bockelmann U., Abstreiter G., Walther M., Böhm G., Tränkle G., Weimann G.: Photoluminescence from Single GaAs/AlGaAs Quantum Dot, Phys. Rev. Lett. **69**, 3216 (1992)
25. Cantrell C.D., Scully M.O.: The EPR-Paradox Revisited, Phys. Rept. C **43**, 499 (1978)
26. Cohen-Tannoudji C.: in Frontiers in Laser Spectroscopy, ed. by R. Balian, S. Haroche, S. Liberman (North Holland, Amsterdam 1977)
27. Cohen-Tannoudji C., Dupont-Roc J., Grynberg G.: Atom-Photon-Interactions (J.Wiley, 1992)
28. Coleman A.J.: Structure of Fermion Density Matrices, Rev. Mod. Phys. **35**, 668 (1963)
29. Cook R.J., Kimble H.J.: Possibility of Direct Observation of Quantum Jumps, Phys. Rev. Lett. **54**, 1023 (1985)
30. Crommie M.F., Lutz C.P., Eigler D.M.: Confinement of Electrons to Quantum Corrals on a Metal Surface, Science **262**, 218 (1993)
31. Dalibard J., Castin Y., Mølmer K.: Wave Function Approach to Dissipative Processes in Quantum Optics, Phys. Rev. Lett. **68**, 580 (1992)
32. Datta S.: Superlattices + Microstructures **6**, 83 (1989)
33. Davies E.B.: Quantum Theory of Open Systems (Academic Press, London 1976)
34. d'Espagnat B.: Conceptual Foundations of Quantum Mechanics, (Benjamin-Cummings, Menlo Park 1971)
35. Deutsch D.: Quantum Computational Networks, Proc. Roy. Soc. London A **425** (1989)
36. Deutsch D., Josza R.: Rapid Solution of Problems by Quantum Computation, Proc. Roy. Soc. London A **439**, 553 (1992)
37. Dicke R.H.: Coherence in Spontaneous Radiation Processes, Phys. Rev. **93**, 99 (1954)
38. DiVincenzo D.: Two-Bit Gates are Universal for Quantum Computation, Phys. Rev. A, in press (1994)
39. Eberly J.H., Hioe F.T.: Nonlinear Constants of Motion for 3-level Quantum Systems, Phys Rev. A **25**, 2168 (1982)
40. Einstein A., Podolski B., Rosen N.: Can Quantum Mechanical Description of Physical Reality be Considered Complete?, Physical Review **47**, 777 (1935)
41. Ekert A.K.: Quantum Cryptography Based on Bell's Theorem, Phys. Rev. Lett. **67**, 661 (1991)
42. Elgin J.N.: Semiclassical Formalism for the Treatment of 3-Level Systems, Phys. Lett. A **80**, 140 (1980)
43. Faist J., Capasso F., et al.: Quantum Cascade Laser, Science **264**, 553 (1994)
44. Fano U.: Description of Quantum Mechanics by Density Matrix and Operator Techniques, Rev. mod. Phys. **29**, 74 (1957)

45. Fano U.: Pairs of 2-Level Systems, Rev. mod. Phys. **55**, 855 (1983)
46. Farmer K.R., Rogers C.T., Buhrman R.A.: Nature of Single Localized Electron States derived from Tunneling Measurements, Phys. Rev. Lett. **58**, 2255 (1987)
47. Feynman R.P.: There is Plenty of Room at the Bottom, in: Miniaturization, p. 282 (Reinhold, New York, 1961)
48. Fick E., Sauermann G.: Quantenstatistik dynamischer Prozesse, Band I (Harri Deutsch, Frankfurt am Main 1983)
49. Fick E., Sauermann G.: Quantenstatistik dynamischer Prozesse, Band IIa: Antwort und Relaxationstheorie (Harri Deutsch, Frankfurt am Main 1983)
50. Finney G.A., Gea-Banacloche J.: Quasi-Classical Approximation for the Spin-Boson Hamiltonian with Counterrotating Terms, Phys. Rev. A **50**, 2040 (1994)
51. Frenkel A.: Spontaneous Localizations of the Wave Function and Classical Behavior, Found. Phys. **20**, 159 (1989)
52. Frerichs V., Schenzle A.: Quantum Zenon Effect without Collapse of the Wave Packet, Phys. Rev. A **44**, 1962 (1991)
53. Gardiner C.W.: Quantum Noise, (Springer, Berlin Heidelberg 1991)
54. Garraway B.M., Knight P.L.: Evolution of Quantum Superpositions in Open Environments, Phys. Rev. A **50**, 2548 (1994)
55. Gell-Mann M., Néeman Y.: The Eightfold Way, (Benjamin, New York 1964)
56. Ghirardi G.C., Omero C., Rimini A., Weber T.: The Stochastic Interpretation of Quantum Mechanics: A Critical Review, Riv. del. Nuovo Cim. **1** (1978)
57. Gisin N., Percival I.C.: Quantum State Diffusion Model Applied to Open Systems, J. Phys. A **25**, 5677 (1992)
58. Glauber R.J.: Time-Dependent Statistics of the Ising Model, J. Math. Phys. **4**, 294 (1963)
59. Gorini V., Kossakowski A., Sudarshan E.C.G.: Completely Positive Dynamical Semigroups of $N$-Level Systems, J. Math. Phys. **17**, 821 (1976)
60. Greiner W., Müller B.: Quantum Mechanics: Symmetries (Springer, Berlin Heidelberg 1994)
61. Greenberger D.M., Horne M.H., Zeilinger A.: Multiparticle Interferometry and the Superposition Principle, Physics Today, Aug. 1993, p.22
62. Haake F.: Statistical Treatment of Open Systems by Generalized Master Equations, Springer Tracts in Modern Physics Vol. **66** (Springer, Berlin Heidelberg 1973)
63. Haken H.: Laser Theory, in Encyclopedia of Physics **XXV**/2c, ed. by S. Flügge (Springer, Berlin Heidelberg 1970)
64. Haken H.: Synergetics. An Introduction, 3rd. edition (Springer, Berlin Heidelberg 1983)
65. Haken H., Wolf H.C.: The Physics of Atoms and Quanta (Springer, Berlin Heidelberg 1994)
66. Quantum Mechanics, Local Realistic Theories, and Lorentz-Invariant Local Theories, Phys. Rev. Lett. **68**, 2981 (1992)
67. Hegerfeldt G.C.: How to Reset an Atom after a Photon Detection: Application to Photon-Counting Processes, Phys. Rev. A **47**, 449 (1993)
68. Hioe F.T., Eberly J.H.: Dynamic Symmetries in Quantum Electronics, Phys Rev. A **28**, 879 (1981)
69. Hioe F.T., Eberly J.H.: N-Level Coherence Vector and Higher Conservation Laws in Quantum Optics and Quantum Mechanics, Phys Rev. Lett. **47**, 838 (1981)
70. Hioe F.T., Eberly J.H.: Nonlinear Constants of Motion for 3-Level Quantum Systems, Phys. Rev. A **25**, 2168 (1982)

71. Ho T., Chu S.: Semiclassical Many-Mode Floquet Theory: SU(3), PRA **31**, 659 (1985)
72. Itano W.M., Bergquist J.C., Wineland D.J.: Science **237**, 612 (1987)
73. Itano W.M., Heinzen D.J., Bollinger J.J., Wineland D.J.: Phys. Rev. A **41**, 2295 (1990)
74. Joos E., Zeh H.D.: The Emergence of Classical Properties through Interaction with the Environment, Z. Phys. B B **59**, 223 (1985)
75. Jordan: Quantum Mechanics in Simple Matrix Form (J.Wiley, New York 1986)
76. Keller M.: Quanteninformation und Physik des Entanglement, Dissertation (Ph.D.-thesis) (Universität Stuttgart 1995)
77. Keller M., Mahler G.: Nanostructures, Entanglement, and the Physics of Quantum Control, J. mod. Optics (1994)
78. Kimble H.J., Cook R.J., Walls A.L.: Intermittent Atomic Fluorescence, Phys. Rev. A **34**, 3190 (1986)
79. Kirk W.P., Reed M.A.: Nanostructures and Mesoscopic Systems (Academic Press, London 1992)
80. Körner H., Mahler G.: Cooperative Few-Level Fluctuations in Coupled Quantum Systems, Phys. Rev. E **47**, 3206 (1993)
81. Körner H., Mahler G.: Optically Driven Quantum Networks: Applications in Molecular Electronics, Phys. Rev. B **48**, 2335 (1993)
82. Körner H.: Stochastische Dynamik optisch gesteuerter Quantennetzwerke mit Anwendungen in der Molekularelektronik, Dissertation (Ph.D.-Thesis), (Universität Stuttgart 1993)
83. Lamb W.: An Operational Interpretation of Nonrelativistic Quantum Mechanics, Phys. Today **22**, 23 (April 1969)
84. Lee C.T.: Bayes Theorem and Continuous Photodectection, Quantum Optics **6**, 397 (1994)
85. Lindberg M., Binder P., Koch S.W.: Phys. Rev. A **45**, 1865 (1992)
86. Lindblad G.: On the Generators of Quantum Dynamical Semigroups, Commun. Math. Phys. **48**, 119 (1976)
87. Lloyd S.: A Potentially Realizable Quantum Computer, Science **261**, 1569 (1993)
88. Ludwig Ch., Gompf B., Glatz W., Petersen J., Eisenmenger W., et al.: Video-STM, LEED and X-Ray Diffraction Investigations of PTCDA on Graphite, Z. Phys. B – Condensed Matter 86 (1992) 397–404
89. Ludwig Ch., Eberle G., Gompf B., Petersen J., Eisenmenger W.: Thermal Motion of One-Dimensional Domain Walls in Monolayers of a Polar Polymer observed by Video-STM, Ann. Physik 2 (1993) 323–329
90. Mabuchi M., Kimble H.J.: Atom Galleries for Whispering Atoms, Optics Letters **19**, 749 (1994)
91. Mahler G., Körner H., Teich, W.: Optical Properties of Quasi-Molecular Structures: From Single Atoms to Quantum Dots, in Festkörperprobleme/Advances in Solid State Physics, Vol. 31, p.357, edited by U. Rössler (Vieweg, Wiesbaden 1991)
92. Mehring M., Wolff E.K., Stoll M.E.: Exploration of the Eight-Dimensional Spin Space of a Spin-1-Particle by NMR, J. Magnetic Resonance **37**, 475 (1980)
93. Meijer P.H.E., Bauer E.: Group Theory (North Holland, Amsterdam 1962)
94. Messiah A.: Quantenmechanik (Walter de Gruyter, Berlin, New York 1979)
95. Meystre P., Sargent M.: Elements of Quantum Optics (Springer, Berlin Heidelberg 1991)

96. Meystre P.: Cavity Quantum Optics and the Quantum Measurement Process, Progr. in Optics XXX (ed. by E. Wolf), p. 261 (North Holland, Amsterdam 1992)

97. Mielnik B.: The Paradox of Two Bottles in Quantum Mechanics, Found. Phys. **20**, 745 (1990)

98. Misra B., Sudershan E.C.G.: The Zeno's Paradox in Quantum Theory, Journal of Mathematical Physics **18** (4), 756–763 (1977)

99. Moerner W.E., Kador L.: Optical Detection and Spectroscopy of Single Molecules in a Solid, Phys. Rev. Lett. **62**, 2535 (1989)

100. Mølmer K., Castin Y., Dalibard J.: Monte Carlo Wavefunction Method in Quantum Optics, J. Opt. Soc. Am. B **10**, 524 (1993)

101. Mollow B.R.: Pure-State Analysis of Resonant Light-Scattering: Radiative Damping, Saturation, and Multiphoton Effects, Phys. Rev. A **12**, 1919 (1975)

102. Nagourney W., Sandberg J., Dehmelt H.: Shelved Optical Electron Amplifier: Observation of Quantum Jumps, Phys. Rev. Lett. **56**, 2797 (1986)

103. Nakajima S., Toyozawa Y., Abe R.: The Physics of Elementary Excitations (Springer, Berlin Heidelberg 1980))

104. Obermayer K., Teich W.G., Mahler G.: Structural Basis of Multistationary Quantum Systems, Phys. Rev. B **37** (14), 8111–8121 (1988)

105. Omnes R.: Consistent Interpretations of Quantum Mechanics, Rev. mod. Phys. **64**, 339 (1992)

106. O'Neill E.L.: Introduction to Statistical Optics (Dover, New York 1992)

107. Oreg J., Hio F.T., Eberly J.H.: Adiabatic Following in Multilevel Systems, Phys. Rev. A **29**, 690 (1984)

108. Orrit M., Bernard J.: Single Pentacene Molecules Detected by Fluorescence Excitation in a p-Terphenyl Crystal, Phys. Rev. Lett. **65**, 2716 (1990)

109. Paz J.P., Habib S., Zurek W.H.: Reduction of the Wave Packet, Phys. Rev. D **47**, 488 (1993)

110. Paz J.P., Mahler G.: Proposed Test for Temporal Bell Inequalities, Phys. Rev. Lett. **71**, 3235 (1993)

111. Pearle P.: Ways to Describe Dynamical State Vector Reduction, Phys. Rev. A **48**, 913 (1993)

112. Pegg D.T., Barnett S.M.: Phase Properties of the Quantized Single-Mode Electromagnetic Field, Phys. Rev. A **39**, 1665 (1989)

113. Peres A., Wooters W.K.: Optimal Detection of Quantum Information, Phys. Rev. Lett. **66**, 1119 (1991)

114. Porrati M., Putterman S.: Coherent Intermittency in the Resonance Fluorescence of a Multilevel Atom, Phys. Rev. A **39**, 3010 (1989)

115. Pöttner J., Lendi K.: Generalized Bloch Equations for Decaying Systems, Phys. Rev. A **31**, 1299 (1985)

116. Prasad S., Scully M.O., Martienssen W.: A Quantum Description of the Beam Splitter, Opt. Commun. **62**, 139 (1987)

117. Rabi I.I., Ramsey J.F., Schwinger J.: Use of Rotating Coordinates in Magnetic Resonance Problems, Rev. Mod. Phys. **26**, 167 (1954)

118. Ralls K.S., Skocpol W.J., Jackel L.D., Howard R.E., Fetter L.A., Epworth R.W., Temnant D.M.: Discrete Resistance Switching in Submicrometer Si-Inversion Layers: Individual Interface Traps and Low-Frequency Noise, Phys. Rev. Lett. **52**, 228 (1984)

119. Rao K.S., Rajeswari V.: Quantum Theory of Angular Momentum (Narosa, New Delhi 1993)

120. Reed M.A., Randall J.N., Aggarwal R.J., Matyi R.J., Moore T.M., Wetsel A.E.: Observation of Discrete Electronic States in a Zero-Dimensional Semiconductor Nanostructure, Phys. Rev. Lett. **60**, 535 (1988)

121. Rhim W.K., Pines A., Waugh J.S.: Time-Reversal Experiments in Dipolar Coupled Spin-Systems, Phys. Rev. B **3**, 684 (1981)
122. Roy S.M., Singh V.: Tests of Signal Locality and Einstein-Bell Locality for Multiparticle Systems, Phys. Rev. Lett. **67**, 2761 (1991)
123. Sardesaro N.B., Stark R.W.: Macroscopic Quantum Coherence and Localization for Normal State Electrons in Mg, Phys. Rev. Lett. **53**, 1681 (1984)
124. Sauter Th., Neuhauser W., Blatt R., Toschek P.E.: Observation of Quantum Jumps, Phys. Rev. Lett. **57**, 1696 (1986)
125. Schenzle A., DeVoe R.G., Brewer R.G.: Possibility of Quantum Jumps, Phys. Rev. A **33**, 2127 (1986)
126. Schenzle A.: Classical and Quantum Noise in Nonlinear Optical Systems, J. Stat. Phys. **54**, 1243 (1989)
127. Schiff L.I.: Quantum Mechanics, 3rd edition (McGraw-Hill, New York 1968)
128. Schulz M., Karmann A.: Single Individual Traps in MOSFETs, Physica Scripta T **35**, 273 (1991)
129. Scully M.O., Englert B., Walther H.: Quantum Optical Tests of Complementarity, Nature **351**, 111 (1991)
130. Selleri F.: Die Debatte um die Quantentheorie (Vieweg, Wiesbaden 1983)
131. Senatore G., March, N.H.: Recent Progress in the Field of Electron Correlations, Rev. mod. Phys. **66**, 445 (1994)
132. Sikorski Ch., Merkl U.: Spectroscopy of Electronic States in InSb Quantum Dots, Phys. Rev. Lett. **62**, 2164 (1989)
133. Spreeuw R.S.C., Woerdman J.P.: Optical Atoms, Progr. in Optics XXXI (ed. by E. Wolf), p. 263 (North Holland, Amsterdam 1993)
134. Stanley H.E.: Introduction to Phase Transitions And Critical Phenomena (Clarendon Press, Oxford 1971)
135. Stenholm S.: The Theory of Quantum Amplifiers, Physica Scripta T **12**, 56 (1986)
136. Sutcliff B.T.: The Chemical Bond and Molecular Structure, J. Molecular Structure **259**, 29 (1992)
137. Tapster P.R., Ravity J.G., Owens P.C.M.: Violation of Bell's Inequality over 4 km of Optical Fiber, Phys. Rev. Lett. **73**, 1923 (1994)
138. Teich W., Mahler G.: Stochastic Dynamics of Individual Quantum Systems: Stationary Rate Equations, Phys. Rev. A **45**, 3300 (1992)
139. Thompson R.S., Rempe G., Kimble H.J.: Observation of Normal Mode Splitting for an Atom in an Optical Cavity, Phys. Rev. Lett. **68**, 1132 (1992)
140. Tougaw P.D., Lent C.S., Porod W.: Bistable Saturation in Coupled Quantum-Dot Cells, J. Appl. Phys. **74**, 3558 (1993)
141. Ueda M., Imoto N., Ogawa T.: Quantum Theory of Continuous Photodetection Process, Phys. Rev. A **41**, 3891 (1990)
142. Universität Stuttgart, Sonderforschungsbereich 329 (Physikalische und chemische Grundlagen der Molekularelektronik): Arbeits- und Ergebnisbericht a) 1986–1988, b) 1989–1991, c) 1992–1994
143. Unruh W.G., Zurek W.H.: Reduction of a Wave Packet in Quantum Brownian Motion, Phys. Rev. D **40**, 1071 (1989)
144. Uren M.J., Kirton M.J., Collins S.: Anomalous Telegraph Noise in Small Area Silicon-Metal-Oxide Semiconductor FET, Phys. Rev. B **37**, 8346 (1988)
145. Vaidman L.: Lorentz-Invariant Elements of Reality and the Joint measurability of Commuting Observables, Phys. Rev. Lett. **70**, 3369 (1993)
146. van Kampen N.G.: Stochastic Processes in Physics and Chemistry (North Holland, Amsterdam 1983)
147. von Neumann J.: Mathematical Foundation of Quantum Mechanics (Princeton University Press, Princeton 1955)

148. Walther H.: Experiments on Cavity Quantum Electrodynamics, Phys. Reports **219**, 263 (1992)
149. Wawer R.: Quantendynamik in SU(N)-Darstellung , Diplomarbeit (Diploma-thesis) (Universität Stuttgart 1994)
150. Weberruß V. A.: Universality in Statistical Physics and Synergetics. A Comprehensive Approach to Modern Theoretical Physics (Vieweg, Wiesbaden 1993)
151. Weller H.: Quantized Semiconductor Particles: A Novel State of Matter for Material Science, Adv. Mater, **5**, 88 (1993)
152. Wilkens M., Meystre P.: Spectrum of Spontaneous Emission in a Fabry-Pérot Cavity: The Effect of Atomic Motion, Optics Commun. **94**, 66 (1992)
153. Wooters W.K., Zurek W.: Nature **299**, 802 (1982)
154. Yao H.I., Eberly J.H.: Dynamical Theory of an Atom interacting with Quantized Cavity Fields, Phys. Rev. Rep. **118**, 239 (1985)
155. Zurek W.: Decoherence and the Transition from Quantum to Classical, Phys. Today, p. 36, Oct. 1991

# Index

# About the Authors

Günter Mahler received his scientific education from the Universities of Frankfurt, Munich, and Regensburg (Germany). In 1972 he earned his Ph.D. in Physics and has been Professor of Theoretical Physics at the Universität Stuttgart since 1978. He has been visiting professor of the Universities of Strasbourg (France), the Arizona State University (USA), and the Santa Fe Institute (USA). He has made numerous contributions to the fields of phonon physics, high-excitation phenomena in semiconductors, quantum transport theory, molecular electronics, and quantum computation.

Until 1992 Volker A. Weberruß worked at the 1. Institut für Theoretische Physik und Synergetik, Universität Stuttgart. In 1992 he earned his Ph.D. in Physics. During this time, the director of the institute, Prof. Dr. Dr. h.c. mult. H. Haken, made it possible for him to participate in a project of the Deutsche Forschungsgemeinschaft (physikalische und chemische Grundlagen der Molekularelektronik). In the course of this project V. A. Weberruß worked on the mathematic modelling of many-component systems of laser theory and nonlinear physics. He summarized the results of this research work in his first book "Universality in Statistical Physics and Synergetics" (1993). Since that time he has been offering scientists in research institutes and university lecturers the possibility of cooperation for the joint production of scientific literature (V.A.W. scientific consultation). In this way the scientists and lecturers are able to put their knowledge at the disposal of a wide public yet without investing a lot of time. This book was produced within the framework of this activity.

# Springer-Verlag and the Environment

We at Springer-Verlag firmly believe that an international science publisher has a special obligation to the environment, and our corporate policies consistently reflect this conviction.

We also expect our business partners – paper mills, printers, packaging manufacturers, etc. – to commit themselves to using environmentally friendly materials and production processes.

The paper in this book is made from low- or no-chlorine pulp and is acid free, in conformance with international standards for paper permanency.